ELEMENTARY LINEAR PROGRAMMING WITH APPLICATIONS

This is a volume in
COMPUTER SCIENCE
AND APPLIED MATHEMATICS

A Series of Monographs and Textbooks
Editor: Werner Rheinboldt

A complete list of titles in this series appears
at the end of this volume.

ELEMENTARY LINEAR PROGRAMMING WITH APPLICATIONS

BERNARD KOLMAN
Drexel University

ROBERT E. BECK
Villanova University

ACADEMIC PRESS
New York   San Francisco   London
A Subsidiary of Harcourt Brace Jovanovich, Publishers

Academic Press, Inc.
111 Fifth Avenue, New York, New York 10003

United Kingdom Edition published by
Academic Press, Inc. (London) Ltd.
24/28 Oval Road, London NW1 7DX

ISBN: 0-12-417860-X
Library of Congress Catalog Card Number: 79-50381

Printed in the United States of America

To Our Families

CONTENTS

Preface	xi
Acknowledgments	xiii

PROLOGUE

INTRODUCTION TO OPERATIONS RESEARCH
INTRODUCTION TO OPERATIONS RESEARCH	1
Further Readings	5

CHAPTER 0

REVIEW OF LINEAR ALGEBRA (OPTIONAL)
REVIEW OF LINEAR ALGEBRA (OPTIONAL)	7
0.1 Matrices	7
0.2 Gauss–Jordan Reduction	15
0.3 The Inverse of a Matrix	24
0.4 Subspaces	29
0.5 Linear Independence and Basis	34
Further Readings	43

CHAPTER 1

GEOMETRY IN R^n
GEOMETRY IN R^n	45
1.1 Hyperplanes	45
1.2 Convex Sets	50
Further Reading	57

CHAPTER **2**

INTRODUCTION TO LINEAR PROGRAMMING   59

   2.1 The Linear Programming Problem   59
   2.2 Matrix Notation; Geometric Solutions   78
   2.3 The Simplex Method   99
   2.4 Degeneracy and Cycling (Optional)   115
   2.5 Artificial Variables   121
       Further Readings   142

CHAPTER **3**

FURTHER TOPICS IN LINEAR PROGRAMMING   143

   3.1 Duality   143
   3.2 Computational Relations between the Primal and Dual
       Problems   164
   3.3 The Dual Simplex Method   190
   3.4 The Revised Simplex Method   196
   3.5 Sensitivity Analysis   206
   3.6 Computer Aspects (Optional)   216
       Further Readings   226

CHAPTER **4**

INTEGER PROGRAMMING   227

   4.1 Examples   227
   4.2 Cutting Plane Methods   240
   4.3 Branch and Bound Methods   254
   4.4 Computer Aspects (Optional)   267
       Further Readings   270

CHAPTER **5**

SPECIAL TYPES OF LINEAR PROGRAMMING
PROBLEMS   271

   5.1 The Transportation Problem   271
       Further Readings   300
   5.2 The Assignment Problem   300
       Further Readings   313

5.3 Graphs and Networks. Basic Definitions   314
Further Reading   319
5.4 The Maximal Flow Problem   319
Further Readings   337
5.5 The Shortest Route Problem   338
Further Readings   349
5.6 The Critical Path Method   349
Further Readings   359
5.7 Computer Aspects (Optional)   359

Solutions to Odd-Numbered Exercises   367
Index   395

PREFACE

Classical optimization techniques have been widely used in engineering and the physical sciences for a long time. They arose from attempts to determine the "best" or "most desirable" solution to a problem. Toward the end of World War II, models for many problems in the management sciences were formulated and algorithms for their solutions were developed. In particular, the new areas of linear, integer, and nonlinear programming and network flows were developed. These new areas of applied mathematics have succeeded in saving billions of dollars by enabling the model builder to find optimal solutions to large and complex applied problems. Of course, the success of these modern optimizations techniques for real problems is due primarily to the rapid development of computer capabilities in the past 30 years and to the consequent decreasing cost of doing computations.

With the increasing emphasis in mathematics on relevance, some of the above-mentioned areas of modern optimization are rapidly becoming part of the undergraduate curricula for business, engineering, computer science, and mathematics students.

This book presents a survey of the basic ideas in linear programming and related areas and is designed for a one-semester or one-quarter course which can be taken by business, engineering, computer science, and mathematics majors. In their professional careers many of these students will work with real applied problems; they will have to formulate models for these problems and obtain numerical answers. Our purpose is to provide such students with an impressive illustration of how simple mathematics can be used to solve difficult problems which arise in real situations and to give them some tools which will prove useful in their professional work.

xi

The Prologue gives a brief survey of operations research and discusses the different steps in solving an operations research problem. Although we assume that most readers have already had some exposure to linear algebra, Chapter 0 provides a quick review of the necessary linear algebra. The linear algebra requirements for the rest of this book are kept to a minimum. It is also possible to learn the required linear algebra from this chapter. Chapter 1 deals with the basic necessary geometric ideas in R^n. Chapter 2 introduces linear programming. Here we give examples of the problems to be considered and present the simplex method as an algorithm for solving linear programming problems. Chapter 3 covers further topics in linear programming, including duality theory and sensitivity analysis. Chapter 4 presents an introduction to integer programming; Chapter 5 covers a few of the more important topics in network flows.

An important feature of this book is a final section in each of Chapters 3, 4, and 5 which discusses computer aspects of the material in the chapter. In solving real optimization problems one almost always must use a computer and make decisions about which particular code to use. These sections will provide some guidance for making such decisions.

The approach in this book is not a rigorous one, and proofs have been kept to a minimum. Each idea is motivated, discussed, and carefully illustrated with examples.

The exercises are of three different types. First, we give routine exercises designed to reinforce the mechanical aspects of the material under study. Second, we present realistic problems requiring the student to formulate a model and obtain solutions to it. Third, we offer projects, some of which ask the student to familiarize himself or herself with those journals that publish papers in the subject under study. Most of the projects are realistic problems, and they will often have some vagueness in their statement; this vagueness must be resolved by the student as he or she formulates a model.

Although access to a computer would further the sense of realism and applicability being stressed by this book, availability of such equipment is not essential to the use of the book. However, problems are included which require the use of a computer. These can be assigned if the appropriate equipment is available.

This book is based on a course developed by one of us (Bernard Kolman) in 1973 at Drexel University under a College Science Improvement Program grant from the National Science Foundation. A prepublication version of this book has been used at both Drexel and Villanova Universities for five years. The published version of the book reflects our classroom experience and student comments.

Bernard Kolman
Robert E. Beck

ACKNOWLEDGMENTS

We gratefully acknowledge the contributions of the following people whose incisive comments greatly helped to improve the manuscript:

IHOR B. CHYZOWYCH—*Sperry UNIVAC*
JOSEPH G. ECKER—*Rensselaer Polytechnic Institute*
MICHAEL LEVITAN—*Villanova University*
GIDEON LIDOR—*City College of New York*
ROGER Y. S. LYNN—*Villanova University*
ARNOLD SHAPIRO—*Pennsylvania State University, Ogontz Campus*

We thank:

Our typists Amelia Maurizio, Catherine McGinnis, and Cornelia Mulhern for skillfully and cheerfully typing the manuscript and showing infinite patience as we revised the material.

Daniel Joyce and Michael Dellipriscoli for providing assistance with solutions to the problems.

Sara Winter for her careful and critical proofreading.

The National Science Foundation for its support in the form of a College Science Improvement Program grant to Drexel University. This book grew out of a course developed under this grant. Dale W. Lick and Chris Rorres provided encouragement and administrative assistance during the period of the grant.

Sperry UNIVAC for providing sample outputs from their mathematical programming codes.

The students at Drexel and Villanova Universities for their many suggestions while using the prepublication version of this book.

Finally, thanks are also due to the staff of Academic Press for their interest, encouragement, and cooperation.

PROLOGUE

INTRODUCTION TO OPERATIONS RESEARCH

What Is Operations Research?

Many definitions of *operations research* (frequently called OR) have been given. A common thread in these definitions is that OR is a *scientific method* for providing a quantitative basis for decision making which can be used in almost any field of endeavor. The techniques of OR give a logical and systematic way of formulating a problem so that the tools of mathematics can be applied to find a solution. However, OR differs from mathematics in the following sense. Most often mathematics problems can be clearly stated and have a specific answer. OR problems are frequently poorly posed: they arise when someone has the vague feeling that the established way of doing things can be improved. Engineering, which is also engaged in solving problems, frequently uses the methods of OR. A central problem in OR is the allocation of scarce resources. In this context, scarce resources include raw materials, labor, capital, energy, and processing time. For example, a manufacturer could consult an operations research analyst to determine which combination of production techniques should be used to meet market demands and minimize costs. In fact, the 1975 Nobel Prize in Economics was awarded to T. C. Koopmans and L. V. Kantorovich for their contributions to the theory of optimum allocation of resources.

Development of Operations Research

The use of scientific methods as an aid in decision making goes back a long time, but the discipline which is now called *operations research*

1

had its birth during World War II. Great Britain, which was struggling for its very existence, gathered a number of its top scientists and mathematicians to study the problem of allocating the country's dwindling resources. The United States Air Force became interested in this new approach to the analysis of military operations and organized a research group. In 1947 George B. Dantzig, a member of this group, developed the simplex algorithm for solving linear programming problems. At approximately the same time the programmable digital computer was developed, giving a means of solving large-scale linear programming problems. The first solution of a linear programming problem on a computer occurred in 1952 on the National Bureau of Standards SEAC machine. The rapid development of mathematical programming techniques has paralleled the rapid growth of computing power. The ability to analyze large and complicated problems with operations research techniques has resulted in savings of *billions* of dollars to industry and government. It is remarkable that a newly developed discipline such as operations research has had such an impact on the science of decision making in such a short time.

Phases of an Operations Research Study

We now look at the steps an operations analyst uses in determining information for decision making. In most cases the analyst is employed as a consultant, so that management has to first recognize the need for the study to be carried out. The consultant can now begin work using the following sequence of steps.

Step 1.   **Problem definition and formulation.** In this phase the goal of the study is defined. The consultant's role at this point is one of helping management to clarify its objectives in undertaking the study. Once an acceptable statement of the goals has been made, the consultant must now identify the decision alternatives. It is likely that there are some options that management will refuse to pursue; thus, the consultant will consider only the alternatives acceptable to management. Attention must also be paid to the limitations, restrictions, and requirements of the various alternatives. For example, management might have to abide by fair employment laws or antipollution laws. Some alternatives may be limited by the available capital, labor, or technology.

Step 2.   **Model construction.** The consultant now develops the appropriate mathematical description of the problem. The limitations, restrictions, and requirements must be translated into mathematical terms, which then give rise to the constraints of the problem. In many cases the goal of the study can be quantified as an expression which is to be maximized or minimized. The decision alternatives are represented by the variables in the problem. Often the mathematical model developed is one which has a familiar form and for which methods of solution are available.

Step 3. **Solution of the model.** The mathematical model developed in Step 2 must now be solved. The method of solution may be as simple as providing the input data for an available computer program or it may call for an investigation of an area of mathematics which so far has not been studied. There may be no method of finding a solution to the mathematical model. In this case the consultant may use heuristic methods or approximate methods, or it may be necessary to go back to Step 2 and modify the model. It should be noted that the solution to the model need not be the solution to the original problem. This will be further discussed below.

Step 4. **Sensitivity analysis.** Frequently the numbers which are given to the consultant are approximate. Obviously, the solution depends on the values which are specified for the model, and since these are subject to variation, it is important to know how the solution will vary with the variation in the input data. For standard types of models these questions have been investigated, and techniques are available for determining the sensitivity of the solution to changes in the input data.

Step 5. **Model evaluation.** At this point the consultant has obtained a solution to the model, and often this solution will represent a solution to the given problem. The consultant must determine whether the answers produced by the model are realistic, acceptable to management, and capable of implementation by management. As in Step 1, the consultant now needs a thorough understanding of the nature of the client's business.

Step 6. **Implementation of the study.** Management must now decide how to implement the recommendations of the consultant.

The Structure of Mathematical Models

When a technical person discusses a model of a situation which is being studied, he or she is referring to some idealized representation of a real-life system. The model may simply involve a change in scale, such as the hobbyist's HO railroad or the architect's display of a newly planned community.

Engineers often use analogue models in which electrical properties substitute for mechanical properties. Usually the electrical analogues are much easier to deal with than the real objects. For example, resetting a dial will change the analogue of the mass of an object. Without the analogue one might have to saw off part of the object.

Mathematical models represent objects by symbols. The variables in the model represent the decision alternatives or items which can be varied in the real-life situation. There are two types of mathematical models: deterministic and probabilistic. Suppose the process described by the model is repeated many times. A *deterministic* model will always yield the same set of output values for a given set of input values, while a *probabilistic* model will typically yield many different sets of output values according to some probability distribution. In this book we will discuss only deterministic models.

The mathematical models which will be considered in this book are structured to include the following four basic components:

(a) **Decision variables or unknowns.** Typically we are seeking values for these unknowns, which describe an optimal allocation of the scarce resources represented by the model. For example, decision variables might represent purchase lot size, number of hours to operate a machine, or which of several alternatives to choose.

(b) **Parameters.** These are inputs which may or may not be adjustable by the analyst, but are known either exactly or approximately. For example, purchase price, rate of consumption, and amount of spoilage could all be parameters.

(c) **Constraints.** These are conditions which limit the values that the decision variables can assume. For example, a variable measuring units of output cannot be negative; a variable measuring the amount to be stored cannot have a value greater than the available capacity.

(d) **Objective function.** This expression measures the effectiveness of the system as a function of the decision variables. The decision variables are to be determined so that the objective function will be optimized. It is sometimes difficult to determine a quantitative measure of the performance of a system. Consequently, several objective functions may be tried before choosing one which will reflect the goals of the client.

Mathematical Techniques in Operations Research

The area of *mathematical programming* plays a prominent role in OR. It consists of a variety of techniques and algorithms for solving certain kinds of mathematical models. These models call for finding values of the decision variables which maximize or minimize the objective function subject to a system of inequality and equality constraints. Mathematical programming is divided into several areas depending on the nature of the constraints, the objective function, and the decision variables. Linear programming deals with those models in which the constraints and the objective function are linear expressions in the decision variables. Integer programming deals with the special linear programming situation in which the decision variables are constrained to take nonnegative integer values. In stochastic programming the parameters do not have fixed values but are described by probability distributions. In nonlinear programming some or all of the constraints and the objective function are nonlinear expressions in the decision variables. Special linear programming problems such as optimally assigning workers to jobs or optimally choosing routes for shipments between plants and warehouses have individually tailored algorithms for obtaining their solutions. These algorithms make use of the techniques of network flow analysis.

Special models, other than mathematical programming techniques, have been developed to handle several important OR problems. These include models for inventory analysis to determine how much of each item to keep on hand, for analysis of waiting-line situations such as checkout counters and tollbooths, and for competitive situations with conflicting goals as would arise in a game.

There are standard techniques available to solve many of the usual models in OR. Some of these methods are iterative, obtaining a better solution at each successive iteration. Some will produce the optimal solution after a finite number of steps. Others converge only after an infinite number of steps and consequently must be truncated. Some models do not lend themselves to the standard approaches, and thus *heuristic* techniques—that is, techniques improvised for the particular problem and without firm mathematical basis—must be used.

Further Readings

Gale, David. *The Theory of Linear Economic Models*. McGraw-Hill, New York, 1960.

Maki, Daniel P., and Maynard Thompson. *Mathematical Models and Applications*. Prentice-Hall, Englewood Cliffs, New Jersey, 1973.

Roberts, Fred S. *Discrete Mathematical Models with Applications to Social, Biological, and Environmental Problems*. Prentice-Hall, Englewood Cliffs, New Jersey, 1976.

Journals

Computer and Information Systems Abstracts
Computer Journal
Decision Science

European Journal of Operational Research
International Abstracts in Operations Research
Journal of Computer and Systems Sciences
Journal of Research of the National Bureau of Standards
Journal of the Canadian Operational Research Society
Naval Research Logistics Quarterly (published by the Office of Naval Research—ONR)
Management Science (published by The Institute for Management Science—TIMS)
Mathematical Programming
Mathematics in Operations Research
Operational Research Quarterly
Operations Research (published by the Operations Research Society of America—ORSA)
Transportation Science

CHAPTER

012345

REVIEW OF LINEAR ALGEBRA

We assume most readers of this book have already had some exposure to linear algebra. We expect that they have learned what a matrix is, how to multiply matrices, and how to tell whether a set of n-tuples is linearly independent. This chapter provides a quick review of the necessary linear algebra material for those readers who wish it. The chapter can also serve as a reference for the linear algebra encountered later in the text. Exercises are included in this chapter to give the student an opportunity to test his or her comprehension of the material.

0.1 MATRICES

Definition. An $m \times n$ **matrix** **A** is a rectangular array of mn numbers (usually real numbers) arranged in m horizontal rows and n vertical columns:

$$\mathbf{A} = \begin{bmatrix} a_{11} & a_{12} & \cdots & a_{1n} \\ a_{21} & a_{22} & \cdots & a_{2n} \\ \vdots & \vdots & & \vdots \\ a_{m1} & a_{m2} & \cdots & a_{mn} \end{bmatrix} \tag{1}$$

The *i*th **row** of **A** is

$$[a_{i1} \quad a_{i2} \cdots a_{in}] \qquad (1 \le i \le m)$$

The *j*th **column** of **A** is

$$\begin{bmatrix} a_{1j} \\ a_{2j} \\ \vdots \\ a_{mj} \end{bmatrix} \qquad (1 \le j \le n)$$

7

The number in the ith row and jth column of $\mathbf{A}$ is denoted by a_{ij}, and is called the **ijth element of A**, or the **(i, j) entry of A**, and we often write (1) as

$$\mathbf{A} = [a_{ij}]$$

The matrix $\mathbf{A}$ is said to be **square of order** n if $m = n$. In this case, the numbers $a_{11}, a_{22}, \ldots, a_{nn}$ form the **main diagonal** of $\mathbf{A}$.

EXAMPLE 1.   If

$$\mathbf{A} = \begin{bmatrix} 1 & -2 \\ 3 & 4 \\ 4 & 2 \end{bmatrix}, \qquad \mathbf{B} = \begin{bmatrix} 3 & -2 & 1 \\ 4 & 3 & 5 \end{bmatrix}, \qquad \text{and} \qquad \mathbf{C} = \begin{bmatrix} -1 & 2 \\ -2 & 4 \end{bmatrix}$$

then $\mathbf{A}$ is 3×2, $\mathbf{B}$ is 2×3, and $\mathbf{C}$ is square of order 2. Moreover, $a_{21} = 3$, $a_{32} = 2$, $b_{23} = 5$, and $c_{21} = -2$.

Definition.   Two $m \times n$ matrices $\mathbf{A} = [a_{ij}]$ and $\mathbf{B} = [b_{ij}]$ are said to be **equal** if $a_{ij} = b_{ij}$ for each choice of i and j where $1 \le i \le m$, $1 \le j \le n$.

We now turn to the definition of several operations on matrices. These operations will enable us to use matrices as we discuss linear programming problems.

Definition.   If $\mathbf{A} = [a_{ij}]$ and $\mathbf{B} = [b_{ij}]$ are $m \times n$ matrices, then the **sum** of $\mathbf{A}$ and $\mathbf{B}$ is the matrix $\mathbf{C} = [c_{ij}]$ defined by

$$c_{ij} = a_{ij} + b_{ij} \qquad (1 \le i \le m, \quad 1 \le j \le n)$$

That is, $\mathbf{C}$ is obtained by adding corresponding entries of $\mathbf{A}$ and $\mathbf{B}$.

EXAMPLE 2.   Let

$$\mathbf{A} = \begin{bmatrix} 2 & -3 & 4 \\ 5 & 1 & -2 \end{bmatrix} \quad \text{and} \quad \mathbf{B} = \begin{bmatrix} 3 & 3 & 2 \\ -2 & 2 & 4 \end{bmatrix}$$

Then

$$\mathbf{A} + \mathbf{B} = \begin{bmatrix} 2+3 & -3+3 & 4+2 \\ 5+(-2) & 1+2 & -2+4 \end{bmatrix} = \begin{bmatrix} 5 & 0 & 6 \\ 3 & 3 & 2 \end{bmatrix}$$

Properties of Matrix Addition

(a) $\mathbf{A} + \mathbf{B} = \mathbf{B} + \mathbf{A}$

(b) $\mathbf{A} + (\mathbf{B} + \mathbf{C}) = (\mathbf{A} + \mathbf{B}) + \mathbf{C}$

(c) There is a unique $m \times n$ matrix $\mathbf{0}$, called the $m \times n$ **zero matrix**, such that

$$\mathbf{A} + \mathbf{0} = \mathbf{A} \quad \text{for any } m \times n \text{ matrix } \mathbf{A}$$

(d) For each $m \times n$ matrix $\mathbf{A}$, there is a unique matrix, denoted by $-\mathbf{A}$, such that

$$\mathbf{A} + (-\mathbf{A}) = \mathbf{0}$$

The matrix $-\mathbf{A}$ is called the **negative** of $\mathbf{A}$. The ijth element of $-\mathbf{A}$ is $-a_{ij}$, where $\mathbf{A} = [a_{ij}]$.

Definition. If $\mathbf{A} = [a_{ij}]$ is an $m \times p$ matrix and $\mathbf{B} = [b_{ij}]$ is a $p \times n$ matrix, then the **product** of $\mathbf{A}$ and $\mathbf{B}$ is the $m \times n$ matrix $\mathbf{C} = [c_{ij}]$ defined by

$$c_{ij} = a_{i1}b_{1j} + a_{i2}b_{2j} + \cdots + a_{ip}b_{pj} \qquad (1 \le i \le m, \quad 1 \le j \le n) \quad (2)$$

EXAMPLE 3.   If

$$\mathbf{A} = \begin{bmatrix} 1 & 3 & -2 \\ 2 & 4 & 3 \end{bmatrix} \quad \text{and} \quad \mathbf{B} = \begin{bmatrix} -2 & 4 \\ 3 & -3 \\ 2 & 1 \end{bmatrix}$$

then

$$\mathbf{AB} = \begin{bmatrix} 1(-2) + 3\cdot3 + (-2)\cdot2 & 1\cdot4 + 3(-3) + (-2)\cdot1 \\ 2(-2) + 4\cdot3 + 3\cdot2 & 2\cdot4 + 4(-3) + 3\cdot1 \end{bmatrix} = \begin{bmatrix} 3 & -7 \\ 14 & -1 \end{bmatrix}$$

Matrix multiplication requires more care than matrix addition. The product $\mathbf{AB}$ can be formed only if the number of columns of $\mathbf{A}$ is the same as the number of rows of $\mathbf{B}$. Also, unlike multiplication of real numbers, we may have $\mathbf{AB} = \mathbf{0}$, the zero matrix, with neither $\mathbf{A} = \mathbf{0}$ nor $\mathbf{B} = \mathbf{0}$, and we may have $\mathbf{AB} = \mathbf{AC}$ without $\mathbf{B} = \mathbf{C}$. Also, if both $\mathbf{A}$ and $\mathbf{B}$ are square of order n, it is possible that $\mathbf{AB} \ne \mathbf{BA}$.

We digress for a moment to recall the summation notation. When we write

$$\sum_{i=1}^{n} a_i$$

we mean

$$a_1 + a_2 + \cdots + a_n$$

The letter i is called the **index of summation**; any other letter can be used in place of it. Thus,

$$\sum_{i=1}^{n} a_i = \sum_{j=1}^{n} a_j = \sum_{r=1}^{n} a_r$$

The summation notation satisfies the following properties:

(i) $\displaystyle\sum_{i=1}^{n} (r_i + s_i)a_i = \sum_{i=1}^{n} r_i a_i + \sum_{i=1}^{n} s_i a_i$

(ii) $\displaystyle\sum_{i=1}^{n} ca_i = c \sum_{i=1}^{n} a_i$

(iii) $\displaystyle\sum_{i=1}^{m} \sum_{j=1}^{n} a_{ij} = \sum_{j=1}^{n} \sum_{i=1}^{m} a_{ij}$

Using the summation notation, Equation (2) for the (i, j) entry of the product $\mathbf{AB}$ can be written as

$$c_{ij} = \sum_{k=1}^{p} a_{ik} b_{kj} \qquad (1 \le i \le m, \quad 1 \le j \le n)$$

Properties of Matrix Multiplication

(a) $\mathbf{A(BC)} = \mathbf{(AB)C}$
(b) $\mathbf{A(B + C)} = \mathbf{AB} + \mathbf{AC}$
(c) $\mathbf{(A + B)C} = \mathbf{AC} + \mathbf{BC}$

Definition. The $n \times n$ matrix $\mathbf{I}_n$, all of whose diagonal elements are 1 and the rest of whose entries are zero, is called the **identity matrix** of order n.

If $\mathbf{A}$ is an $m \times n$ matrix, then

$$\mathbf{I}_m\mathbf{A} = \mathbf{AI}_n = \mathbf{A}$$

Linear Systems

The linear system of m equations in n unknowns

$$\left.\begin{aligned}
a_{11}x_1 + a_{12}x_2 + \cdots + a_{1n}x_n &= b_1 \\
a_{21}x_1 + a_{22}x_2 + \cdots + a_{2n}x_n &= b_2 \\
\vdots \qquad \vdots \qquad\qquad \vdots \quad & \vdots \\
a_{m1}x_1 + a_{m2}x_2 + \cdots + a_{mn}x_n &= b_m
\end{aligned}\right\} \tag{3}$$

can be written in matrix form as follows. Let

$$\mathbf{A} = \begin{bmatrix} a_{11} & a_{12} & \cdots & a_{1n} \\ a_{21} & a_{22} & \cdots & a_{2n} \\ \vdots & \vdots & & \vdots \\ a_{m1} & a_{m2} & \cdots & a_{mn} \end{bmatrix}, \quad \mathbf{x} = \begin{bmatrix} x_1 \\ x_2 \\ \vdots \\ x_n \end{bmatrix}, \quad \mathbf{b} = \begin{bmatrix} b_1 \\ b_2 \\ \vdots \\ b_m \end{bmatrix}$$

Then (3) can be written as

$$\mathbf{Ax} = \mathbf{b}$$

The matrix **A** is called the **coefficient matrix** of the linear system (3), and the matrix

$$[A \mid b] = \begin{bmatrix} a_{11} & a_{12} & \cdots & a_{1n} & \vdots & b_1 \\ a_{21} & a_{22} & \cdots & a_{2n} & \vdots & b_2 \\ \vdots & \vdots & & \vdots & \vdots & \vdots \\ a_{m1} & a_{m2} & \cdots & a_{mn} & \vdots & b_m \end{bmatrix}$$

obtained by adjoining **b** to **A** is called the **augmented matrix** of (3).

EXAMPLE 4. Consider the linear system

$$3x - 2y + 4z + 5w = 6$$
$$2x + 3y - 2z + w\ = 7$$
$$x - 5y + 2z\quad\ = 8$$

Letting

$$A = \begin{bmatrix} 3 & -2 & 4 & 5 \\ 2 & 3 & -2 & 1 \\ 1 & -5 & 2 & 0 \end{bmatrix}, \quad x = \begin{bmatrix} x \\ y \\ z \\ w \end{bmatrix}, \quad b = \begin{bmatrix} 6 \\ 7 \\ 8 \end{bmatrix}$$

we can write the given linear system in matrix form as

$$Ax = b$$

Scalar Multiplication

Definition. If $A = [a_{ij}]$ is an $m \times n$ matrix and r is a real number, then the **scalar multiple** of **A** by r, r**A**, is the $m \times n$ matrix $B = [b_{ij}]$, where $b_{ij} = r a_{ij}$ $(1 \le i \le m, \ 1 \le j \le n)$.

EXAMPLE 5. If $r = -2$ and

$$A = \begin{bmatrix} 2 & -3 & 5 \\ 2 & 4 & 3 \\ 0 & 6 & -3 \end{bmatrix}$$

then

$$rA = \begin{bmatrix} -4 & 6 & -10 \\ -4 & -8 & -6 \\ 0 & -12 & 6 \end{bmatrix}$$

Properties of Scalar Multiplication

(a) $r(s\mathbf{A}) = (rs)\mathbf{A}$
(b) $(r + s)\mathbf{A} = r\mathbf{A} + s\mathbf{A}$
(c) $r(\mathbf{A} + \mathbf{B}) = r\mathbf{A} + r\mathbf{B}$
(d) $\mathbf{A}(r\mathbf{B}) = r(\mathbf{AB})$

The Transpose of a Matrix

Definition. If $\mathbf{A} = [a_{ij}]$ is an $m \times n$ matrix, then the $n \times m$ matrix $\mathbf{A}^\mathrm{T} = [b_{ij}]$, where

$$b_{ij} = a_{ji} \qquad (1 \le i \le m, \quad 1 \le j \le n)$$

is called the **transpose** of $\mathbf{A}$. Thus, the transpose of $\mathbf{A}$ is obtained by merely interchanging the rows and columns of $\mathbf{A}$.

EXAMPLE 6. If

$$\mathbf{A} = \begin{bmatrix} 1 & 3 & 2 \\ -2 & 6 & 5 \end{bmatrix}$$

then

$$\mathbf{A}^\mathrm{T} = \begin{bmatrix} 1 & -2 \\ 3 & 6 \\ 2 & 5 \end{bmatrix}$$

Properties of the Transpose

If r is a scalar and $\mathbf{A}$ and $\mathbf{B}$ are matrices, then
(a) $(\mathbf{A}^\mathrm{T})^\mathrm{T} = \mathbf{A}$
(b) $(\mathbf{A} + \mathbf{B})^\mathrm{T} = \mathbf{A}^\mathrm{T} + \mathbf{B}^\mathrm{T}$
(c) $(\mathbf{AB})^\mathrm{T} = \mathbf{B}^\mathrm{T}\mathbf{A}^\mathrm{T}$ (Note that the order changes.)
(d) $(r\mathbf{A})^\mathrm{T} = r\mathbf{A}^\mathrm{T}$

If we cross out some rows, or columns, of a given matrix $\mathbf{A}$, we obtain a **submatrix** of $\mathbf{A}$.

EXAMPLE 7. Let

$$\mathbf{A} = \begin{bmatrix} 2 & 3 & 5 & -1 \\ 3 & 4 & 2 & 7 \\ 8 & 2 & 6 & 1 \end{bmatrix}$$

If we cross out the second row and third column, we obtain the submatrix

$$\begin{bmatrix} 2 & 3 & -1 \\ 8 & 2 & 1 \end{bmatrix}$$

We can now view a given matrix $\mathbf{A}$ as being partitioned into submatrices. Moreover, the partitioning can be carried out in many different ways.

EXAMPLE 8. The matrix

$$
A = \left[\begin{array}{ccc:cc}
a_{11} & a_{12} & a_{13} & a_{14} & a_{15} \\
a_{21} & a_{22} & a_{23} & a_{24} & a_{25} \\
\hdashline
a_{31} & a_{32} & a_{33} & a_{34} & a_{35} \\
a_{41} & a_{42} & a_{43} & a_{44} & a_{45}
\end{array}\right]
$$

is partitioned as

$$
A = \begin{bmatrix} A_{11} & A_{12} \\ A_{21} & A_{22} \end{bmatrix}
$$

Another partitioning of A is

$$
A = \left[\begin{array}{cc:cc:c}
a_{11} & a_{12} & a_{13} & a_{14} & a_{15} \\
a_{21} & a_{22} & a_{23} & a_{24} & a_{25} \\
\hdashline
a_{31} & a_{32} & a_{33} & a_{34} & a_{35} \\
a_{41} & a_{42} & a_{43} & a_{44} & a_{45}
\end{array}\right]
$$

Another example of a partitioned matrix is the augmented matrix $[A \vdots b]$ of a linear system $Ax = b$. Partitioned matrices can be multiplied by multiplying the corresponding submatrices. This idea is illustrated in the following example.

EXAMPLE 9. Consider the partitioned matrices

$$
A = \left[\begin{array}{cc:cc:c}
a_{11} & a_{12} & a_{13} & a_{14} & a_{15} \\
a_{21} & a_{22} & a_{23} & a_{24} & a_{25} \\
\hdashline
a_{31} & a_{32} & a_{33} & a_{34} & a_{35} \\
a_{41} & a_{42} & a_{43} & a_{44} & a_{45}
\end{array}\right] = \begin{bmatrix} A_{11} & A_{12} & A_{13} \\ A_{21} & A_{22} & A_{23} \end{bmatrix}
$$

and

$$
B = \left[\begin{array}{cc:cc}
b_{11} & b_{12} & b_{13} & b_{14} \\
b_{21} & b_{22} & b_{23} & b_{24} \\
\hdashline
b_{31} & b_{32} & b_{33} & b_{34} \\
b_{41} & b_{42} & b_{43} & b_{44} \\
\hdashline
b_{51} & b_{52} & b_{53} & b_{54}
\end{array}\right] = \begin{bmatrix} B_{11} & B_{12} \\ B_{21} & B_{22} \\ B_{31} & B_{32} \end{bmatrix}
$$

We then find, as the reader should verify, that

$$\mathbf{AB} = \left[\begin{array}{c|c} \mathbf{A}_{11}\mathbf{B}_{11} + \mathbf{A}_{12}\mathbf{B}_{21} + \mathbf{A}_{13}\mathbf{B}_{31} & \mathbf{A}_{11}\mathbf{B}_{12} + \mathbf{A}_{12}\mathbf{B}_{22} + \mathbf{A}_{13}\mathbf{B}_{32} \\ \hline \mathbf{A}_{21}\mathbf{B}_{11} + \mathbf{A}_{22}\mathbf{B}_{21} + \mathbf{A}_{23}\mathbf{B}_{31} & \mathbf{A}_{21}\mathbf{B}_{12} + \mathbf{A}_{22}\mathbf{B}_{22} + \mathbf{A}_{23}\mathbf{B}_{32} \end{array}\right]$$

Addition of partitioned matrices is carried out in the obvious manner.

0.1 Exercises

1. If

$$\begin{bmatrix} a+b & c+d \\ c-d & a-b \end{bmatrix} = \begin{bmatrix} 6 & 8 \\ 10 & 2 \end{bmatrix}$$

find a, b, c, and d.

In Exercises 2–4, let

$$\mathbf{A} = \begin{bmatrix} 2 & 3 & 1 \\ 3 & 1 & 2 \end{bmatrix}, \quad \mathbf{B} = \begin{bmatrix} 2 & 0 \\ 3 & 2 \\ 1 & 2 \end{bmatrix}, \quad \mathbf{C} = \begin{bmatrix} 2 & -1 & 3 \\ 4 & 2 & 6 \\ 3 & 2 & 1 \end{bmatrix},$$

$$\mathbf{D} = \begin{bmatrix} 3 & -1 \\ 2 & -3 \end{bmatrix}, \quad \mathbf{E} = \begin{bmatrix} 2 & 3 & -2 \\ 0 & 2 & 5 \\ 1 & 2 & 3 \end{bmatrix} \quad \text{and} \quad \mathbf{F} = \begin{bmatrix} 2 & -3 \\ 4 & 5 \end{bmatrix}$$

Compute, if possible:

2. (a) $\mathbf{C} + \mathbf{E}$
   (b) $\mathbf{AB}$ and $\mathbf{BA}$
   (c) $\mathbf{AB} + \mathbf{DF}$
   (d) $\mathbf{A}(\mathbf{BD})$

3. (a) $\mathbf{A}(\mathbf{B} + \mathbf{D})$
   (b) $3\mathbf{B} - 2\mathbf{F}$
   (c) $\mathbf{A}^T$
   (d) $(\mathbf{C} + \mathbf{E})^T$

4. (a) $(\mathbf{AB})^T$
   (b) $(\mathbf{B}^T + \mathbf{A})\mathbf{C}$
   (c) $\mathbf{A}^T(\mathbf{D} + \mathbf{F})$
   (d) $(2\mathbf{C} - 3\mathbf{E})^T\mathbf{B}$

5. Let

$$\mathbf{A} = \begin{bmatrix} 1 & 3 \\ 2 & 2 \end{bmatrix} \quad \text{and} \quad \mathbf{B} = \begin{bmatrix} 3 & -4 \\ 2 & 5 \end{bmatrix}$$

Show that $\mathbf{AB} \neq \mathbf{BA}$.

6. If

$$\mathbf{A} = \begin{bmatrix} 1 & -2 \\ -1 & 2 \end{bmatrix} \quad \text{and} \quad \mathbf{B} = \begin{bmatrix} 2 & 6 \\ 1 & 3 \end{bmatrix}$$

show that $\mathbf{AB} = \mathbf{0}$.

7. If

$$A = \begin{bmatrix} 5 & 1 \\ 4 & -2 \end{bmatrix}, \quad B = \begin{bmatrix} 1 & -1 \\ -2 & -5 \end{bmatrix}, \quad \text{and} \quad C = \begin{bmatrix} -1 & -4 \\ 2 & 8 \end{bmatrix}$$

show that $AC = BC$.

8. Consider the following linear system:

$$\begin{aligned} 3x \qquad\quad + 2z + 2w &= -8 \\ 2x + 3y + 5z - \quad w &= 4 \\ 3x + 2y + 4z \qquad\quad &= 6 \\ x \qquad\quad + z + w &= -6 \end{aligned}$$

(a) Find the coefficient matrix.
(b) Write the linear system in matrix form.
(c) Find the augmented matrix.

9. If A is an $m \times n$ matrix, show that

$$AI_n = I_m A = A$$

10. Show that $(-1)A = -A$.

11. Consider the matrices

$$A = \begin{bmatrix} 3 & 1 & 2 & -1 & 2 \\ 3 & 2 & 1 & 2 & -1 \\ 3 & 4 & 2 & 1 & 5 \\ 2 & -1 & 2 & 3 & 1 \\ 2 & -1 & 1 & 4 & 2 \\ -1 & 2 & 3 & 4 & 5 \end{bmatrix} \quad \text{and} \quad B = \begin{bmatrix} 1 & 2 & -1 & 3 & 2 \\ 2 & 3 & 2 & 4 & 5 \\ 3 & 4 & 3 & 3 & 2 \\ -1 & 2 & -3 & 1 & 1 \\ 2 & 1 & 2 & 4 & 3 \end{bmatrix}$$

Find AB by partitioning A and B in two different ways.

12. (a) Prove that if A has a row of zeros, then AB has a row of zeros.
    (b) Prove that if B has a column of zeros, then AB has a column of zeros.

13. Show that the jth column of the matrix product AB is equal to the matrix product AB_j, where B_j is the jth column of B.

14. Show that if $Ax = b$ has more than one solution, then it has infinitely many solutions. (*Hint:* If x_1 and x_2 are solutions, consider $x_3 = rx_1 + sx_2$, where $r + s = 1$.)

0.2 GAUSS–JORDAN REDUCTION

The reader has undoubtedly solved linear systems of three equations in three unknowns by the method of elimination of variables. We now discuss a systematic way of eliminating variables which will allow the reader to solve larger systems of equations. It is not difficult to see that it is more efficient to carry out the operations on the augmented matrix of the given linear system instead of performing them on the equations of the linear system. Thus, we start with the augmented matrix of the given linear system and transform it to a matrix of a certain special form. This new matrix represents a linear system that has exactly the same solutions as the given system. However, this new linear system can be solved very easily. This method is called **Gauss–Jordan reduction**.

Definition.   An $m \times n$ matrix **A** is said to be in **reduced row echelon form** when it satisfies the following properties:

(a) All rows consisting entirely of zeros, if any, are at the bottom of the matrix.
(b) The first nonzero entry in each row that does not consist entirely of zeros is a 1, called the **leading entry** of its row.
(c) If rows i and $i + 1$ are two successive rows that do not consist entirely of zeros, then the leading entry of row $i + 1$ is to the right of the leading entry of row i.
(d) If a column contains a leading entry of some row, then all other entries in that column are zero.

   Notice that a matrix in reduced row echelon form might not have any rows that consist entirely of zeros.

EXAMPLE 1.   The following matrices are in reduced row echelon form:

$$
\begin{bmatrix} 1 & 0 & 0 & 2 \\ 0 & 1 & 0 & -5 \\ 0 & 0 & 1 & 3 \end{bmatrix}
\quad
\begin{bmatrix} 1 & 0 & 0 & 3 & 0 & 1 \\ 0 & 0 & 1 & 2 & 0 & 2 \\ 0 & 0 & 0 & 0 & 1 & 3 \end{bmatrix}
\quad
\begin{bmatrix} 1 & 0 & 3 & 0 & 4 \\ 0 & 1 & 2 & 0 & 1 \\ 0 & 0 & 0 & 1 & 2 \\ 0 & 0 & 0 & 0 & 0 \\ 0 & 0 & 0 & 0 & 0 \end{bmatrix}
$$

EXAMPLE 2.   The following matrices are not in reduced row echelon form. (Why not?)

$$
\begin{bmatrix} 1 & 3 & 0 & 3 \\ 0 & 0 & 0 & 0 \\ 0 & 0 & 1 & 2 \end{bmatrix}
\quad
\begin{bmatrix} 1 & 0 & -2 & 1 \\ 0 & 4 & 2 & 2 \\ 0 & 0 & 1 & 3 \end{bmatrix}
$$

$$
\begin{bmatrix} 1 & 0 & 2 & 1 \\ 0 & 1 & 3 & 2 \\ 0 & 1 & 2 & 3 \\ 0 & 0 & 0 & 0 \end{bmatrix}
\quad
\begin{bmatrix} 1 & 2 & 5 & -2 \\ 0 & 1 & 3 & 2 \\ 0 & 0 & 1 & 2 \\ 0 & 0 & 0 & 0 \end{bmatrix}
$$

   We now define three operations on the rows of a matrix which can be used to transform it to reduced row echelon form.

Definition.   An **elementary row operation** on an $m \times n$ matrix $\mathbf{A} = [a_{ij}]$ is one of the following operations:

(a) Interchange two rows of $\mathbf{A}$.
(b) Multiply a row of $\mathbf{A}$ by a nonzero constant.
(c) Add a multiple of one row of $\mathbf{A}$ to another row of $\mathbf{A}$.

EXAMPLE 3.   Let

$$\mathbf{A} = \begin{bmatrix} 1 & 2 & 0 & 3 \\ 3 & -2 & 1 & 5 \\ 4 & 2 & 3 & -4 \end{bmatrix}$$

If we interchange the first and third rows of $\mathbf{A}$, we obtain

$$\mathbf{B} = \begin{bmatrix} 4 & 2 & 3 & -4 \\ 3 & -2 & 1 & 5 \\ 1 & 2 & 0 & 3 \end{bmatrix}$$

If we multiply the third row of $\mathbf{A}$ by -2, we obtain

$$\mathbf{C} = \begin{bmatrix} 1 & 2 & 0 & 3 \\ 3 & -2 & 1 & 5 \\ -8 & -4 & -6 & 8 \end{bmatrix}$$

If we add (-3) times the first row of $\mathbf{A}$ to the second row of $\mathbf{A}$, we obtain

$$\mathbf{D} = \begin{bmatrix} 1 & 2 & 0 & 3 \\ 0 & -8 & 1 & -4 \\ 4 & 2 & 3 & -4 \end{bmatrix}$$

Theorem 0.1.   Every $m \times n$ matrix can be transformed to reduced row echelon form by a finite sequence of elementary row operations.

We omit the proof of this theorem and illustrate the method with the following example.

EXAMPLE 4.   Let

$$\mathbf{A} = \begin{bmatrix} 0 & 2 & 5 & -2 & 1 \\ 0 & 0 & 2 & 1 & 3 \\ 2 & -4 & -7 & 8 & -7 \\ 2 & 0 & 3 & 4 & -5 \end{bmatrix}$$

Step 1. Find the first column in **A** which does not consist entirely of zeros; this column is called the **pivotal column**.

$$\mathbf{A} = \begin{bmatrix} 0 & 2 & 5 & -2 & 1 \\ 0 & 0 & 2 & 1 & 3 \\ 2 & -4 & -7 & 8 & -7 \\ 2 & 0 & 3 & 4 & -5 \end{bmatrix}$$

$\uparrow$
pivotal column
of **A**

Step 2. Find the first nonzero entry in the pivotal column. This element, called the **pivot**, is circled.

$$\mathbf{A} = \begin{bmatrix} 0 & 2 & 5 & -2 & 1 \\ 0 & 0 & 2 & 1 & 3 \\ ② & -4 & -7 & 8 & -7 \\ 2 & 0 & 3 & 4 & -5 \end{bmatrix}$$

Step 3. Interchange, if necessary, the first row of **A** with the row where the pivot is located and call the new matrix $\mathbf{A}_1$. Thus, the pivot now has been moved to position (1, 1) in $\mathbf{A}_1$:

$$\mathbf{A}_1 = \begin{bmatrix} 2 & -4 & -7 & 8 & -7 \\ 0 & 0 & 2 & 1 & 3 \\ 0 & 2 & 5 & -2 & 1 \\ 2 & 0 & 3 & 4 & -5 \end{bmatrix}$$

Step 4. Divide the first row of $\mathbf{A}_1$ by the entry in position (1, 1). That is, divide the first row of $\mathbf{A}_1$ by the pivot. The matrix thus obtained is denoted by $\mathbf{A}_2$:

$$\mathbf{A}_2 = \begin{bmatrix} 1 & -2 & -\tfrac{7}{2} & 4 & -\tfrac{7}{2} \\ 0 & 0 & 2 & 1 & 3 \\ 0 & 2 & 5 & -2 & 1 \\ 2 & 0 & 3 & 4 & -5 \end{bmatrix}$$

Step 5. Add suitable multiples of the first row of $\mathbf{A}_2$ to all its other rows so that all entries in the pivotal column, except for the entry where the pivot was located, become zero. Thus, all entries in the pivotal column and rows 2, 3, . . . , m are zero. Call the new matrix $\mathbf{A}_3$:

$$\mathbf{A}_3 = \begin{bmatrix} 1 & -2 & -\tfrac{7}{2} & 4 & -\tfrac{7}{2} \\ 0 & 0 & 2 & 1 & 3 \\ 0 & 2 & 5 & -2 & 1 \\ 0 & 4 & 10 & -4 & 2 \end{bmatrix}$$

-2 times the first row of $\mathbf{A}_2$ was added to its fourth row.

Step 6. Ignore, but do not erase, the first row of $\mathbf{A}_3$ and denote the resulting $(m-1) \times n$ matrix by $\mathbf{B}$. Now repeat Steps 1–5 on $\mathbf{B}$:

$$
\mathbf{B} = \begin{array}{ccccc} 1 & -2 & -\tfrac{7}{2} & 4 & -\tfrac{7}{2} \\ \left[\begin{array}{ccccc} 0 & 0 & 2 & 1 & 3 \\ 0 & ② & 5 & -2 & 1 \\ 0 & 4 & 10 & -4 & 2 \end{array}\right] \end{array}
$$

$$\uparrow$$
pivotal column
of $\mathbf{B}$

$$
\mathbf{B}_1 = \begin{array}{ccccc} 1 & -2 & -\tfrac{7}{2} & 4 & -\tfrac{7}{2} \\ \left[\begin{array}{ccccc} 0 & 2 & 5 & -2 & 1 \\ 0 & 0 & 2 & 1 & 3 \\ 0 & 4 & 10 & -4 & 2 \end{array}\right] \end{array}
$$

The first and second rows of $\mathbf{B}$ were interchanged.

$$
\mathbf{B}_2 = \begin{array}{ccccc} 1 & -2 & -\tfrac{7}{2} & 4 & -\tfrac{7}{2} \\ \left[\begin{array}{ccccc} 0 & 1 & \tfrac{5}{2} & -1 & \tfrac{1}{2} \\ 0 & 0 & 2 & 1 & 3 \\ 0 & 4 & 10 & -4 & 2 \end{array}\right] \end{array}
$$

The first row of $\mathbf{B}_1$ was divided by 2.

$$
\mathbf{B}_3 = \begin{array}{ccccc} 1 & -2 & -\tfrac{7}{2} & 4 & -\tfrac{7}{2} \\ \left[\begin{array}{ccccc} 0 & 1 & \tfrac{5}{2} & -1 & \tfrac{1}{2} \\ 0 & 0 & 2 & 1 & 3 \\ 0 & 0 & 0 & 0 & 0 \end{array}\right] \end{array}
$$

-4 times the first row of $\mathbf{B}_2$ was added to its third row.

Step 7. Add multiples of the first row of $\mathbf{B}_3$ to all the rows of $\mathbf{A}_3$ above the first row of $\mathbf{B}_3$ so that all the entries in the pivotal column, except for the entry where the pivot was located, become zero:

$$
\mathbf{B}_3 = \begin{array}{ccccc} 1 & 0 & \tfrac{3}{2} & 2 & -\tfrac{5}{2} \\ \left[\begin{array}{ccccc} 0 & 1 & \tfrac{5}{2} & -1 & \tfrac{1}{2} \\ 0 & 0 & 2 & 1 & 3 \\ 0 & 0 & 0 & 0 & 0 \end{array}\right] \end{array}
$$

2 times the first row of $\mathbf{B}_3$ was added to the shaded row.

Step 8. Ignore, but do not erase, the first row of $\mathbf{B}_3$ and denote the resulting $(m - 2) \times n$ matrix by $\mathbf{C}$. Repeat Steps 1–7 on $\mathbf{C}$.

$$\mathbf{C} = \begin{bmatrix} 1 & 0 & \frac{3}{2} & 2 & -\frac{5}{2} \\ 0 & 1 & \frac{5}{2} & -1 & \frac{1}{2} \\ 0 & 0 & ② & 1 & 3 \\ 0 & 0 & 0 & 0 & 0 \end{bmatrix}$$

$$\uparrow$$
pivotal column
of $\mathbf{C}$

$$\mathbf{C}_1 = \mathbf{C}_2 = \mathbf{C}_3 = \begin{bmatrix} 1 & 0 & \frac{3}{2} & 2 & -\frac{5}{2} \\ 0 & 1 & \frac{5}{2} & -1 & \frac{1}{2} \\ 0 & 0 & 1 & \frac{1}{2} & \frac{3}{2} \\ 0 & 0 & 0 & 0 & 0 \end{bmatrix}$$

No rows of $\mathbf{C}$ had to be interchanged. The first row of $\mathbf{C}$ was divided by 2. No multiple of the first row of $\mathbf{C}$ had to be added to its second row.

$$\mathbf{C}_3 = \begin{bmatrix} 1 & 0 & 0 & \frac{5}{4} & -\frac{19}{4} \\ 0 & 1 & 0 & -\frac{9}{4} & -\frac{13}{4} \\ 0 & 0 & 1 & \frac{1}{2} & \frac{3}{2} \\ 0 & 0 & 0 & 0 & 0 \end{bmatrix}$$

$-\frac{3}{2}$ times the first row of $\mathbf{C}_3$ was added to the first shaded row; $-\frac{5}{2}$ times the first row of $\mathbf{C}_3$ was added to the second shaded row.

The final matrix

$$\begin{bmatrix} 1 & 0 & 0 & \frac{5}{4} & -\frac{19}{4} \\ 0 & 1 & 0 & -\frac{9}{4} & -\frac{13}{4} \\ 0 & 0 & 1 & \frac{1}{2} & \frac{3}{2} \\ 0 & 0 & 0 & 0 & 0 \end{bmatrix}$$

is in reduced row echelon form.

The **Gauss–Jordan reduction** for solving a given linear system consists of transforming the augmented matrix of the linear system to a matrix in reduced row echelon form. This latter matrix is the augmented matrix of a linear system whose solutions are exactly the same as those of the given linear system. The new linear system can be solved very easily. We illustrate the method with several examples.

EXAMPLE 5. Consider the linear system

$$\begin{aligned} x + 3y + 2z &= 5 \\ 3x + y - z &= -8 \\ 2x + 2y + 3z &= 1 \end{aligned}$$

The augmented matrix of this linear system can be transformed to the following matrix in reduced row echelon form (verify):

$$\left[\begin{array}{ccc|c} 1 & 0 & 0 & -3 \\ 0 & 1 & 0 & 2 \\ 0 & 0 & 1 & 1 \end{array} \right]$$

which represents the linear system

$$x \qquad\qquad = -3$$
$$y \qquad = 2$$
$$z = 1$$

Thus, the unique solution to the given linear system is

$$x = -3$$
$$y = 2$$
$$z = 1$$

EXAMPLE 6.   Consider the linear system

$$x + \ y + 2z + 3w = 13$$
$$x - 2y + \ z + \ w = 8$$
$$3x + \ y + \ z - \ w = 1$$

The augmented matrix of this linear system can be transformed to the following matrix in reduced row echelon form (verify):

$$\begin{bmatrix} 1 & 0 & 0 & -1 & -2 \\ 0 & 1 & 0 & 0 & -1 \\ 0 & 0 & 1 & 2 & 8 \\ 0 & 0 & 0 & 0 & 0 \end{bmatrix}$$

which represents the linear system

$$x \qquad\qquad -w = -2$$
$$y \qquad\qquad = -1$$
$$z + 2w = 8$$

This linear system can be solved, obtaining

$$x = -2 + r$$
$$y = -1$$
$$z = 8 - 2r$$
$$w = r$$

where r is any real number. This solution may be written in matrix form as

$$\begin{bmatrix} x \\ y \\ z \\ w \end{bmatrix} = \begin{bmatrix} -2 \\ -1 \\ 8 \\ 0 \end{bmatrix} + r \begin{bmatrix} 1 \\ 0 \\ -2 \\ 1 \end{bmatrix}$$

The situation in this example is typical of what occurs in linear programming problems in that the number of unknowns generally exceeds the number of variables. As this example shows, there may be infinitely many solutions to such a problem. In linear programming we study how to choose a "best" solution from among these.

EXAMPLE 7.   Consider the linear system

$$x + 2y - 3z = 2$$
$$x + 3y + z = 7$$
$$x + y - 7z = 3$$

The augmented matrix of this linear system can be transformed to the following matrix in reduced row echelon form (verify):

$$\begin{bmatrix} 1 & 0 & -11 & \vdots & 0 \\ 0 & 1 & 4 & \vdots & 0 \\ 0 & 0 & 0 & \vdots & 1 \end{bmatrix}$$

which represents the linear system

$$x \quad - 11z = 0$$
$$y + 4z = 0$$
$$0 = 1$$

Since this last system obviously has no solution, neither does the given system.

The last example is the way in which we recognize that a linear system has no solution. That is, its augmented matrix in reduced row echelon form has a row whose first n entries are zero and whose $(n + 1)$th entry is 1.

0.2   Exercises

In Exercises 1–4 transform the given matrix to a matrix in reduced row echelon form.

1. $\begin{bmatrix} 1 & 2 & 3 & -1 \\ 2 & 3 & 5 & 2 \\ 1 & 4 & 2 & 3 \end{bmatrix}$

2. $\begin{bmatrix} 1 & 3 & -1 & 2 & 1 \\ 1 & 2 & -3 & 2 & 2 \\ 1 & 4 & 1 & 2 & -1 \\ 1 & 5 & 3 & 2 & -1 \end{bmatrix}$

3. $\begin{bmatrix} 0 & 1 & 2 & -1 & 3 \\ 3 & 4 & 1 & 2 & -1 \\ 6 & 9 & 4 & 3 & 1 \\ -3 & -1 & 5 & -5 & 10 \end{bmatrix}$

4. $\begin{bmatrix} 1 & 0 & 2 & 3 & 4 \\ 4 & -4 & 1 & 1 & 3 \\ -2 & 4 & 3 & 5 & 1 \end{bmatrix}$

In Exercises 5–10, find all solutions to the given linear systems.

5. (a)
$$x + y + z = 1$$
$$2x - y + 2z = -4$$
$$3x + y + z = 3$$

(b)
$$x + y + 2z = -1$$
$$2x - y + 3z = -6$$
$$5x + 2y + 9z = 2$$

6. (a)
$$x + y + 2z = -3$$
$$2x + 2y - 5z = 15$$
$$3x + y - z = 10$$
$$2x + y + 2z = 5$$

(b)
$$x + y + 2z + w = 4$$
$$2x - 2y + 3z - 2w = 5$$
$$x + 7y + 3z + 5w = 7$$

7. (a)
$$x + y + 2z + w = 5$$
$$3x + y - 2z + 2w = 8$$
$$-x + y + 6z = 4$$

(b)
$$x + 2y + z - w = -2$$
$$2x + y + z + 2w = 6$$
$$3x + 2y + w = 6$$
$$x + 3y + z + 2w = -1$$
$$2x - 5y + z - 2w = 14$$

8. (a)
$$x + 3y + z + w = 4$$
$$2x + y + 2z - 2w = 8$$
$$x + 8y + z + 5w = 4$$
$$3x - y + 3z - 5w = 12$$

(b)
$$x + 2y + z = -1$$
$$2x + 3y - 5z = -7$$
$$4x + 5y + 7z = 5$$

9. (a)
$$2x + y + 2z + w = 2$$
$$x + 3y - 2z - 3w = -4$$
$$4x + 2y + z = 2$$
$$-2x - 6y + z + 4w = 2$$

(b)
$$x + 2y - z = 5$$
$$4x - y + 4z = 8$$
$$-x + 7y - 7z = 7$$

10. (a)
$$x - y - z = 2$$
$$2x + y + 3z = -5$$
$$3x + 4y - z = 14$$

(b)
$$x + 2y - 3z = 4$$
$$2x + y - 4z = 5$$
$$x + 5y - 5z = 6$$

In Exercises 11 and 12 find all values of a for which the resulting linear system has (1) no solution, (2) a unique solution, and (3) infinitely many solutions.

11.
$$x + y = 3$$
$$x + (a^2 - 8)y = a$$

12.
$$x + 2y - 2z = 4$$
$$-y + 5z = 2$$
$$x + y + (a^2 - 13)z = a + 2$$

13. Let $\mathbf{A}$ be an $n \times n$ matrix in reduced row echelon form. Show that if $\mathbf{A} \neq \mathbf{I}_n$, then $\mathbf{A}$ has a row of zeros.

14. Consider the linear system $\mathbf{Ax} = \mathbf{0}$. Show that if $\mathbf{x}_1$ and $\mathbf{x}_2$ are solutions, then $\mathbf{x} = r\mathbf{x}_1 + s\mathbf{x}_2$ is a solution for any real numbers r and s.

0.3   THE INVERSE OF A MATRIX

In this section we restrict our attention to square matrices.

Definition.   An $n \times n$ matrix A is called **nonsingular** or **invertible** if there exists an $n \times n$ matrix B such that

$$AB = BA = I_n$$

The matrix B is called the **inverse** of A. If no such matrix B exists, then A is called **singular** or **noninvertible**. If the inverse of A exists, we shall write it as A^{-1}. Thus

$$AA^{-1} = A^{-1}A = I_n$$

EXAMPLE 1.   Let

$$A = \begin{bmatrix} 1 & 2 \\ 3 & 4 \end{bmatrix} \quad \text{and} \quad B = \begin{bmatrix} -2 & 1 \\ \frac{3}{2} & -\frac{1}{2} \end{bmatrix}$$

Since

$$AB = BA = I_2$$

it follows that B is the inverse of A or $B = A^{-1}$ and that A is nonsingular.

EXAMPLE 2.   Let

$$A = \begin{bmatrix} 1 & 3 \\ 2 & 6 \end{bmatrix}$$

Does A have an inverse? If it does, let such an inverse be denoted by

$$B = \begin{bmatrix} x & y \\ z & w \end{bmatrix}$$

Then

$$AB = \begin{bmatrix} 1 & 3 \\ 2 & 6 \end{bmatrix} \begin{bmatrix} x & y \\ z & w \end{bmatrix} = I_2 = \begin{bmatrix} 1 & 0 \\ 0 & 1 \end{bmatrix}$$

Thus,

$$\begin{bmatrix} x + 3z & y + 3w \\ 2x + 6z & 2y + 6w \end{bmatrix} = \begin{bmatrix} 1 & 0 \\ 0 & 1 \end{bmatrix}$$

which means that we must have

$$x + 3z = 1$$

$$2x + 6z = 0$$

Since this linear system has no solution, we conclude that A has no inverse.

Theorem 0.2.   (Properties of Inverses)
   (a) If $\mathbf{A}$ is nonsingular, then $\mathbf{A}^{-1}$ is nonsingular and

$$(\mathbf{A}^{-1})^{-1} = \mathbf{A}$$

   (b) If $\mathbf{A}$ and $\mathbf{B}$ are nonsingular, then $\mathbf{AB}$ is nonsingular and

$$(\mathbf{AB})^{-1} = \mathbf{B}^{-1}\mathbf{A}^{-1} \qquad \text{(Note that the order changes.)}$$

   (c) If $\mathbf{A}$ is nonsingular, then $\mathbf{A}^{\mathrm{T}}$ is nonsingular and

$$(\mathbf{A}^{\mathrm{T}})^{-1} = (\mathbf{A}^{-1})^{\mathrm{T}}$$

We now develop a method for calculating $\mathbf{A}^{-1}$. Suppose $\mathbf{A} = [a_{ij}]$ is a given $n \times n$ matrix. We want to find $\mathbf{A}^{-1}$ so that

$$\mathbf{A}\mathbf{A}^{-1} = \mathbf{I}_n$$

Let $\mathbf{A}_1, \mathbf{A}_2, \ldots, \mathbf{A}_n$ denote the columns of $\mathbf{A}^{-1}$, and let $\mathbf{e}_1, \mathbf{e}_2, \ldots, \mathbf{e}_n$ denote the columns of $\mathbf{I}_n$. It follows from Exercise 13 of Section 0.1 that the jth column of $\mathbf{A}\mathbf{A}^{-1}$ is $\mathbf{A}\mathbf{A}_j$. Since equal matrices must agree in their respective columns, we conclude that the problem of finding $\mathbf{A}^{-1}$ can be reduced to the problem of finding n column vectors $\mathbf{A}_1, \mathbf{A}_2, \ldots, \mathbf{A}_n$ such that

$$\mathbf{A}\mathbf{A}_j = \mathbf{e}_j \qquad (1 \le j \le n)$$

Thus, to compute $\mathbf{A}^{-1}$ we need to solve n linear systems, each having n equations in n unknowns. Each of these linear systems can be solved by Gauss–Jordan reduction. Thus, we form the augmented matrices $[\mathbf{A} \mid \mathbf{e}_1]$, $[\mathbf{A} \mid \mathbf{e}_2], \ldots, [\mathbf{A} \mid \mathbf{e}_n]$ and put each one into reduced row echelon form. However, if we notice that the coefficient matrix of each of these linear systems is $\mathbf{A}$, we can solve them all simultaneously. Thus, we form the $n \times 2n$ matrix

$$[\mathbf{A} \mid \mathbf{e}_1 \quad \mathbf{e}_2 \cdots \mathbf{e}_n] = [\mathbf{A} \mid \mathbf{I}_n]$$

and transform it to reduced row echelon form $[\mathbf{C} \mid \mathbf{D}]$. Of course, $\mathbf{C}$ is the reduced row echelon form of $\mathbf{A}$. The matrix $[\mathbf{C} \mid \mathbf{D}]$ represents the n linear systems

$$\mathbf{C}\mathbf{A}_j = \mathbf{d}_j \qquad (1 \le j \le n)$$

where $\mathbf{d}_1, \mathbf{d}_2, \ldots, \mathbf{d}_n$ are the columns of $\mathbf{D}$, or the matrix equation

$$\mathbf{C}\mathbf{A}^{-1} = \mathbf{D}$$

There are now two possibilities:

1. $\mathbf{C}$ has no row consisting entirely of zeros. Then from Exercise 13 in Section 0.2, it follows that $\mathbf{C} = \mathbf{I}_n$, so $\mathbf{D} = \mathbf{A}^{-1}$, and we have obtained the inverse of $\mathbf{A}$.
2. $\mathbf{C}$ has a row consisting entirely of zeros. It can then be shown that $\mathbf{A}$ is singular and has no inverse.

The above method for calculating A^{-1} can be efficiently organized as follows. Form the $n \times 2n$ matrix $[A \mid I_n]$ and perform elementary row operations to transform this matrix to $[I_n \mid A^{-1}]$. Every elementary row operation that is performed on a row of A is also performed on the corresponding row of I_n.

EXAMPLE 3.   Let

$$A = \begin{bmatrix} 1 & 1 & 2 \\ 0 & 2 & 5 \\ 3 & 2 & 2 \end{bmatrix}$$

Then

$$[A \mid I_3] = \begin{bmatrix} 1 & 1 & 2 & 1 & 0 & 0 \\ 0 & 2 & 5 & 0 & 1 & 0 \\ 3 & 2 & 2 & 0 & 0 & 1 \end{bmatrix}$$

Thus,

$$\begin{bmatrix} 1 & 1 & 2 & 1 & 0 & 0 \\ 0 & 2 & 5 & 0 & 1 & 0 \\ 3 & 2 & 2 & 0 & 0 & 1 \end{bmatrix}$$

Add -3 times the first row to the third row, obtaining

$$\begin{bmatrix} 1 & 1 & 2 & 1 & 0 & 0 \\ 0 & 2 & 5 & 0 & 1 & 0 \\ 0 & -1 & -4 & -3 & 0 & 1 \end{bmatrix}$$

Divide the second row by 2, obtaining

$$\begin{bmatrix} 1 & 1 & 2 & 1 & 0 & 0 \\ 0 & 1 & \frac{5}{2} & 0 & \frac{1}{2} & 0 \\ 0 & -1 & -4 & -3 & 0 & 1 \end{bmatrix}$$

Add the second row to the third row, obtaining

$$\begin{bmatrix} 1 & 1 & 2 & 1 & 0 & 0 \\ 0 & 1 & \frac{5}{2} & 0 & \frac{1}{2} & 0 \\ 0 & 0 & -\frac{3}{2} & -3 & \frac{1}{2} & 1 \end{bmatrix}$$

Multiply the third row by $-\frac{2}{3}$, obtaining

$$\begin{bmatrix} 1 & 1 & 2 & 1 & 0 & 0 \\ 0 & 1 & \frac{5}{2} & 0 & \frac{1}{2} & 0 \\ 0 & 0 & 1 & 2 & -\frac{1}{3} & -\frac{2}{3} \end{bmatrix}$$

Add -1 times the second row to the first row, obtaining

$$\begin{bmatrix} 1 & 0 & -\frac{1}{2} & \vdots & 1 & -\frac{1}{2} & 0 \\ 0 & 1 & \frac{5}{2} & \vdots & 0 & \frac{1}{2} & 0 \\ 0 & 0 & 1 & \vdots & 2 & -\frac{1}{3} & -\frac{2}{3} \end{bmatrix}$$

Add $-\frac{5}{2}$ times the third row to the second row, obtaining

$$\begin{bmatrix} 1 & 0 & -\frac{1}{2} & \vdots & 1 & -\frac{1}{2} & 0 \\ 0 & 1 & 0 & \vdots & -5 & \frac{4}{3} & \frac{5}{3} \\ 0 & 0 & 1 & \vdots & 2 & -\frac{1}{3} & -\frac{2}{3} \end{bmatrix}$$

Add $\frac{1}{2}$ times the third row to the first row, obtaining

$$\begin{bmatrix} 1 & 0 & 0 & \vdots & 2 & -\frac{2}{3} & -\frac{1}{3} \\ 0 & 1 & 0 & \vdots & -5 & \frac{4}{3} & \frac{5}{3} \\ 0 & 0 & 1 & \vdots & 2 & -\frac{1}{3} & -\frac{2}{3} \end{bmatrix}$$

Hence,

$$\mathbf{A}^{-1} = \begin{bmatrix} 2 & -\frac{2}{3} & -\frac{1}{3} \\ -5 & \frac{4}{3} & \frac{5}{3} \\ 2 & -\frac{1}{3} & -\frac{2}{3} \end{bmatrix}$$

If the matrix **C** in reduced row echelon form obtained from **A** has a row of zeros, we conclude that **A** is singular. However, observe that once a matrix obtained from **A** by this process has a row of zeros, then the final matrix **C** in reduced row echelon form will also have a row of zeros. Thus, we can stop our calculations as soon as we get a matrix from **A** with a row of zeros, and conclude that $\mathbf{A}^{-1}$ does not exist.

EXAMPLE 4.   Let

$$\mathbf{A} = \begin{bmatrix} 1 & 2 & 1 \\ 2 & 3 & -4 \\ 1 & 3 & 7 \end{bmatrix}$$

To find $\mathbf{A}^{-1}$ we proceed as above:

$$\begin{bmatrix} 1 & 2 & 1 & \vdots & 1 & 0 & 0 \\ 2 & 3 & -4 & \vdots & 0 & 1 & 0 \\ 1 & 3 & 7 & \vdots & 0 & 0 & 1 \end{bmatrix}$$

Add -2 times the first row to the second row, obtanining

$$\begin{bmatrix} 1 & 2 & 1 & \vdots & 1 & 0 & 0 \\ 0 & -1 & -6 & \vdots & -2 & 1 & 0 \\ 1 & 3 & 7 & \vdots & 0 & 0 & 1 \end{bmatrix}$$

Add -1 times the first row to the third row, obtaining

$$\begin{bmatrix} 1 & 2 & 1 & | & 1 & 0 & 0 \\ 0 & -1 & -6 & | & -2 & 1 & 0 \\ 0 & 1 & 6 & | & -1 & 0 & 1 \end{bmatrix}$$ Add the second row to the third row, obtaining

$$\begin{bmatrix} 1 & 2 & 1 & | & 1 & 0 & 0 \\ 0 & -1 & -6 & | & -2 & 1 & 0 \\ 0 & 0 & 0 & | & -3 & 1 & 1 \end{bmatrix}$$

Since we have a matrix with a row of zeros under $\mathbf{A}$, we stop and conclude that $\mathbf{A}$ is singular.

The method just presented for calculating $\mathbf{A}^{-1}$ has the nice feature that we do not have to know, in advance, whether or not $\mathbf{A}^{-1}$ exists. That is, we set out to find $\mathbf{A}^{-1}$, and we either find it or determine that it does not exist.

Suppose now that $\mathbf{A}$ is an $n \times n$ matrix, and consider the linear system

$$\mathbf{A}\mathbf{x} = \mathbf{b} \tag{1}$$

If $\mathbf{A}$ is nonsingular, then $\mathbf{A}^{-1}$ exists. Multiplying both sides of (1) by $\mathbf{A}^{-1}$ on the left, we have

$$\mathbf{A}^{-1}\mathbf{A}\mathbf{x} = \mathbf{A}^{-1}\mathbf{b}$$

or

$$\mathbf{I}_n\mathbf{x} = \mathbf{x} = \mathbf{A}^{-1}\mathbf{b}$$

Thus, if $\mathbf{A}^{-1}$ is known, then the solution to (1) can be simply calculated as $\mathbf{A}^{-1}\mathbf{b}$.

0.3 Exercises

1. If

$$\mathbf{A}^{-1} = \begin{bmatrix} 2 & 1 \\ 3 & 4 \end{bmatrix}$$

find $\mathbf{A}$.

2. If

$$\mathbf{A}^{-1} = \begin{bmatrix} 3 & 2 \\ 1 & 3 \end{bmatrix} \quad \text{and} \quad \mathbf{B}^{-1} = \begin{bmatrix} 2 & 3 \\ 4 & 1 \end{bmatrix}$$

find $(\mathbf{AB})^{-1}$.

In Exercises 3–8 find the inverses of the given matrices, if possible.

3. (a) $\begin{bmatrix} 1 & 3 \\ 4 & 5 \end{bmatrix}$ **(b)** $\begin{bmatrix} 1 & -1 & 2 \\ -3 & -3 & -3 \\ 2 & -2 & 4 \end{bmatrix}$ **(c)** $\begin{bmatrix} 1 & 0 & 1 \\ 1 & -1 & 4 \\ 2 & 3 & 2 \end{bmatrix}$

4. (a) $\begin{bmatrix} 2 & 5 \\ 3 & 2 \end{bmatrix}$ **(b)** $\begin{bmatrix} 2 & 3 & 4 \\ 0 & 1 & 2 \\ -2 & 5 & 1 \end{bmatrix}$ **(c)** $\begin{bmatrix} 2 & 1 & 3 \\ 4 & 6 & 2 \\ -1 & -6 & 4 \end{bmatrix}$

5. (a) $\begin{bmatrix} 1 & 3 \\ -2 & 4 \end{bmatrix}$ **(b)** $\begin{bmatrix} 1 & 2 & 3 \\ 2 & 0 & 1 \\ -1 & 2 & 3 \end{bmatrix}$ **(c)** $\begin{bmatrix} 1 & 5 & 3 \\ 2 & 5 & 1 \\ -1 & 1 & 3 \end{bmatrix}$

6. (a) $\begin{bmatrix} 2 & -2 \\ 4 & 3 \end{bmatrix}$ **(b)** $\begin{bmatrix} 1 & 2 & 1 \\ 1 & 3 & -1 \\ 1 & 2 & 2 \end{bmatrix}$ **(c)** $\begin{bmatrix} 1 & 0 & 2 \\ 2 & 1 & -1 \\ 0 & 4 & 4 \end{bmatrix}$

7. (a) $\begin{bmatrix} 3 & 3 \\ 2 & 2 \end{bmatrix}$ **(b)** $\begin{bmatrix} 2 & 1 & 3 \\ 5 & 8 & 3 \\ -2 & -8 & 4 \end{bmatrix}$ **(c)** $\begin{bmatrix} 1 & 2 & 5 \\ 2 & 4 & 10 \\ -4 & 3 & -9 \end{bmatrix}$

8. (a) $\begin{bmatrix} 2 & 3 \\ 4 & -3 \end{bmatrix}$ **(b)** $\begin{bmatrix} 1 & 2 & 3 \\ 2 & 0 & 0 \\ 6 & 3 & 4 \end{bmatrix}$ **(c)** $\begin{bmatrix} 1 & 7 & 3 \\ 2 & -6 & -2 \\ 5 & 3 & 4 \end{bmatrix}$

9. Consider the matrix

$$\mathbf{A} = \begin{bmatrix} a & b \\ c & d \end{bmatrix}$$

Show that if $ad - bc \neq 0$, then

$$\mathbf{A}^{-1} = \begin{bmatrix} \dfrac{d}{ad-bc} & \dfrac{-b}{ad-bc} \\ \dfrac{-c}{ad-bc} & \dfrac{a}{ad-bc} \end{bmatrix} = \dfrac{1}{ad-bc} \begin{bmatrix} d & -b \\ -c & a \end{bmatrix}$$

0.4 SUBSPACES

Definition. An *n*-vector (or *n*-tuple) is an $n \times 1$ matrix

$$\mathbf{x} = \begin{bmatrix} x_1 \\ x_2 \\ \vdots \\ x_n \end{bmatrix}$$

whose entries are real numbers, called the **components** of **x**.

The set of all *n*-vectors is denoted by R^n, and is called ***n*-space**. When we do not need to specify the value of *n*, we merely refer to an *n*-vector as a **vector**. As a space-saving device, at times some authors write the vector x as $(x_1, x_2, \ldots, x_n)$. However, it is thought of as a column whenever operations, and especially matrix multiplication, are performed on it. Throughout this book the vector x will be written as a column or as $[x_1 \quad x_2 \cdots x_n]^T$.

Since R^n is the set of $n \times 1$ matrices, the operations of addition and scalar multiplication, which we discussed in Section 0.1, are defined in this set.

EXAMPLE 1. Consider the vectors

$$x = \begin{bmatrix} -4 \\ 2 \\ 3 \\ 4 \end{bmatrix} \quad \text{and} \quad y = \begin{bmatrix} 3 \\ -2 \\ 5 \\ 3 \end{bmatrix}$$

in R^4 and let $c = -3$. Then

$$x + y = \begin{bmatrix} -4 + 3 \\ 2 + (-2) \\ 3 + 5 \\ 4 + 3 \end{bmatrix} = \begin{bmatrix} -1 \\ 0 \\ 8 \\ 7 \end{bmatrix}$$

and

$$-3x = \begin{bmatrix} -3(-4) \\ -3(2) \\ -3(3) \\ -3(4) \end{bmatrix} = \begin{bmatrix} 12 \\ -6 \\ -9 \\ -12 \end{bmatrix}$$

The reader will recall that R^3 is the world we live in, and we can thus visualize vectors in R^3. For example, in Figure 0.1 we show the vectors

$$x = \begin{bmatrix} 1 \\ 2 \\ 3 \end{bmatrix}, \quad y = \begin{bmatrix} 1 \\ -\frac{3}{2} \\ 2 \end{bmatrix}, \quad \text{and} \quad z = \begin{bmatrix} 0 \\ 1 \\ 0 \end{bmatrix}$$

We now turn our attention to certain subsets of R^n, which are used in applications.

Definition. A nonempty subset V of R^n is called a **subspace** if the following properties are satisfied:

(a) If x and y are any vectors in V, then $x + y$ is in V.
(b) If r is any real number and x is any vector in V, then rx is in V.

EXAMPLE 2. The simplest example of a subspace of R^n is R^n itself.

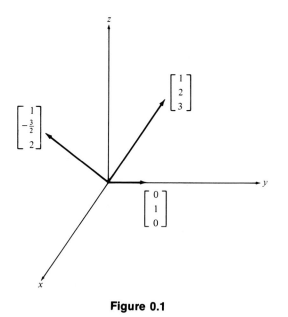

Figure 0.1

Another example is the subset consisting only of the zero vector. These are called the **trivial** subspaces.

EXAMPLE 3. Consider the subset V of R^3 consisting of all vectors of the form

$$\begin{bmatrix} x_1 \\ x_2 \\ x_1 + x_2 \end{bmatrix}$$

Show that V is a subspace.

Solution. Let r be a real number and let

$$\mathbf{x} = \begin{bmatrix} x_1 \\ x_2 \\ x_1 + x_2 \end{bmatrix} \quad \text{and} \quad \mathbf{y} = \begin{bmatrix} y_1 \\ y_2 \\ y_1 + y_2 \end{bmatrix}$$

be vectors in V. Then we must show that $\mathbf{x} + \mathbf{y}$ and $r\mathbf{x}$ are in V. A vector is in V if its third component is the sum of the first two components. We now have

$$\mathbf{x} + \mathbf{y} = \begin{bmatrix} x_1 + y_1 \\ x_2 + y_2 \\ (x_1 + x_2) + (y_1 + y_2) \end{bmatrix} = \begin{bmatrix} x_1 + y_1 \\ x_2 + y_2 \\ (x_1 + y_1) + (x_2 + y_2) \end{bmatrix}$$

so that $\mathbf{x} + \mathbf{y}$ is in V. Also,

$$r\mathbf{x} = \begin{bmatrix} rx_1 \\ rx_2 \\ r(x_1 + x_2) \end{bmatrix} = \begin{bmatrix} rx_1 \\ rx_2 \\ rx_1 + rx_2 \end{bmatrix}$$

is in V. Hence, V is a subspace.

EXAMPLE 4.   Consider the subset V of R^3 consisting of all vectors of the form

$$\begin{bmatrix} x_1 \\ x_2 \\ 1 \end{bmatrix}$$

Is V a subspace of R^3?

Solution.   Let r be a real number and let

$$\mathbf{x} = \begin{bmatrix} x_1 \\ x_2 \\ 1 \end{bmatrix} \quad \text{and} \quad \mathbf{y} = \begin{bmatrix} y_1 \\ y_2 \\ 1 \end{bmatrix}$$

be vectors in V. Then

$$\mathbf{x} + \mathbf{y} = \begin{bmatrix} x_1 + y_1 \\ x_2 + y_2 \\ 1 + 1 \end{bmatrix} = \begin{bmatrix} x_1 + y_1 \\ x_2 + y_2 \\ 2 \end{bmatrix}$$

which is not in V, since its third component is 2, while the third component of every vector in V must be 1. Hence, V is *not* a subspace.

EXAMPLE 5.   Let A be an $m \times n$ matrix and consider the linear system $A\mathbf{x} = \mathbf{0}$. The set V of all solutions to this system is a subset of R^n. We now show that V is a subspace of R^n. Thus, let $\mathbf{x}$ and $\mathbf{y}$ be in V. Then $\mathbf{x}$ and $\mathbf{y}$ are solutions, so that

$$A\mathbf{x} = \mathbf{0} \quad \text{and} \quad A\mathbf{y} = \mathbf{0}$$

We have

$$A(\mathbf{x} + \mathbf{y}) = A\mathbf{x} + A\mathbf{y} = \mathbf{0} + \mathbf{0} = \mathbf{0}$$

which implies that $\mathbf{x} + \mathbf{y}$ is a solution, or $\mathbf{x} + \mathbf{y}$ is in V. Also, if r is any real number, then

$$A(r\mathbf{x}) = r(A\mathbf{x}) = r\mathbf{0} = \mathbf{0}$$

so that $r\mathbf{x}$ is in V. Thus, V is a subspace of R^n.

0.4 Exercises

1. Verify that $\{0\}$ is a subspace of R^n.

2. Which of the following subsets of R^3 are subspaces? The set of all vectors of the form

(a) $\begin{bmatrix} x_1 \\ x_2 \\ x_3 \end{bmatrix}$ where $x_3 = x_1 + 2x_2$

(b) $\begin{bmatrix} x_1 \\ x_2 \\ x_3 \end{bmatrix}$ where $x_3 \neq 0$

(c) $\begin{bmatrix} x_1 \\ x_2 \\ x_1 + 1 \end{bmatrix}$

3. Which of the following subsets of R^4 are subspaces? The set of all vectors of the form

(a) $\begin{bmatrix} x_1 \\ x_2 \\ x_3 \\ x_4 \end{bmatrix}$ where $x_1 = 0$ and $x_3 = x_1 + 2$

(b) $\begin{bmatrix} x_1 \\ x_2 \\ x_3 \\ x_4 \end{bmatrix}$ where $x_1 = 0$ and $x_2 = -2x_3$

(c) $\begin{bmatrix} x_1 \\ x_2 \\ x_3 \\ x_4 \end{bmatrix}$ where $x_1 + x_2 = 0$

4. Which of the following subsets of R^3 are subspaces? The set of all vectors of the form

(a) $\begin{bmatrix} 2a + b \\ b - c \\ a \end{bmatrix}$

(b) $\begin{bmatrix} a \\ b + a \\ a - 3 \end{bmatrix}$

(c) $\begin{bmatrix} a + b + c \\ a - c \\ b + c \end{bmatrix}$

5. Which of the following subsets of R^4 are subspaces? The set of all vectors of the form

(a) $\begin{bmatrix} a + b \\ 0 \\ b - a \\ a + 2b \end{bmatrix}$

(b) $\begin{bmatrix} a + 2b \\ -b \\ 0 \\ a + b \end{bmatrix}$

(c) $\begin{bmatrix} 0 \\ 1 \\ -3a \\ 4a + 3b \end{bmatrix}$

6. Show that $0\mathbf{x} = \mathbf{0}$ for any vector $\mathbf{x}$.

7. Show that $r\mathbf{0} = \mathbf{0}$ for any real number r.

8. Show that $-(-\mathbf{x}) = \mathbf{x}$.

9. Show that if $r\mathbf{x} = \mathbf{0}$, then $r = 0$ or $\mathbf{x} = \mathbf{0}$.

10. Show that $(-1)\mathbf{x} = -\mathbf{x}$.

11. Show that if $\mathbf{x} + \mathbf{y} = \mathbf{x} + \mathbf{z}$, then $\mathbf{y} = \mathbf{z}$.

12. Let $\mathbf{x}$ and $\mathbf{y}$ be fixed vectors in R^n. Show that the set of all vectors $r\mathbf{x} + s\mathbf{y}$, where r and s are any real numbers, is a subspace of R^n.

0.5 LINEAR INDEPENDENCE AND BASIS

In this section we examine the structure of a subspace of R^n.

Definition. A vector $\mathbf{v}$ in R^n is said to be a **linear combination** of the vectors $\mathbf{v}_1, \mathbf{v}_2, \ldots, \mathbf{v}_k$ if it can be written as

$$\mathbf{v} = c_1\mathbf{v}_1 + c_2\mathbf{v}_2 + \cdots + c_k\mathbf{v}_k$$

where $c_1, c_2, \ldots, c_k$ are real numbers.

EXAMPLE 1. Let

$$\mathbf{v} = \begin{bmatrix} 4 \\ 7 \\ 2 \end{bmatrix}, \quad \mathbf{v}_1 = \begin{bmatrix} 1 \\ 2 \\ -1 \end{bmatrix}, \quad \text{and} \quad \mathbf{v}_2 = \begin{bmatrix} 2 \\ 3 \\ 4 \end{bmatrix}$$

The vector $\mathbf{v}$ is a linear combination of $\mathbf{v}_1$ and $\mathbf{v}_2$ if we can find constants c_1 and c_2 such that

$$c_1\mathbf{v}_1 + c_2\mathbf{v}_2 = \mathbf{v}$$

or

$$c_1 \begin{bmatrix} 1 \\ 2 \\ -1 \end{bmatrix} + c_2 \begin{bmatrix} 2 \\ 3 \\ 4 \end{bmatrix} = \begin{bmatrix} 4 \\ 7 \\ 2 \end{bmatrix}$$

which leads to the linear system

$$c_1 + 2c_2 = 4$$

$$2c_1 + 3c_2 = 7$$

$$-c_1 + 4c_2 = 2$$

Using Gauss–Jordan reduction, we obtain the solution $c_1 = 2$ and $c_2 = 1$.

Definition.   Let $S = \{v_1, v_2, \ldots, v_k\}$ be a set of vectors in a subspace V of R^n. The set S **spans** V, or V is **spanned by** S, if every vector in V is a linear combination of the vectors in S.

EXAMPLE 2.   Let $S = \{v_1, v_2, v_3\}$, where

$$v_1 = \begin{bmatrix} 1 \\ 2 \\ 1 \end{bmatrix}, \quad v_2 = \begin{bmatrix} 2 \\ 1 \\ 0 \end{bmatrix}, \quad \text{and} \quad v_3 = \begin{bmatrix} 0 \\ 1 \\ 1 \end{bmatrix}$$

Determine whether S spans R^3.

Solution.   Let

$$x = \begin{bmatrix} x_1 \\ x_2 \\ x_3 \end{bmatrix}$$

be any vector in R^3. The set S spans R^3 if we can find constants $c_1, c_2,$ and c_3 such that

$$c_1 v_1 + c_2 v_2 + c_3 v_3 = x$$

We then have the linear system

$$c_1 + 2c_2 + 0c_3 = x_1$$

$$2c_1 + c_2 + c_3 = x_2$$

$$c_1 + 0c_2 + c_3 = x_3$$

A solution to this linear system can be easily obtained for any choice of $x_1, x_2,$ and x_3:

$$c_1 = -x_1 + 2x_2 - 2x_3, \qquad c_2 = x_1 - x_2 + x_3, \qquad c_3 = x_1 - 2x_2 + 3x_3$$

Thus, S spans R^3.

EXAMPLE 3. Let $S = \{v_1, v_2\}$, where

$$v_1 = \begin{bmatrix} 1 \\ 2 \\ 1 \end{bmatrix} \quad \text{and} \quad v_2 = \begin{bmatrix} 1 \\ 1 \\ 1 \end{bmatrix}$$

Determine whether S spans R^3.

Solution. Let

$$x = \begin{bmatrix} x_1 \\ x_2 \\ x_3 \end{bmatrix}$$

be any vector in R^3. We need to find constants c_1 and c_2 such that

$$c_1 v_1 + c_2 v_2 = x$$

If we transform the augmented matrix of the resulting linear system to reduced row echelon form, we obtain

$$\begin{bmatrix} 1 & 0 & | & x_2 - x_1 \\ 0 & 1 & | & 2x_1 - x_2 \\ 0 & 0 & | & x_3 - x_1 \end{bmatrix}$$

Thus, a solution exists only when $x_3 - x_1 = 0$. Since we must find constants for any choice of x_1, x_2, and x_3, we conclude that S does not span R^3.

Definition. Let $S = \{v_1, v_2, \ldots, v_k\}$ be a set of distinct vectors in a subspace V of R^n. The set S is said to be **linearly dependent** if we can find constants $c_1, c_2, \ldots, c_k$ not all zero such that

$$c_1 v_1 + c_2 v_2 + \cdots + c_k v_k = 0 \tag{1}$$

Otherwise, S is said to be **linearly independent**. That is, S is linearly independent if Equation (1) can be satisfied only with

$$c_1 = c_2 = \cdots = c_k = 0$$

Of course, (1) always holds when

$$c_1 = c_2 = \cdots = c_k = 0$$

The essential point in the above definition is whether the equation can hold with not all of the constants $c_1, c_2, \ldots, c_k$ being zero.

EXAMPLE 4. Consider the vectors

$$v_1 = \begin{bmatrix} 1 \\ 2 \\ 0 \\ 1 \end{bmatrix}, \quad v_2 = \begin{bmatrix} 0 \\ 1 \\ 1 \\ 2 \end{bmatrix}, \quad \text{and} \quad v_3 = \begin{bmatrix} 1 \\ 1 \\ -1 \\ 1 \end{bmatrix}$$

Determine whether $S = \{v_1, v_2, v_3\}$ is linearly dependent or linearly independent.

Solution.   From Equation (1) we have

$$c_1 v_1 + c_2 v_2 + c_3 v_3 = 0$$

which yields the linear system

$$
\begin{aligned}
c_1 \qquad\quad + c_3 &= 0 \\
2c_1 + \quad c_2 + c_3 &= 0 \\
c_2 - c_3 &= 0 \\
c_1 + 2c_2 + c_3 &= 0
\end{aligned}
$$

Since this linear system has only the trivial solution $c_1 = c_2 = c_3 = 0$, we conclude that S is linearly independent.

EXAMPLE 5.   Consider the vectors

$$
v_1 = \begin{bmatrix} 1 \\ -1 \\ -3 \end{bmatrix}, \qquad
v_2 = \begin{bmatrix} 1 \\ 0 \\ -2 \end{bmatrix}, \qquad \text{and} \quad
v_3 = \begin{bmatrix} 1 \\ -2 \\ -4 \end{bmatrix}
$$

Determine whether $S = \{v_1, v_2, v_3\}$ is linearly dependent or linearly independent.

Solution.   From Equation (1) we have

$$c_1 v_1 + c_2 v_2 + c_3 v_3 = 0$$

which yields the linear system

$$
\begin{aligned}
c_1 + \quad c_2 + \quad c_3 &= 0 \\
-c_1 \qquad\quad - 2c_3 &= 0 \\
-3c_1 - 2c_2 - 4c_3 &= 0
\end{aligned}
$$

which has the solution (verify)

$$c_1 = 2, \qquad c_2 = -1, \quad \text{and} \quad c_3 = -1$$

Hence, S is linearly dependent.

Theorem 0.3.   The set $S = \{v_1, v_2, \ldots, v_k\}$ of vectors in R^n is linearly dependent if and only if one of the vectors in S is a linear combination of the other vectors in S.

Proof.   Suppose S is linearly dependent. Then we can write

$$c_1 v_1 + c_2 v_2 + \cdots + c_k v_k = 0$$

where not all the constants $c_1, c_2, \ldots, c_k$ are zero. Suppose $c_j \neq 0$. Then

$$\mathbf{v}_j = -\frac{c_1}{c_j}\mathbf{v}_1 - \frac{c_2}{c_j}\mathbf{v}_2 - \cdots - \frac{c_{j-1}}{c_j}\mathbf{v}_{j-1} - \frac{c_{j+1}}{c_j}\mathbf{v}_{j+1} - \cdots - \frac{c_k}{c_j}v_k$$

Conversely, suppose one of the vectors—say $\mathbf{v}_j$—is a linear combination of the other vectors in S:

$$\mathbf{v}_j = a_1\mathbf{v}_1 + a_2\mathbf{v}_2 + \cdots + a_{j-1}\mathbf{v}_{j-1} + a_{j+1}\mathbf{v}_{j+1} + \cdots + a_k\mathbf{v}_k$$

Then

$$a_1\mathbf{v}_1 + a_2\mathbf{v}_2 + \cdots + a_{j-1}\mathbf{v}_{j-1} - 1\mathbf{v}_j + a_{j+1}\mathbf{v}_{j+1} + \cdots + a_k\mathbf{v}_k = 0 \quad (2)$$

Since at least one of the coefficients in (2) is nonzero, we conclude that S is linearly dependent.

Thus, in Example 5 we found that

$$2\mathbf{v}_1 - \mathbf{v}_2 - \mathbf{v}_3 = 0$$

so that we could solve for any one vector in terms of the other two. However, the theorem does *not* say that in general *every* vector in S is a linear combination of the other vectors in S. For suppose we have

$$5\mathbf{v}_1 + 0\mathbf{v}_2 - 2\mathbf{v}_3 + 5\mathbf{v}_4 = 0$$

Then we cannot solve for $\mathbf{v}_2$ in terms of $\mathbf{v}_1$, $\mathbf{v}_3$, and $\mathbf{v}_4$.

Theorem 0.4. Let A be an $n \times n$ matrix. Then A is nonsingular if and only if the columns of A (as vectors in R^n) form a linearly independent set.

EXAMPLE 6. Let

$$A = \begin{bmatrix} 1 & 2 & 0 \\ -1 & 1 & 1 \\ 2 & 3 & 2 \end{bmatrix}$$

Since A is nonsingular (verify), the columns of A form the linearly independent set

$$\begin{bmatrix} 1 \\ -1 \\ 2 \end{bmatrix}, \quad \begin{bmatrix} 2 \\ 1 \\ 3 \end{bmatrix}, \quad \begin{bmatrix} 0 \\ 1 \\ 2 \end{bmatrix}$$

in R^3.

Definition. A set of vectors $S = \{\mathbf{v}_1, \mathbf{v}_2, \ldots, \mathbf{v}_k\}$ in a subspace V of R^n is called a **basis** for V if S spans V and S is linearly independent.

EXAMPLE 7. The sets

$$S = \left\{ \begin{bmatrix} 1 \\ 0 \\ 0 \end{bmatrix}, \begin{bmatrix} 0 \\ 1 \\ 0 \end{bmatrix}, \begin{bmatrix} 0 \\ 0 \\ 1 \end{bmatrix} \right\} \quad \text{and} \quad T = \left\{ \begin{bmatrix} 1 \\ 0 \\ 1 \end{bmatrix}, \begin{bmatrix} 0 \\ 1 \\ 1 \end{bmatrix}, \begin{bmatrix} 1 \\ 1 \\ 0 \end{bmatrix} \right\}$$

are bases for R^3.

Theorem 0.5. If $S = \{v_1, v_2, \ldots, v_k\}$ is a basis for a subspace V of R^n and $T = \{w_1, w_2, \ldots, w_r\}$ is a linearly independent set of vectors in V, then $r \le k$.

Corollary. If $S = \{v_1, v_2, \ldots, v_k\}$ and $T = \{w_1, w_2, \ldots, w_r\}$ are bases for a subspace V of R^n, then $k = r$.

Definition. The **dimension** of a subspace V of R^n is the number of vectors in a basis for V.

Thus, the dimension of R^n is n.

Theorem 0.6. Let V be a vector space of dimension k and let $S = \{v_1, v_2, \ldots, v_k\}$ be a set of k vectors in V.

(a) If S is linearly independent, then S is a basis for V.
(b) If S spans V, then S is a basis for V.

Theorem 0.7. If $S = \{v_1, v_2, \ldots, v_k\}$ is a basis for a subspace V of R^n, then every vector x in V can be written in one and only one way as a linear combination of the vectors in S.

Proof. Suppose

$$x = a_1 v_1 + a_2 v_2 + \cdots + a_k v_k$$

and also

$$x = b_1 v_1 + b_2 v_2 + \cdots + b_k v_k$$

Subtracting the second expression from the first, we obtain

$$0 = x - x = (a_1 - b_1)v_1 + (a_2 - b_2)v_2 + \cdots + (a_k - b_k)v_k \tag{3}$$

Since S is a basis, it is a linearly independent set of vectors. Equation (3) is the expression which must be tested for linear independence. Therefore, $a_j - b_j = 0$ $(1 \le j \le k)$ or $a_j = b_j$ and the expression for x is unique.

Definition. Suppose that $S = \{v_1, v_2, \ldots, v_k\}$ is a basis for a subspace V of R^n and let x be a vector in V. Then

$$x = c_1 v_1 + c_2 v_2 + \cdots + c_k v_k$$

The vector

$$[\mathbf{x}]_S = \begin{bmatrix} c_1 \\ c_2 \\ \vdots \\ c_k \end{bmatrix}$$

is called the **coordinate vector** of **x** with respect to S.

EXAMPLE 8. Consider the bases

$$S = \left\{ \begin{bmatrix} 1 \\ 0 \\ 0 \end{bmatrix}, \begin{bmatrix} 0 \\ 1 \\ 0 \end{bmatrix}, \begin{bmatrix} 0 \\ 0 \\ 1 \end{bmatrix} \right\} \quad \text{and} \quad T = \left\{ \begin{bmatrix} 1 \\ 0 \\ 1 \end{bmatrix}, \begin{bmatrix} 0 \\ 1 \\ 1 \end{bmatrix}, \begin{bmatrix} 1 \\ 1 \\ 1 \end{bmatrix} \right\}$$

for R^3. Let

$$\mathbf{x} = \begin{bmatrix} 0 \\ 5 \\ 3 \end{bmatrix}$$

Expressing **x** in terms of the given bases, we obtain (verify)

$$\mathbf{x} = 0 \begin{bmatrix} 1 \\ 0 \\ 0 \end{bmatrix} + 5 \begin{bmatrix} 0 \\ 1 \\ 0 \end{bmatrix} + 3 \begin{bmatrix} 0 \\ 0 \\ 1 \end{bmatrix}$$

and

$$\mathbf{x} = -2 \begin{bmatrix} 1 \\ 0 \\ 1 \end{bmatrix} + 3 \begin{bmatrix} 0 \\ 1 \\ 1 \end{bmatrix} + 2 \begin{bmatrix} 1 \\ 1 \\ 1 \end{bmatrix}$$

so that

$$[\mathbf{x}]_S = \begin{bmatrix} 0 \\ 5 \\ 3 \end{bmatrix} \quad \text{and} \quad [\mathbf{x}]_T = \begin{bmatrix} -2 \\ 3 \\ 2 \end{bmatrix}$$

Consider the matrix equation

$$\begin{bmatrix} a_{11} & a_{12} & \dots & a_{1n} \\ a_{21} & a_{22} & \dots & a_{2n} \\ \vdots & \vdots & & \vdots \\ a_{n1} & a_{n2} & \dots & a_{nn} \end{bmatrix} \begin{bmatrix} x_1 \\ x_2 \\ \vdots \\ x_n \end{bmatrix} = \begin{bmatrix} b_1 \\ b_2 \\ \vdots \\ b_n \end{bmatrix} \tag{4}$$

Multiplying out the left-hand side, we obtain

$$\begin{bmatrix} a_{11}x_1 + a_{12}x_2 + \cdots + a_{1n}x_n \\ a_{21}x_1 + a_{22}x_2 + \cdots + a_{2n}x_n \\ \vdots \\ a_{n1}x_1 + a_{n2}x_2 + \cdots + a_{nn}x_n \end{bmatrix} = \begin{bmatrix} b_1 \\ b_2 \\ \vdots \\ b_n \end{bmatrix} \qquad (5)$$

Using the properties of matrix addition and scalar multiplication, we may rewrite (5) as

$$x_1 \begin{bmatrix} a_{11} \\ a_{21} \\ \vdots \\ a_{n1} \end{bmatrix} + x_2 \begin{bmatrix} a_{21} \\ a_{22} \\ \vdots \\ a_{n2} \end{bmatrix} + \cdots + x_n \begin{bmatrix} a_{n1} \\ a_{n2} \\ \vdots \\ a_{nn} \end{bmatrix} = \begin{bmatrix} b_1 \\ b_2 \\ \vdots \\ b_n \end{bmatrix}$$

Thus, if (4) is written in compact form as

$$\mathbf{Ax} = \mathbf{b}$$

and $\mathbf{A}_1, \mathbf{A}_2, \ldots, \mathbf{A}_n$ are the columns of $\mathbf{A}$, then we have shown that (4) can be written as

$$x_1\mathbf{A}_1 + x_2\mathbf{A}_2 + \cdots + x_n\mathbf{A}_n = \mathbf{b}$$

That is, writing the matrix equation $\mathbf{Ax} = \mathbf{b}$ is equivalent to writing $\mathbf{b}$ as a linear combination of the columns of $\mathbf{A}$. Conversely, if $\mathbf{x}$ is a solution to $\mathbf{Ax} = \mathbf{b}$, then $\mathbf{b}$ is a linear combination of the columns of $\mathbf{A}$.

An important application of this discussion occurs in the case when $\mathbf{A}$ is nonsingular. In this case, the columns of $\mathbf{A}$ form a linearly independent set of n vectors in R^n. Thus, this set is a basis for R^n, and $\mathbf{b}$ can be uniquely written as a linear combination of the columns of $\mathbf{A}$. The solution to $\mathbf{Ax} = \mathbf{b}$ gives the coordinate vector of $\mathbf{b}$ with respect to this basis.

0.5   Exercises

1. Let

$$S = \left\{ \begin{bmatrix} -2 \\ 1 \\ 2 \end{bmatrix}, \begin{bmatrix} 0 \\ -1 \\ -2 \end{bmatrix}, \begin{bmatrix} -3 \\ 2 \\ 4 \end{bmatrix} \right\}$$

Which of the following vectors are linear combinations of the vectors in S?

(a) $\begin{bmatrix} 1 \\ -1 \\ -2 \end{bmatrix}$ (b) $\begin{bmatrix} -6 \\ 2 \\ 4 \end{bmatrix}$ (c) $\begin{bmatrix} 3 \\ 2 \\ -1 \end{bmatrix}$ (d) $\begin{bmatrix} 1 \\ 1 \\ 1 \end{bmatrix}$

2. Let

$$S = \left\{ \begin{bmatrix} 1 \\ -2 \\ 3 \\ 4 \end{bmatrix}, \begin{bmatrix} 3 \\ -1 \\ -2 \\ -2 \end{bmatrix}, \begin{bmatrix} 1 \\ 3 \\ -8 \\ -10 \end{bmatrix}, \begin{bmatrix} 0 \\ 1 \\ 1 \\ 0 \end{bmatrix} \right\}$$

Which of the following vectors are linear combinations of the vectors in S?

(a) $\begin{bmatrix} -2 \\ 0 \\ 6 \\ 6 \end{bmatrix}$ (b) $\begin{bmatrix} 0 \\ 0 \\ 0 \\ 1 \end{bmatrix}$ (c) $\begin{bmatrix} 0 \\ 0 \\ 1 \\ 0 \end{bmatrix}$ (d) $\begin{bmatrix} 3 \\ 1 \\ 0 \\ -2 \end{bmatrix}$

3. Which of the following sets of vectors span R^2?

(a) $\left\{ \begin{bmatrix} 2 \\ -1 \end{bmatrix}, \begin{bmatrix} 0 \\ 1 \end{bmatrix} \right\}$ (b) $\left\{ \begin{bmatrix} 1 \\ 0 \end{bmatrix}, \begin{bmatrix} 0 \\ 1 \end{bmatrix}, \begin{bmatrix} 3 \\ 4 \end{bmatrix} \right\}$

(c) $\left\{ \begin{bmatrix} 2 \\ 3 \end{bmatrix} \right\}$ (d) $\left\{ \begin{bmatrix} -1 \\ 3 \end{bmatrix}, \begin{bmatrix} 2 \\ -6 \end{bmatrix} \right\}$

4. Which of the following sets of vectors span R^3?

(a) $\left\{ \begin{bmatrix} 3 \\ 2 \\ 2 \end{bmatrix}, \begin{bmatrix} 0 \\ 1 \\ 0 \end{bmatrix}, \begin{bmatrix} 1 \\ -2 \\ -1 \end{bmatrix} \right\}$ (b) $\left\{ \begin{bmatrix} 1 \\ 1 \\ 0 \end{bmatrix}, \begin{bmatrix} 3 \\ -1 \\ 2 \end{bmatrix} \right\}$

(c) $\left\{ \begin{bmatrix} 1 \\ 0 \\ 1 \end{bmatrix}, \begin{bmatrix} 0 \\ 0 \\ 1 \end{bmatrix}, \begin{bmatrix} 3 \\ 0 \\ 1 \end{bmatrix}, \begin{bmatrix} 1 \\ 1 \\ 1 \end{bmatrix} \right\}$

5. Which of the following sets of vectors in R^3 are linearly dependent? For those that are, express one vector as a linear combination of the rest.

(a) $\left\{ \begin{bmatrix} 3 \\ 2 \\ -1 \end{bmatrix}, \begin{bmatrix} -1 \\ 0 \\ -1 \end{bmatrix}, \begin{bmatrix} 1 \\ 1 \\ 1 \end{bmatrix} \right\}$ (b) $\left\{ \begin{bmatrix} 3 \\ 2 \\ 1 \end{bmatrix}, \begin{bmatrix} 1 \\ 2 \\ -1 \end{bmatrix} \right\}$

(c) $\left\{ \begin{bmatrix} 1 \\ 1 \\ 2 \end{bmatrix}, \begin{bmatrix} 1 \\ 2 \\ -1 \end{bmatrix}, \begin{bmatrix} 3 \\ 4 \\ -2 \end{bmatrix}, \begin{bmatrix} 5 \\ 7 \\ -1 \end{bmatrix} \right\}$ (d) $\left\{ \begin{bmatrix} 2 \\ 1 \\ 3 \end{bmatrix}, \begin{bmatrix} 4 \\ 5 \\ 1 \end{bmatrix}, \begin{bmatrix} 1 \\ 2 \\ -1 \end{bmatrix} \right\}$

6. Follow the directions of Exercise 5 for the following sets of vectors in R^4.

(a) $\left\{ \begin{bmatrix} -1 \\ 2 \\ 4 \\ 3 \end{bmatrix}, \begin{bmatrix} 3 \\ -1 \\ 2 \\ 5 \end{bmatrix}, \begin{bmatrix} -5 \\ 5 \\ 6 \\ 1 \end{bmatrix} \right\}$ (b) $\left\{ \begin{bmatrix} 2 \\ 1 \\ 1 \\ 1 \end{bmatrix}, \begin{bmatrix} 0 \\ 0 \\ 1 \\ 2 \end{bmatrix}, \begin{bmatrix} 8 \\ 6 \\ 4 \\ 6 \end{bmatrix}, \begin{bmatrix} 3 \\ 2 \\ 0 \\ 1 \end{bmatrix} \right\}$

(c) $\left\{ \begin{bmatrix} 3 \\ 1 \\ 2 \\ 2 \end{bmatrix}, \begin{bmatrix} 2 \\ 1 \\ 3 \\ 1 \end{bmatrix}, \begin{bmatrix} 2 \\ 1 \\ 2 \\ 1 \end{bmatrix}, \begin{bmatrix} 1 \\ 1 \\ 1 \\ 1 \end{bmatrix} \right\}$ (d) $\left\{ \begin{bmatrix} 0 \\ 0 \\ 1 \\ 2 \end{bmatrix}, \begin{bmatrix} 1 \\ 1 \\ 1 \\ 1 \end{bmatrix}, \begin{bmatrix} 0 \\ 1 \\ 0 \\ 2 \end{bmatrix} \right\}$

7. Which of the sets in Exercise 3 are bases for R^2?

8. Which of the sets in Exercise 4 are bases for R^3?

9. Which of the sets in Exercise 5 are bases for R^3?

10. Which of the sets in Exercise 6 are bases for R^4?

11. Which of the following sets form a basis for R^3? Express the vector

$$\begin{bmatrix} -3 \\ 0 \\ 1 \end{bmatrix}$$

as a linear combination of the vectors in each set that is a basis.

(a) $\left\{ \begin{bmatrix} 2 \\ 3 \\ 4 \end{bmatrix}, \begin{bmatrix} 4 \\ -2 \\ 1 \end{bmatrix}, \begin{bmatrix} 0 \\ -8 \\ -7 \end{bmatrix} \right\}$ (b) $\left\{ \begin{bmatrix} -3 \\ -1 \\ 2 \end{bmatrix}, \begin{bmatrix} 3 \\ 4 \\ 1 \end{bmatrix}, \begin{bmatrix} -3 \\ -2 \\ 1 \end{bmatrix} \right\}$

(c) $\left\{ \begin{bmatrix} 0 \\ 1 \\ 0 \end{bmatrix}, \begin{bmatrix} 3 \\ 2 \\ 1 \end{bmatrix}, \begin{bmatrix} 1 \\ 1 \\ 1 \end{bmatrix} \right\}$

12. Let

$$S = \left\{ \begin{bmatrix} 2 \\ 3 \end{bmatrix}, \begin{bmatrix} 1 \\ 1 \end{bmatrix} \right\}$$

be a basis for R^2. Find the coordinate vector $[\mathbf{x}]_S$ of the indicated vector $\mathbf{x}$ with respect to S.

(a) $\mathbf{x} = \begin{bmatrix} 3 \\ 4 \end{bmatrix}$ (b) $\mathbf{x} = \begin{bmatrix} -2 \\ 1 \end{bmatrix}$ (c) $\mathbf{x} = \begin{bmatrix} 1 \\ 1 \end{bmatrix}$ (d) $\mathbf{x} = \begin{bmatrix} 1 \\ 2 \end{bmatrix}$

13. Let

$$S = \left\{ \begin{bmatrix} 1 \\ 0 \\ 1 \end{bmatrix}, \begin{bmatrix} 0 \\ 1 \\ 1 \end{bmatrix}, \begin{bmatrix} 1 \\ 0 \\ 0 \end{bmatrix} \right\}$$

be a basis for R^3. Find the coordinate vector $[\mathbf{x}]_S$ of the indicated vector $\mathbf{x}$ with respect to S.

(a) $\mathbf{x} = \begin{bmatrix} 1 \\ 0 \\ 1 \end{bmatrix}$ (b) $\mathbf{x} = \begin{bmatrix} -1 \\ 2 \\ 3 \end{bmatrix}$ (c) $\mathbf{x} = \begin{bmatrix} 2 \\ 1 \\ 1 \end{bmatrix}$ (d) $\mathbf{x} = \begin{bmatrix} 3 \\ 2 \\ 1 \end{bmatrix}$

14. Suppose that S_1 and S_2 are finite sets of vectors in R^n and that S_1 is a subset of S_2. Prove:
(a) If S_1 is linearly dependent, so is S_2.
(b) If S_2 is linearly independent, so is S_1.

15. Show that any set of vectors in R^n which includes the zero vector must be linearly dependent.

16. Show that any set of $n + 1$ vectors in R^n must be linearly dependent.

17. Show that R^n cannot be spanned by a set containing fewer than n vectors.

Further Readings

Kolman, Bernard. *Introductory Linear Algebra with Applications*. Second edition. Macmillan, New York, 1980.

Kolman, Bernard. *Elementary Linear Algebra*. Second edition. Macmillan, New York, 1977.

CHAPTER

012345

GEOMETRY IN R^n

1.1 HYPERPLANES

Recall that the equation of a line in R^2 is

$$ax + by = k \tag{1}$$

If we define the column vectors (or 2×1 matrices)

$$\mathbf{c} = \begin{bmatrix} a \\ b \end{bmatrix} \quad \text{and} \quad \mathbf{x} = \begin{bmatrix} x \\ y \end{bmatrix}$$

then we can write (1) as

$$\mathbf{c}^T\mathbf{x} = k$$

We also recall that the equation of a plane in R^3 is

$$ax + by + cz = k \tag{2}$$

If we define the vectors

$$\mathbf{c} = \begin{bmatrix} a \\ b \\ c \end{bmatrix} \quad \text{and} \quad \mathbf{x} = \begin{bmatrix} x \\ y \\ z \end{bmatrix}$$

then we can write (2) as

$$\mathbf{c}^T\mathbf{x} = k$$

We can define the analogous algebraic notion in R^n as follows.

Definition. Let $\mathbf{c}$ be a given nonzero vector in R^n and let k be a

45

constant. **A hyperplane H in R^n is the set**

$$H = \{x \in R^n \mid c^T x = k\}$$

EXAMPLE 1. If

$$c = \begin{bmatrix} 1 \\ -2 \\ 3 \\ 4 \end{bmatrix}$$

is a vector in R^4, then the set

$$H = \{x \in R^4 \mid c^T x = -2\}$$

which is the set of solutions to the linear equation

$$x - 2y + 3z + 4w = -2 \tag{3}$$

is a hyperplane in R^4.

Of course, a hyperplane in R^2 is merely a line, and a hyperplane in R^3 is a plane. Geometrically, a hyperplane contains all but one of the mutually perpendicular directions. For example, in R^3 there are three mutually perpendicular directions, and any hyperplane contains two of them.

A hyperplane H defined as

$$H = \{x \in R^n \mid c^T x = k\}$$

divides R^n into two subsets:

$$H_1 = \{x \in R^n \mid c^T x \leq k\}$$

and

$$H_2 = \{x \in R^n \mid c^T x \geq k\}$$

The sets H_1 and H_2 are called **closed half-spaces**. We see that $H_1 \cap H_2$ is the original hyperplane H. The space R^n can also be divided into three disjoint parts using H. Let

$$\hat{H}_1 = \{x \in R^n \mid c^T x < k\}$$

and

$$\hat{H}_2 = \{x \in R^n \mid c^T x > k\}$$

Then R^n is the union of $\hat{H}_1$, $\hat{H}_2$, and H. The sets $\hat{H}_1$ and $\hat{H}_2$ are called **open half-spaces**.

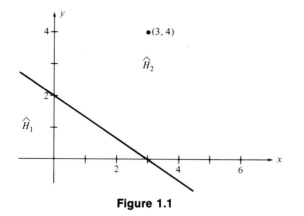

Figure 1.1

EXAMPLE 2. Consider the hyperplane in R^2 given by

$$H = \left\{ \begin{bmatrix} x \\ y \end{bmatrix} \,\middle|\, [2 \quad 3] \begin{bmatrix} x \\ y \end{bmatrix} = 6 \right\}$$

which is a straight line (Figure 1.1).

To find out which set is $\hat{H}_1$ (the other side of H will be $\hat{H}_2$), we select an arbitrary point not on the hyperplane (line in this case) and calculate the value of $c^T x$. Suppose our test point is $(3, 4)$. Then

$$[2 \quad 3] \begin{bmatrix} 3 \\ 4 \end{bmatrix} = 6 + 12 = 18 > 6$$

Thus $\begin{bmatrix} 3 \\ 4 \end{bmatrix} \in \hat{H}_2$ and the region in which $\begin{bmatrix} 3 \\ 4 \end{bmatrix}$ lies on Figure 1.1 is labeled $\hat{H}_2$. Often it is convenient to choose $(0, 0)$ as the test point.

Each of the equations (1), (2), and (3) is a linear equation and defines a hyperplane. Thus, a linear equation in n unknowns is the equation of a hyperplane in R^n. Combining several linear equations each in n unknowns into a system of linear equations defines a collection of hyperplanes. The solution set to m linear equations in n unknowns is the set of points in R^n which lie in the intersection of the m hyperplanes.

EXAMPLE 3. The hyperplanes in R^2 defined by the linear equations

$$x \qquad\quad = 0$$
$$x + \quad y = 2$$
$$x - 2y = -4$$

intersect in the point $(0, 2)$.

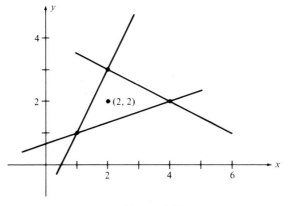

Figure 1.2

EXAMPLE 4. The hyperplanes in R^2 defined by the linear equations

$$x - 3y = -2$$
$$2x - y = 1$$
$$x + 2y = 8$$

do not all intersect. However, they enclose a region in R^2 (Figure 1.2). The point $(2, 2)$ is in this region. Testing this point in each of the equations as in Example 2, we find that

$$2 - 3 \cdot 2 = -4 < -2$$
$$2 \cdot 2 - 2 = \quad 2 > 1$$
$$2 + 2 \cdot 2 = \quad 6 < 8$$

Thus, the region is defined by the inequalities

$$x - 3y \leq -2$$
$$2x - y \geq 1$$
$$x + 2y \leq 8$$

EXAMPLE 5. To sketch the region in R^2 defined by the inequalities

$$x + y \leq 8$$
$$-2x + 3y \leq 9$$

we first sketch the two hyperplanes defined by the corresponding equalities (Figure 1.3a). By checking the test point $(0, 0)$ in each of the inequalities, we determine that the region defined by these inequalities is the shaded region in Figure 1.3b.

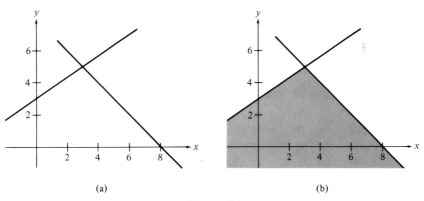

(a)

(b)

Figure 1.3

EXAMPLE 6. Using the same technique as in the previous example, we find the sketch of the region in R^3 defined by the inequalities

$$x \geq 0$$

$$y \geq 0$$

$$z \geq 0$$

$$5x + 3y + 5z \leq 15$$

$$10x + 4y + 5z \leq 20$$

The first three inequalities limit us to the first octant of R^3. The other two inequalities define certain closed half-spaces. The region, which is the intersection of these two half-spaces in the first octant, is shown in Figure 1.4.

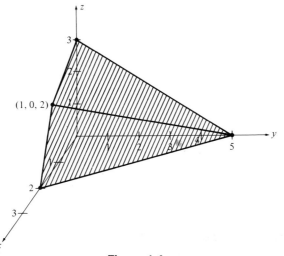

Figure 1.4

1.2 CONVEX SETS

Definition. Let x_1 and x_2 be two distinct points (or vectors) in R^n. The **line** determined by x_1 and x_2 is the set of points

$$\{x \in R^n \mid x = \lambda x_1 + (1 - \lambda)x_2, \quad \lambda \text{ real}\}$$

EXAMPLE 1. Let

$$x_1 = \begin{bmatrix} 2 \\ 3 \end{bmatrix} \quad \text{and} \quad x_2 = \begin{bmatrix} 1 \\ 5 \end{bmatrix}$$

be two points in R^2. Then

$$x = \begin{bmatrix} x \\ y \end{bmatrix}$$

is on the line determined by the given points if and only if

$$\begin{bmatrix} x \\ y \end{bmatrix} = \lambda \begin{bmatrix} 2 \\ 3 \end{bmatrix} + (1 - \lambda)\begin{bmatrix} 1 \\ 5 \end{bmatrix} = \begin{bmatrix} 2\lambda + (1 - \lambda) \\ 3\lambda + 5(1 - \lambda) \end{bmatrix}$$

Thus,

$$x = 2\lambda + (1 - \lambda) = \lambda + 1$$

$$y = 3\lambda + 5(1 - \lambda) = -2\lambda + 5$$

To obtain the usual linear equation for the line, we solve the first equation for λ in terms of x and substitute into the second equation. Thus,

$$y = -2(x - 1) + 5 = -2x + 7$$

or

$$y + 2x = 7$$

The reader can verify this equation by calculating the slope from the two points and using one of the points and the slope to obtain the equation in the usual manner.

The previous definition is useful because it gives a convenient way of specifying parts of the line.

Definition. The **line segment** joining the distinct points x_1 and x_2 of R^n is the set

$$\{x \in R^n \mid x = \lambda x_1 + (1 - \lambda)x_2, \quad 0 \le \lambda \le 1\}$$

Observe that if $\lambda = 0$, we get x_2, and if $\lambda = 1$, we get x_1. The points of the line segment where $0 < \lambda < 1$ are called the **interior points** of the line segment.

We can now introduce the geometric object which will play a crucial role in our subsequent work—a convex set.

Definition. A subset S of R^n is called **convex** if for any two distinct points $\mathbf{x}_1$ and $\mathbf{x}_2$ in S the line segment joining $\mathbf{x}_1$ and $\mathbf{x}_2$ lies in S. That is, S is convex if whenever $\mathbf{x}_1$ and $\mathbf{x}_2 \in S$ so does

$$\mathbf{x} = \lambda\mathbf{x}_1 + (1 - \lambda)\mathbf{x}_2 \quad \text{for} \quad 0 \le \lambda \le 1$$

EXAMPLE 2. The sets in R^2 in Figures 1.5 and 1.6 are convex. The sets in R^2 in Figure 1.7 are not convex.

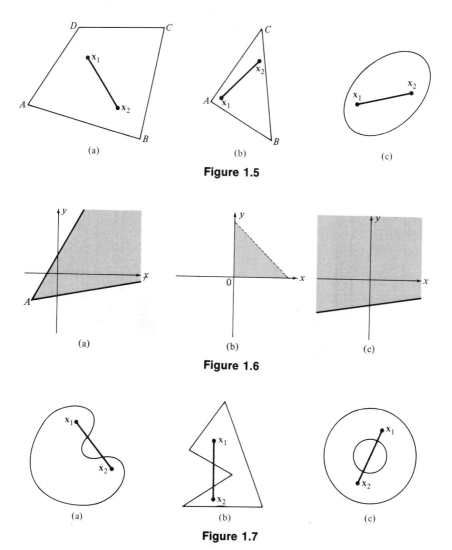

(a)   (b)   (c)

Figure 1.5

(a)   (b)   (c)

Figure 1.6

(a)   (b)   (c)

Figure 1.7

Theorem 1.1. A closed (open) half-space is a convex set.

Proof. We consider only the case where the half-space H_1 is closed and defined by $\mathbf{c}^T\mathbf{x} \leq k$. The remaining cases are left as exercises.

Let $\mathbf{x}_1$ and $\mathbf{x}_2 \in H_1$ and consider $\mathbf{x} = \lambda\mathbf{x}_1 + (1 - \lambda)\mathbf{x}_2$ $(0 \leq \lambda \leq 1)$. Then

$$\mathbf{c}^T\mathbf{x} = \mathbf{c}^T[\lambda\mathbf{x}_1 + (1 - \lambda)\mathbf{x}_2]$$
$$= \lambda\mathbf{c}^T\mathbf{x}_1 + (1 - \lambda)\mathbf{c}^T\mathbf{x}_2$$

Since $\lambda \geq 0$ and $1 - \lambda \geq 0$, we obtain

$$\mathbf{c}^T\mathbf{x} \leq \lambda k + (1 - \lambda)k = k$$

Thus,

$$\mathbf{c}^T\mathbf{x} \leq k$$

so that $\mathbf{x} \in H_1$.

Theorem 1.2. A hyperplane is a convex set.

Proof. Exercise.

Theorem 1.3. The intersection of a finite collection of convex sets is convex.

Proof. Exercise.

Theorem 1.4. Let $\mathbf{A}$ be an $m \times n$ matrix and $\mathbf{b}$ a vector in R^m. The set of solutions to the system of linear equations $\mathbf{Ax} = \mathbf{b}$ is a convex set.

Proof. Exercise.

We now give a sequence of definitions and theorems which will lead to a neat characterization of a certain kind of convex set. The theorem which gives this characterization is called the Krein–Milman theorem (Theorem 1.9). It was proved for more general convex sets by the Russian mathematicians M. Krein and D. P. Milman in 1940.

Definition. A point $\mathbf{x} \in R^n$ is a **convex combination** of the points $\mathbf{x}_1$, $\mathbf{x}_2$, ..., $\mathbf{x}_r$ in R^n if for some real numbers $c_1, c_2, \ldots, c_r$ which satisfy

$$\sum_{i=1}^{r} c_i = 1 \quad \text{and} \quad c_i \geq 0, \quad 1 \leq i \leq r$$

we have

$$\mathbf{x} = \sum_{i=1}^{r} c_i\mathbf{x}_i$$

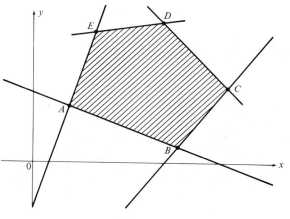

Figure 1.8

Theorem 1.5. The set of all convex combinations of a finite set of points in R^n is a convex set.

Proof. Exercise.

Theorem 1.6. A set S in R^n is convex if and only if every convex combination of a finite number of points of S is in S.

Proof. Exercise.

Definition. A point **u** in a convex set S is called an **extreme point** of S if it is not an interior point of any line segment in S. That is, **u** is an extreme point of S if there are no distinct points $\mathbf{x}_1$, $\mathbf{x}_2$ in S such that

$$\mathbf{u} = \lambda\mathbf{x}_1 + (1 - \lambda)\mathbf{x}_2, \qquad 0 < \lambda < 1$$

EXAMPLE 3. The only extreme points of the convex set in Figure 1.8 are A, B, C, D, and E (verify).

EXAMPLE 4. The extreme points of the convex sets shown in Figures 1.5 and 1.6 are given in the following table:

Figure	Extreme points
1.5a	A, B, C, D
1.5b	A, B, C
1.5c	The entire edge of the ellipse
1.6a	A
1.6b	0
1.6c	None

Theorem 1.7.   Let S be a convex set in R^n. A point $\mathbf{u}$ in S is an extreme point of S if and only if $\mathbf{u}$ is not a convex combination of other points of S.

Proof.   Exercise.

Definition.   The **convex hull** $C(T)$ of a given set T of points in R^n is the set of all convex combinations of finite sets of points of T.

It can be shown that $C(T)$ is the smallest convex set containing T. That is, if S is any convex set containing T, then $C(T) \subseteq S$. This statement is sometimes used as the definition of $C(T)$, the convex hull of T. The proof showing the equivalence of these two descriptions of $C(T)$ is left as an exercise to the reader.

Definition.   The convex hull of a finite set of points is called a **convex polytope.**

EXAMPLE 5.

(a) If T consists of two points (Figure 1.9a), then $C(T)$ is the line segment joining the two points (Figure 1.9b).

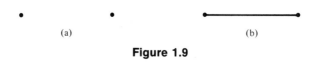

(a)   (b)

Figure 1.9

(b) If T consists of three points (Figure 1.10a), then $C(T)$ is the triangle whose vertices are those three points and its interior (Figure 1.10b). What is $C(T)$ if all three points lie on the same line?

(c) If T is the set of points on the curve in Figure 1.11a, then $C(T)$ is the region bounded by this curve (Figure 1.11b). This region is not a convex polytope since T is an infinite set.

Convex sets are of two types: bounded and unbounded. To define a bounded convex set we first need the concept of a rectangle. A **rectangle**

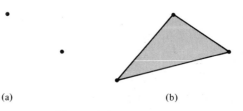

(a)   (b)

Figure 1.10

(a)   (b)

Figure 1.11

in R^n is a set

$$R = \{x \in R^n \mid a_i \le x_i \le b_i\}$$

where $a_i < b_i$, $i = 1, 2, \ldots, n$ are real numbers. A **bounded** convex set is one which can be enclosed in a rectangle in R^n. An **unbounded** convex set cannot be so enclosed. The convex sets in R^2 illustrated in Figure 1.5 are bounded; those in Figure 1.6 are unbounded.

Theorem 1.8. A convex polytope is bounded.

Proof. Let the convex polytope S be the convex hull of the points x_1, $x_2, \ldots, x_r$. These points are contained in a rectangle R, which is necessarily convex. But then $S \subseteq R$, since S is the smallest convex set containing $x_1, x_2, \ldots, x_r$.

A convex set is **closed** if it contains all the points of its boundary. We will not give a precise definition of the boundary of a set in general. However, the boundary of any half-space is the hyperplane which defines it. The boundary of the intersection of a finite number of half-spaces is the union of the segments of the hyperplanes which enclose the intersection. Examples of closed convex sets are a closed half-space, a hyperplane, and the convex hull of a finite number of points.

Theorem 1.9. (Krein–Milman) A closed bounded convex set in R^n is the convex hull of its extreme points.

We will state but not prove the following theorem about convex polytopes. The proof uses the Krein–Milman theorem.

Theorem 1.10. A closed bounded convex set is a convex polytope if and only if it has finitely many extreme points.

Definition. A **convex polyhedron** is the intersection of finitely many closed half-spaces.

Theorem 1.11. A bounded convex polyhedron is a convex polytope.

Theorem 1.12. In a convex polytope, every point is a convex combination of extreme points.

Proof. Exercise.

Convex Functions

A problem of paramount importance in calculus is the determination of the local or relative minima of a function defined over some interval of the real line or over some subset of R^n. In the problems arising in mathematical programming we are not content with a local minimum; we must find the global minimum of a function f defined on a set S contained in R^n. Fortunately, the structure of S and the properties of f are sufficiently nice for a large class of mathematical programming problems, and in these cases a local minimum is, in fact, the global minimum.

Definition. A function f defined on a convex set S in R^n is called a **convex function** if

$$f[\lambda x_1 + (1 - \lambda)x_2] \le \lambda f(x_1) + (1 - \lambda)f(x_2)$$

for $0 \le \lambda \le 1$ and any $x_1, x_2 \in S$. Looking at the graph of f (Figure 1.12), the left-hand side of the inequality represents the graph of f over the line segment in S joining the points x_1 and x_2. The right-hand side represents the line segment joining $[x_1, f(x_1)]$ and $[x_2, f(x_2)]$. Thus, a function f is convex if the line segment joining any two points of the graph of f does not lie below the graph.

Theorem 1.13. If f is a convex function defined on a closed and bounded convex set, then every local minimum of f is a global minimum of f.

Theorem 1.14. Let $f: R^n \to R$ be defined by $f(x) = c^T x$, where c and x are in R^n (f is called **linear**). Then f is a convex function.

Proof. Exercise.

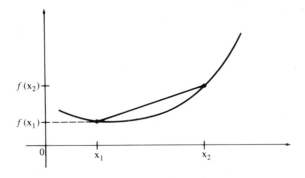

Figure 1.12

1.2 Exercises

1. Let H_2 be the half-space in R^n defined by $\mathbf{c}^T\mathbf{x} \geq k$. Show that H_2 is convex.

2. Show that any open half-space in R^n is convex.

3. Show that a hyperplane H in R^n is convex.

4. Show that the intersection of a finite collection of convex sets is convex.

5. Give two proofs of Theorem 1.4. One proof should use the definition of convex sets and the other should use Theorem 1.3.

6. Prove Theorem 1.5.

7. Prove Theorem 1.6.

8. Prove Theorem 1.7.

9. Sketch the convex set formed by the intersection of the given half-spaces and determine its extreme points.

    (a) $x \geq 0$ (b) $x \geq 0$

               $y \geq 0$ $y \geq 0$

          $x + y \leq 5$ $x - y \leq -2$

        $2x + y \leq 8$ $2x - y \leq 0$

                            $4x - y \leq 12$

    (c) $x \geq 0$ (d) $x \geq 0$

               $y \geq 0$ $y \geq 0$

        $3x + y \leq 6$ $4x + y \geq 8$

       $2x + 3y \geq 4$ $3x + 2y \geq 6$

10. Determine which of the convex sets in Exercise 9 are bounded.

11. Sketch the convex hull of each of the following sets of points:
    (a) $(1, 1), (2, 4), (5, 3)$
    (b) $(0, 0), (4, 0), (0, 3), (1, 1)$
    (c) $(1, 2), (2, 4), (3, 1), (5, 6)$

12. Let $C(T)$ be the convex hull of a set T in R^n. Let S be a convex set which contains T. Show that $C(T) \subseteq S$.

13. Show that a rectangle in R^n is a convex set.

14. Prove Theorem 1.12.

15. Prove Theorem 1.14.

Further Reading

Grünbaum, B. *Convex Polytopes.* Wiley (Interscience), New York, 1967.

CHAPTER

012345

INTRODUCTION TO LINEAR PROGRAMMING

This chapter and the next, which represent the heart of this book, introduce the basic ideas and techniques of linear programming. This area of applied mathematics was developed in the late 1940s to solve a number of resource allocation problems for the federal government. It has become an essential tool in operations research and has been applied to a remarkably varied number of real problems with enormous savings in money and resources. In this chapter we first introduce the linear programming problem and then discuss a simple geometric solution for small problems. The main part of the chapter presents a general algebraic technique for the solution of this problem.

2.1 THE LINEAR PROGRAMMING PROBLEM

We start by giving several examples of linear programming problems.

EXAMPLE 1. (**Activity analysis or product mix**) A lumber mill saws both construction-grade and finish-grade boards from the logs which it receives. Suppose that it takes 2 hours to rough-saw each 1000 board feet of the construction-grade boards and 3 hours to plane each 1000 board feet of these boards. Suppose that it also takes 2 hours to rough-saw each 1000 board feet of the finish-grade boards, but it takes 5 hours to plane each 1000 board feet of these boards. The saw is available 8 hours per day and the plane is available 15 hours per day. If the profit on each 1000 board feet of construction-grade boards is $100 and the profit on each 1000 board feet of finish-grade boards is $120, how many board feet of each type of lumber should be sawed to maximize the profit?

59

Mathematical Model. Let x and y denote the amount of finish-grade and construction-grade lumber, respectively, to be sawed per day. Let the units of x and y be thousands of board feet. The number of hours required daily for the saw is

$$2x + 2y$$

Since only 8 hours are available daily, x and y must satisfy the inequality

$$2x + 2y \leq 8$$

Similarly, the number of hours required for the plane is

$$5x + 3y$$

so x and y must satisfy

$$5x + 3y \leq 15$$

Of course, we must also have

$$x \geq 0 \quad \text{and} \quad y \geq 0$$

The profit (in dollars) to be maximized is given by

$$z = 120x + 100y$$

Thus, our mathematical model is

Find values of x and y which will

$$\text{maximize} \quad z = 120x + 100y$$

subject to the restrictions

$$2x + 2y \leq 8$$
$$5x + 3y \leq 15$$
$$x \geq 0$$
$$y \geq 0$$

EXAMPLE 2. **(The diet problem)** A nutritionist is planning a menu consisting of two main foods A and B. Each ounce of A contains 2 units of fat, 1 unit of carbohydrate, and 4 units of protein. Each ounce of B contains 3 units of fat, 3 units of carbohydrates, and 3 units of protein. The nutritionist wants the meal to provide at least 18 units of fat, at least 12 units of carbohydrate, and at least 24 units of protein. If an ounce of A costs 20 cents and an ounce of B costs 25 cents, how many ounces of each food should be served to minimize the cost of the meal yet satisfy the nutritionist's requirements?

Mathematical Model. Let x and y denote the number of ounces of foods A and B, respectively, which are served. The number of units of fat contained in the meal is

$$2x + 3y$$

so that x and y have to satisfy the inequality

$$2x + 3y \geq 18$$

Similarly, to meet the nutritionist's requirements for carbohydrate and protein, we must have x and y satisfy

$$x + 3y \geq 12$$

and

$$4x + 3y \geq 24$$

Of course, we also require

$$x \geq 0 \quad \text{and} \quad y \geq 0$$

The cost of the meal, which is to be minimized, is

$$z = 20x + 25y$$

Thus, our mathematical model is

Find values of x and y which will

$$\text{minimize} \quad z = 20x + 25y$$

subject to the restrictions

$$2x + 3y \geq 18$$
$$x + 3y \geq 12$$
$$4x + 3y \geq 24$$
$$x \geq 0$$
$$y \geq 0$$

EXAMPLE 3. (The transportation problem) A manufacturer of sheet polyethylene has two plants, one located in Salt Lake City and the other located in Denver. There are three distributing warehouses, one in Los Angeles, another in Chicago, and the third in New York City. The Salt Lake City plant can supply 120 tons of the product per week, whereas the Denver plant can supply 140 tons per week. The Los Angeles warehouse needs 100 tons weekly to meet its demand, the Chicago warehouse needs 60 tons weekly, and the New York City warehouse needs 80 tons weekly. The following table gives the shipping cost (in dollars) per ton of the product:

		To		
		Los Angeles	Chicago	New York City
From	Salt Lake City	5	7	9
	Denver	6	7	10

How many tons of polyethylene should be shipped from each plant to each warehouse to minimize the total shipping cost while meeting the demand?

Mathematical Model. Let P_1 and P_2 denote the plants in Salt Lake City and in Denver, respectively. Let W_1, W_2, and W_3 denote the warehouses in Los Angeles, Chicago, and New York City, respectively. Let

$$x_{ij} = \text{Number of tons shipped from } P_i \text{ to } W_j$$

$$c_{ij} = \text{Cost of shipping 1 ton from } P_i \text{ to } W_j$$

for $i = 1, 2$ and $j = 1, 2, 3$. The total amount of polyethylene sent from P_1 is

$$x_{11} + x_{12} + x_{13}$$

Since P_1 can only supply 120 tons, we must have

$$x_{11} + x_{12} + x_{13} \leq 120$$

Similarly, since P_2 can only supply 140 tons, we must have

$$x_{21} + x_{22} + x_{23} \leq 140$$

The total amount of polyethylene received at W_1 is

$$x_{11} + x_{21}$$

Since the demand at W_1 is 100 tons, we would like to have

$$x_{11} + x_{21} \geq 100$$

Similarly, since the demands at W_2 and W_3 are 60 and 80 tons, respectively, we would like to have

$$x_{12} + x_{22} \geq 60$$

and

$$x_{13} + x_{23} \geq 80$$

Of course, we must also have

$$x_{ij} \geq 0 \quad \text{for} \quad i = 1, 2 \quad \text{and} \quad j = 1, 2, 3$$

The total transportation cost, which we want to minimize, is

$$z = c_{11}x_{11} + c_{12}x_{12} + c_{13}x_{13} + c_{21}x_{21} + c_{22}x_{22} + c_{23}x_{23}$$

Thus, our mathematical model is

Find values of x_{11}, x_{12}, x_{13}, x_{21}, x_{22}, and x_{23} which will

$$\text{minimize} \quad z = \sum_{i=1}^{2} \sum_{j=1}^{3} c_{ij}x_{ij}$$

subject to the restrictions

$$\sum_{j=1}^{3} x_{ij} \leq s_i, \quad i = 1, 2$$

$$\sum_{i=1}^{2} x_{ij} \geq d_j, \quad j = 1, 2, 3$$

$$x_{ij} \geq 0, \quad i = 1, 2 \quad \text{and} \quad j = 1, 2, 3$$

where available supplies are

$$s_1 = 120 \quad \text{and} \quad s_2 = 140$$

and the required demands are

$$d_1 = 100, \quad d_2 = 60, \quad \text{and} \quad d_3 = 80$$

EXAMPLE 4. **(A blending problem)** A manufacturer of artificial sweetener blends 14 kilograms of saccharin and 18 kilograms of dextrose to prepare two new products: SWEET and LO-SUGAR. Each kilogram of SWEET contains 0.4 kilogram of dextrose and 0.2 kilogram of saccharin, while each kilogram of LO-SUGAR contains 0.3 kilogram of dextrose and 0.4 kilogram of saccharin. If the profit on each kilogram of SWEET is 20 cents and the profit on each kilogram of LO-SUGAR is 30 cents, how many kilograms of each product should be made to maximize the profit?

Mathematical Model. Let x and y denote the number of kilograms of SWEET and LO-SUGAR, respectively, being made. The number of kilograms of dextrose being used is

$$.4x + .3y$$

so that we must have

$$.4x + .3y \leq 18$$

Similarly, the number of kilograms of saccharin being used is

$$.2x + .4y$$

so that we must have

$$.2x + .4y \leq 14$$

Of course, we also require

$$x \geq 0 \quad \text{and} \quad y \geq 0$$

The total profit (in cents), which we seek to maximize, is

$$z = 20x + 30y$$

Thus, our mathematical model is

Find values of x and y which will

maximize $\quad z = 20x + 30y$

subject to the restrictions

$$.4x + .3y \leq 18$$
$$.2x + .4y \leq 14$$
$$x \geq 0$$
$$y \geq 0$$

EXAMPLE 5. (**A financial problem**) Suppose that the financial advisor of a university's endowment fund must invest *exactly* $100,000 in two types of securities: bond AAA, paying a dividend of 7%, and stock BB, paying a dividend of 9%. The advisor has been told that no more than $30,000 can be invested in stock BB, while the amount invested in bond AAA must be at least twice the amount invested in stock BB. How much should be invested in each security to maximize the university's return?

Mathematical Model. Let x and y denote the amounts invested in bond AAA and stock BB, respectively. We must then have

$$x + y = 100,000$$
$$x \geq 2y$$
$$y \leq 30,000$$

Of course, we also require

$$x \geq 0 \quad \text{and} \quad y \geq 0$$

The return to the university, which we seek to maximize, is

$$z = .07x + .09y$$

Thus, our mathematical model is

Find values of x and y which will

$$\text{maximize} \quad z = .07x + .09y$$

subject to the restrictions

$$x + y = 100{,}000$$

$$x - 2y \geq 0$$

$$y \leq 30{,}000$$

$$x \geq 0$$

$$y \geq 0$$

Following the form of the previous examples, the **general linear programming problem** can be stated as follows:

Find values of $x_1, x_2, \ldots, x_n$ which will

$$\text{maximize or minimize} \quad z = c_1x_1 + c_2x_2 + \cdots + c_nx_n \tag{1}$$

subject to the restrictions

$$\left. \begin{array}{l} a_{11}x_1 + a_{12}x_2 + \cdots + a_{1n}x_n \leq (\geq)(=) b_1 \\ a_{21}x_1 + a_{22}x_2 + \cdots + a_{2n}x_n \leq (\geq)(=) b_2 \\ \quad \vdots \qquad\quad \vdots \qquad\qquad\quad \vdots \qquad\qquad\quad \vdots \\ a_{m1}x_1 + a_{m2}x_2 + \cdots + a_{mn}x_n \leq (\geq)(=) b_m \end{array} \right\} \tag{2}$$

where in each inequality in (2) one and only one of the symbols $\leq, \geq, =$ occurs. The linear function in (1) is called the **objective function**. The equalities or inequalities in (2) are called **constraints**. Note that the left-hand sides of all the inequalities or equalities in (2) are linear functions of the variables $x_1, x_2, \ldots, x_n$, just as the objective function is. A problem in which not all the constraints or the objective function are linear functions of the variables is a nonlinear programming problem. Such problems are discussed in more advanced texts.

We shall say that a linear programming problem is in **standard form** if it is in the following form:

Find values of $x_1, x_2, \ldots, x_n$ which will

$$\text{maximize} \quad z = c_1x_1 + c_2x_2 + \cdots + c_nx_n \tag{3}$$

subject to the constraints

$$\left.\begin{array}{l} a_{11}x_1 + a_{12}x_2 + \cdots + a_{1n}x_n \le b_1 \\ a_{21}x_1 + a_{22}x_2 + \cdots + a_{2n}x_n \le b_2 \\ \vdots \qquad \vdots \qquad \qquad \vdots \qquad \vdots \\ a_{m1}x_1 + a_{m2}x_2 + \cdots + a_{mn}x_n \le b_m \end{array}\right\} \qquad (4)$$

$$x_j \ge 0, \qquad j = 1, 2, \ldots, n \qquad (5)$$

Examples 1 and 4 are in standard form. The other examples are not. Why?

We shall say that a linear programming problem is in **canonical form** if it is in the following form:

Find values of $x_1, x_2, \ldots, x_s$ which will

$$\text{maximize} \quad z = c_1x_1 + c_2x_2 + \cdots + c_sx_s$$

subject to the constraints

$$\begin{array}{l} a_{11}x_1 + a_{12}x_2 + \cdots + a_{1s}x_s = b_1 \\ a_{21}x_1 + a_{22}x_2 + \cdots + a_{2s}x_s = b_2 \\ \vdots \qquad \vdots \qquad \qquad \vdots \qquad \vdots \\ a_{m1}x_1 + a_{m2}x_2 + \cdots + a_{ms}x_s = b_m \end{array}$$

$$x_j \ge 0, \qquad j = 1, 2, \ldots, s$$

EXAMPLE 6.   The following linear programming problem is in canonical form:

$$\text{Maximize} \quad z = 3x + 2y \quad \text{\textit{not in standard form}}$$

subject to the constraints

$$2x + 4y = 5$$

$$x - 2y = 7$$

$$x \ge 0, \quad y \ge 0$$

Some other authors use different names for what we call standard and canonical linear programming problems. Some also require that all the variables be nonnegative in a linear programming problem. The reader should carefully check the definitions when referring to other books or papers.

EXAMPLE 7.   The following linear programming problems are neither in standard form nor in canonical form. Why?

(a) Minimize $z = 3x + 2y$

subject to the constraints

$$2x + y \le 4$$
$$3x - 2y \le 6$$
$$x \ge 0, \quad y \ge 0$$

(b) Maximize $z = 2x_1 + 3x_2 + 4x_3$

subject to the constraints

$$3x_1 + 2x_2 - 3x_3 \le 4 \quad \text{yes}$$
$$2x_1 + 3x_2 + 2x_3 \le 6 \quad \text{yes}$$
$$3x_1 - x_2 + 2x_3 \ge -8 \quad \text{no (unless multiply it by } -1)$$
$$x_1 \ge 0, \quad x_2 \ge 0, \quad x_3 \ge 0$$

(c) Maximize $z = 3x + 2y$

subject to the constraints

$$2x - 6y = 7$$
$$3x + 2y = 8$$
$$6x + 7y \le 4$$
$$x \ge 0, \quad y \ge 0$$

(d) Minimize $z = 2x + 5y$

subject to the constraints

$$3x + 2y = 4$$
$$4x + 5y = 7$$
$$x \ge 0, \quad y \ge 0$$

(e) Maximize $z = 2x + 5y$

subject to the constraints

$$3x + 2y \le 6$$
$$2x + 9y \le 8$$
$$x \ge 0$$

(f) Minimize $z = 2x_1 + 3x_2 + x_3$

subject to the constraints

$$2x_1 +\ \ x_2 - x_3 = 4$$
$$3x_1 + 2x_2 + x_3 = 8$$
$$x_1 -\ \ x_2\ \ \ \ \ \ = 6$$
$$x_1 \geq 0, \quad x_2 \geq 0$$

We shall now show that every linear programming problem which has unconstrained variables can be solved by solving a corresponding linear programming problem in which all the variables are constrained to be nonnegative. Moreover, we show that every linear programming problem can be formulated as a corresponding standard linear programming problem or as a corresponding canonical linear programming problem. That is, we can show that there is a standard linear programming problem (or canonical linear program problem) whose solution determines a solution to the given arbitrary linear programming problem.

Minimization Problem as a Maximization Problem

Every minimization problem can be viewed as a maximization problem and conversely. This can be seen from the observation that

$$\min \sum_{i=1}^{n} c_i x_i = -\max \left(-\sum_{i=1}^{n} c_i x_i \right)$$

That is, to minimize the objective function we could maximize its negative instead.

Reversing an Inequality

If we multiply the inequality

$$k_1 x_1 + k_2 x_2 + \cdots + k_n x_n \geq b$$

by -1, we obtain the inequality

$$-k_1 x_1 - k_2 x_2 - \cdots - k_n x_n \leq -b$$

EXAMPLE 8. Consider the linear programming problem given in Example 7(b). If we multiply the third constraint

$$3x_1 - x_2 + 2x_3 \geq -8$$

by -1, we obtain the equivalent linear programming problem:

$$\text{Maximize} \quad z = 2x_1 + 3x_2 + 4x_3$$

subject to

$$3x_1 + 2x_2 - 3x_3 \le 4$$
$$2x_1 + 3x_2 + 2x_3 \le 6$$
$$-3x_1 + x_2 - 2x_3 \le 8$$
$$x_1 \ge 0, \quad x_2 \ge 0, \quad x_3 \ge 0$$

which is in standard form.

Changing an Equality to an Inequality

Observe that we can write the equation $x = 6$ as the pair of inequalities $x \le 6$ and $x \ge 6$, and hence as the pair $x \le 6$ and $-x \le -6$.

In the general case the equation

$$\sum_{j=1}^{n} a_{ij}x_j = b_i$$

can be written as the pair of inequalities

$$\sum_{j=1}^{n} a_{ij}x_j \le b_i$$

$$\sum_{j=1}^{n} -a_{ij}x_j \le -b_i$$

EXAMPLE 9.   Consider the linear programming problem given in Example 7(c). It contains the two equality constraints

$$2x - 6y = 7$$
$$3x + 2y = 8$$

These may be written as the equivalent four inequalities

$$2x - 6y \le 7$$
$$2x - 6y \ge 7$$
$$3x + 2y \le 8$$
$$3x + 2y \ge 8$$

Thus, we obtain the equivalent linear programming problem:

$$\text{Maximize} \quad z = 3x + 2y$$

subject to

$$2x - 6y \le 7$$
$$-2x + 6y \le -7$$
$$3x + 2y \le 8$$
$$-3x - 2y \le -8$$
$$6x + 7y \le 4$$
$$x \ge 0, \quad y \ge 0$$

which is in standard form.

We have thus shown that every linear programming problem not in standard form can be transformed into an equivalent linear programming problem which is in standard form.

Unconstrained Variables

The problems in Examples 7(e) and 7(f) have variables that are not constrained to be nonnegative. Suppose that x_j is not constrained to be nonnegative. We replace x_j by two new variables x_j^+ and x_j^-, letting

$$x_j = x_j^+ - x_j^-$$

where $x_j^+ \ge 0$ and $x_j^- \ge 0$. That is, any number is the difference of two nonnegative numbers. In this manner we may introduce constraints on unconstrained variables.

EXAMPLE 10. Consider the problem in Example 7(e). Letting $y = y^+ - y^-$, our problem becomes the following linear programming problem:

$$\text{Maximize} \quad z = 2x + 5y^+ - 5y^-$$

subject to

$$3x + 2y^+ - 2y^- \le 6$$
$$2x + 9y^+ - 9y^- \le 8$$
$$x \ge 0, \quad y^+ \ge 0, \quad y^- \ge 0$$

which is in standard form.

EXAMPLE 11. The problem in Example 7(f) can be converted to a

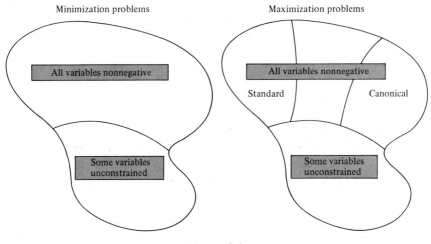

Figure 2.1

maximization problem. We also let $x_3 = x_3^+ - x_3^-$. With these changes we obtain the following linear programming problem:

$$\text{Maximize} \quad z = -2x_1 - 3x_2 - x_3^+ + x_3^-$$

subject to

$$2x_1 + x_2 - x_3^+ + x_3^- = 4$$
$$3x_1 + 2x_2 + x_3^+ - x_3^- = 8$$
$$x_1 - x_2 \qquad\qquad = 6$$
$$x_1 \geq 0, \quad x_2 \geq 0, \quad x_3^+ \geq 0, \quad x_3^- \geq 0$$

which is in canonical form.

A diagrammatic representation of the various types of linear programming problems is given in Figure 2.1.

In Section 2.2 we will show how to convert a linear programming problem in standard form to one in canonical form. Thus, any linear programming problem can be put in either standard form or canonical form.

2.1 Exercises

In Exercises 1–11 set up a linear programming model of the situation described. Determine if the model is in standard form. If it is not, state what must be changed to put the model into standard form.

1. **(Blending problem)** A new rose dust is being prepared by using two available products: PEST and BUG. Each kilogram of PEST contains 30 grams of

carbaryl and 40 grams of Malathion, while each kilogram of BUG contains 40 grams of carbaryl and 20 grams of Malathion. The final blend must contain at least 120 grams of carbaryl and at most 80 grams of Malathion. If each kilogram of PEST costs $3.00 and each kilogram of BUG costs $2.50, how many kilograms of each pesticide should be used to minimize the cost?

2. **(Equipment purchasing problem)** A container manufacturer is considering the purchase of two different types of cardboard-folding machines: model A and model B. Model A can fold 30 boxes per minute and requires 1 attendant, while model B can fold 50 boxes per minute and requires 2 attendants. Suppose the manufacturer must fold at least 320 boxes per minute and cannot afford more than 12 employees for the folding operation. If a model A machine costs $15,000 and a model B machine costs $20,000, how many machines of each type should be bought to minimize the cost?

3. **(Disease treatment problem)** Dr. R. C. McGonigal treats cases of tactutis with a combination of the brand-name compounds Palium and Timade. The Palium costs $.40/pill and the Timade costs $.30/pill. Each compound contains SND plus an activator. The typical dosage requires at least 10 mg of SND per day. Palium contains 4 mg of SND and Timade contains 2 mg of SND. In excessive amounts the activators can be harmful. Consequently Dr. McGonigal limits the total amount of activator to no more than 2 mg per day. Palium and Timade each contain $\frac{1}{2}$ mg of activator per pill. How many of each pill per day should Dr. McGonigal prescribe to minimize the cost of the medication, provide enough SND, and yet not exceed the maximum permissible limit of activator?

4. **(Agricultural problem)** A farmer owns a farm which produces corn, soybeans, and oats. There are 12 acres of land available for cultivation. Each crop which is planted has certain requirements for labor and capital. These data along with the net profit figures are given in the accompanying table.

	Labor (hours)	Capital ($)	Net profit ($)
Corn (per acre)	6	36	40
Soybeans (per acre)	6	24	30
Oats (per acre)	2	18	20

48   360

The farmer has available $360 for capital and knows that there are 48 hours available for working these crops. How much of each crop should be planted to maximize profit?

5. **(Blending problem)** A coffee packer blends Brazilian coffee and Colombian coffee to prepare two products: Super and Deluxe brands. Each kilogram of Super coffee contains 0.5 kilogram of Brazilian coffee and 0.5 kilogram of Colombian coffee, while each kilogram of Deluxe coffee contains 0.25 kilo-

gram of Brazilian coffee and 0.75 kilogram of Colombian coffee. The packer has 120 kilograms of Brazilian coffee and 160 kilograms of Colombian coffee on hand. If the profit on each kilogram of Super coffee is 20 cents and the profit on each kilogram of Deluxe coffee is 30 cents, how many kilograms of each type of coffee should be blended to maximize profit?

6. **(Air pollution problem)** Consider an airshed in which there is one major contributor to air pollution—a cement-manufacturing plant whose annual production capacity is 2,500,000 barrels of cement. Figures are not available to determine whether the plant has been operating at capacity. Although the kilns are equipped with mechanical collectors for air pollution control, the plant still emits 2.0 pounds of dust per barrel of cement produced. There are two types of electrostatic precipitators which can be installed to control dust emission. The four-field type would reduce emissions by 1.5 pounds of dust/barrel and would cost $.14/barrel to operate. The five-field type would reduce emissions by 1.8 pounds of dust/barrel and would cost $.18/barrel to operate. The EPA requires that particulate emissions be reduced by at least 84%. How many barrels of cement should be produced using each new control process to minimize the cost of controls and still meet the EPA requirements? (*Hint:* Let x_1 = fraction of current production to be processed using four-field collectors, and let x_2 = fraction of current production to be processed using five-field collectors.)*

7. **(Mixing problem)** The R. H. Lawn Products Co. has available 80 metric tons of nitrate and 50 metric tons of phosphate to use during the coming week in producing their three types of fertilizer. The mixture ratios and profit figures are given in the accompanying table. Determine how the current inventory should be used to maximize the profit.

	Nitrate	Phosphate	Profit ($/1000 bags)
Regular Lawn (metric tons/1000 bags)	4	2	300
Super Lawn (metric tons/1000 bags)	4	3	500
Garden (metric tons/1000 bags)	2	2	400

8. **(Investment problem)** The administrator of a $200,000 trust fund set up by Mr. Smith's will must adhere to certain guidelines. The total amount of $200,000 need not be fully invested at any one time. The money may be invested in three different types of securities: a utilities stock paying a 9% dividend, an electronics stock paying a 4% dividend, and a bond paying a

---

* Robert E. Kohn, "A Mathematical Programming Model for Air Pollution Control," *School Science and Mathematics,* June 1969.

5% interest. Suppose that the amount invested in the stocks cannot be more than half the total amount invested; the amount invested in the utilities stock cannot exceed $40,000; and the amount invested in the bond must be at least $70,000. What investment policy should be pursued to maximize the return?

9. A book publisher is planning to bind the latest potential bestseller in three different bindings; paperback, book club, and library. Each book goes through a sewing and gluing process. The time required for each process is given in the accompanying table.

	Paperback	Book Club	Library
Sewing (in minutes)	2	2	3
Gluing (in minutes)	4	6	10

Suppose the sewing process is available 7 hours per day and the gluing process 10 hours per day. Assume that the profits are $.50 on a paperback edition, $.80 on a book club edition, and $1.20 on a library edition. How many books will be manufactured in each binding when the profit is maximized?

10. Major oil companies use linear programming to model many phases of their operations. Consider the following simplified version of part of a refinery operation. Two kinds of aviation gasoline, high-octane and low-octane, are made by blending four components from the refinery output. For the low-octane gasoline the components are augmented with a small amount of tetraethyllead (TEL) to give the low-octane ratings shown in the accompanying table. The high-octane gasoline is made from the same components when these have been augmented with a larger amount of TEL giving the high-octane ratings in the table. Assume that the octane rating OR of the mixture is the volumetric average of the octane ratings of the components. That is, letting V denote volume, we have

$$OR_{mix} = \frac{OR_{comp1} \times V_{comp1} + OR_{comp2} \times V_{comp2} + \cdots}{V_{comp1} + V_{comp2} + \cdots}$$

The vapor pressure (a measure of the tendency of the gasoline to evaporate) of both gasolines must be 7. Assume that the vapor pressure of a mixture is the volumetric average of the vapor pressures of the components. Vapor pressure and octane rating are the only two physical properties for which there are constraints.

Data for the components and desired mixtures are given in the accompanying table.

	Vapor pressure	OR High	OR Low	Demand	Supply	Revenue	Cost
Component							
Alkylate	5	108	98		700		7.20
Catalytic cracked	6.5	94	87		600		4.35
Straight-run	4	87	80		900		3.80
Isopentane	18	108	100		500		4.30
Mixture							
High-octane	7	100	—	1300		6.50	
Low-octane	7	—	90	800		7.50	

Assume that the demands must be met exactly. Measure the profit by using revenue less cost. Set up a model of this situation which maximizes this measure of profit.

11. A local health food store packages three types of snack foods—Chewy, Crunchy, and Nutty—by mixing sunflower seeds, raisins, and peanuts. The specifications for each mixture are given in the accompanying table.

Mixture	Sunflower seeds	Raisins	Peanuts	Selling price per kg
Chewy	X_{11}	At least 60% X_{21}	At most 20% X_{31}	$2.00
Crunchy	At least 60% X_{12}	X_{22}	X_{32}	$1.60
Nutty	At most 20% X_{13}	X_{23}	At least 60% X_{33}	$1.20

The suppliers of the ingredients can deliver each week at most 100 kg of sunflower seeds at $1.00/kg, 80 kg of raisins at $1.50/kg, and 60 kg of peanuts at $.80/kg. Determine a mixing scheme that will maximize the store's profit.

2.1 Projects

1. **(Feed problem)** A laboratory needs to supply its research dogs a mixture of commercially available dog foods which meet the National Research Council (NRC) nutrient requirements (Table 2.1). The nutritional composition of each of the eight available foods is given in Table 2.2. Note that these data are given in terms of percentages.

   (a) Set up a constraint for each of the food constituents listed in Table 2.1 based on the NRC requirements.

   (b) An additional constraint must be provided, since the requirements are given in terms of percentages. It must say that the sum of the amounts used is 1. That is, each variable represents a fraction of the total amount to be blended. Write this constraint.

   (c) Set up the objective function using the cost data given in Table 2.2.

Table 2.1 National Research Council Nutrient Requirements

	Food must have at least		Food must have at most	
Protein	20%		Fiber	8%
Fat	5		Moisture	5
Linoleic acid	1.4			
Calcium	1			
Phosphorus	0.8			
Potassium	0.5			
Salt	1			
Magnesium	0.4			
NFE	25			

(d) Consider an arbitrary constraint

$$a_1 x_1 + a_2 x_2 + \cdots + a_8 x_8 \geq b \tag{1}$$

Using the constraint in part b, show that the constraint in (1) is automatically satisfied if for all i, $a_i \geq b$. Similarly, show that (1) is impossible to satisfy if for all i, $a_i < b$. Formulate and prove similar results if the inequality in (1) is reversed.

(e) Using the results in part d, identify the redundant constraints and the impossible constraints in part a. Rewrite the model, eliminating impossible and redundant constraints.

(f) Discuss why $x_1 = x_2 = \cdots = x_7 = 0$, $x_8 = 1$ is an optimal solution.

Table 2.2 Dog Food Constituents (in Percent by Weight)

	Wayne	Wayne TW	Purina Meal	Purina Chow	Purina HP	Gaines	Burgerbits	Agway 2000
Protein	25.0	24.0	27.0	23.8	26.0	21.0	23.0	25.5
Fat	8.0	9.0	10.5	9.4	10.0	8.0	7.0	10.5
Linoleic acid	2.1	1.6	1.6	1.6	1.6	.9	1.5	1.5
Calcium	2.15	1.20	2.50	1.75	1.60	1.0	1.50	1.50
Phosphorus	1.43	1.00	1.40	1.03	1.20	.80	.80	1.70
Potassium	.73	.98	.80	.71	.90	.50	.50	.69
Salt	1.15	1.15	.78	.64	1.10	1.00	1.50	1.00
Magnesium	.170	.220	.290	.270	.150	.036	.050	.230
Fiber	3.5	4.7	4.3	3.7	4.0	5.0	5.0	2.9
NFE	45.77	46.15	41.83	48.10	41.45	51.76	47.15	45.28
Moisture	10.0	10.0	9.0	9.0	12.0	10.0	12.0	9.2
Cost ($/kg)	.17	.17	.17	.16	.21	.20	.17	.16

2. **(Advertising)** The advertising programs of major companies are designed to achieve certain goals in the hope of stimulating sales. There are many media which accept advertising, and the company must decide how to allocate its advertising budget among the different media to achieve the greatest possible benefit. To aid in making this type of decision, there are a number of research firms which collect data concerning audience of each medium. Suppose a car manufacturer, who requires a four-color one-page unit in a weekly magazine, is presented by the research firm with the accompanying table representing readership characteristics and advertising limitations of three different weekly magazines.

	TV Guide	*Newsweek*	*Time*
Cost per four-color one-page unit	$55,000	$35,335	$49,480
Total male readers per unit	19,089,000	11,075,000	10,813,000
Men fifty years or older per unit	4,312,000	2,808,000	2,714,000
Men who are college graduates per unit	2,729,000	3,387,000	3,767,000

The advertising manager has a monthly budget limitation of $200,000 and must decide what amount to spend for each magazine. Since she is worried about the possible duplication of *TV Guide* with her television advertising schedule, she decides to limit *TV Guide* to a maximum of two advertising units. She can use as many as four advertising units per month in each of *Newsweek* and *Time*. Each time a person reads a magazine, it is counted as an exposure to the advertising in the magazine. The advertising manager wants to obtain at least 12,000,000 exposures to men who are college graduates, and since men 50 years and older are not good prospects for her products, she wants to limit the number of exposures to no more than 16,000,000 such men. Set up a linear programming model to determine how many advertising units the advertising manager should buy in each magazine if she wants to keep within her budget and wants to maximize the total number of male readers.

3. **(Construction problem)** A private contractor has five machines available at certain times during the day for a period of one week that are capable of doing excavation work. He wants to determine which combination of machines he should use in order to get the job done the cheapest way. The size of the excavation is 1000 cubic yards of material, and it has to be removed in one week's time.

His machine operators will work at most an 8-hour day, 5 days per week. In the accompanying table is the capacity of each machine, the cycle time for each machine (the time it takes the machine to dig to its capacity and move the excavated material to a truck), the availability, and the cost (which includes wages for the operator).

Machine	Capacity (cu yd)	Rate/hr ($/hr)	Availability (hr/day)	Time needed to excavate 1 unit of capacity (min)
Shovel dozer	2	17.50	6.0	4.25
Large backhoe	$2\frac{1}{2}$	40.00	6.0	1.00
Backhoe A	$1\frac{1}{2}$	27.50	6.0	1.00
Backhoe B	1	22.00	8.0	1.00
Crane with clamshell	$1\frac{1}{2}$	47.00	5.5	2.25

Set up a linear programming problem to determine what combination of machines to use to complete the job at minimum cost. From the data given you will have to compute the number of cubic yards each machine can excavate in 1 hour. Remember to include a constraint which says that the job must be finished. (What would the solution be if this last constraint were not included?)

4. **(Literature search)** Among the many journals which deal with linear programming problems are *Operations Research, Management Science, Naval Logistics Research Quarterly, Mathematics in Operations Research, Operational Research Quarterly,* and the *Journal of the Canadian Operational Research Society*. Write a short report on a paper appearing in one of these journals that describes a real situation. The report should include a description of the situation and a discussion of the assumptions which were made in constructing the model.

2.2 MATRIX NOTATION; GEOMETRIC SOLUTIONS

It is convenient to write linear programming problems in matrix notation. Consider the standard linear programming problem:

$$\text{Maximize} \quad z = c_1 x_1 + c_2 x_2 + \cdots + c_n x_n \tag{1}$$

subject to

$$\left.\begin{array}{l} a_{11}x_1 + a_{12}x_2 + \cdots + a_{1n}x_n \le b_1 \\ a_{21}x_1 + a_{22}x_2 + \cdots + a_{2n}x_n \le b_2 \\ \quad\vdots \qquad\quad \vdots \qquad\qquad\quad \vdots \qquad \vdots \\ a_{m1}x_1 + a_{m2}x_2 + \cdots + a_{mn}x_n \le b_m \end{array}\right\} \tag{2}$$

$$x_j \ge 0, \quad j = 1, 2, \ldots, n \tag{3}$$

Letting

$$
\mathbf{A} = \begin{bmatrix} a_{11} & a_{12} & \dots & a_{1n} \\ a_{21} & a_{22} & \dots & a_{2n} \\ \vdots & \vdots & & \vdots \\ a_{m1} & a_{m2} & \dots & a_{mn} \end{bmatrix}, \qquad \mathbf{x} = \begin{bmatrix} x_1 \\ x_2 \\ \vdots \\ x_n \end{bmatrix},
$$

$$
\mathbf{b} = \begin{bmatrix} b_1 \\ b_2 \\ \vdots \\ b_m \end{bmatrix}, \quad \text{and} \quad \mathbf{c} = \begin{bmatrix} c_1 \\ c_2 \\ \vdots \\ c_n \end{bmatrix}
$$

we can write our given linear programming problem as:

Find a vector $\mathbf{x} \in R^n$ that will

$$
\text{maximize} \quad z = \mathbf{c}^T \mathbf{x} \tag{4}
$$

subject to

$$
\mathbf{A}\mathbf{x} \le \mathbf{b} \tag{5}
$$

$$
\mathbf{x} \ge \mathbf{0} \tag{6}
$$

Here writing $\mathbf{v} \le \mathbf{w}$ for two vectors $\mathbf{v}$ and $\mathbf{w}$ means that each entry of $\mathbf{v}$ is less than or equal to the corresponding entry of $\mathbf{w}$. Specifically, $\mathbf{x} \ge \mathbf{0}$ means that each entry of $\mathbf{x}$ is nonnegative.

EXAMPLE 1. The linear programming problem in Example 1 of Section 2.1 can be written in matrix form as follows:

Find a vector $\mathbf{x}$ in R^2 which will

$$
\text{maximize} \quad z = \begin{bmatrix} 120 & 100 \end{bmatrix} \begin{bmatrix} x \\ y \end{bmatrix}
$$

subject to

$$
\begin{bmatrix} 2 & 2 \\ 5 & 3 \end{bmatrix} \begin{bmatrix} x \\ y \end{bmatrix} \le \begin{bmatrix} 8 \\ 15 \end{bmatrix}
$$

$$
\begin{bmatrix} x \\ y \end{bmatrix} \ge \mathbf{0}
$$

Definition. A vector $\mathbf{x} \in R^n$ satisfying the constraints of a linear programming problem is called a **feasible solution** to the problem. A feasible solution which maximizes or minimizes the objective function of a linear programming problem is called an **optimal solution**.

EXAMPLE 2. Consider the linear programming problem in Example 1. The vectors

$$\mathbf{x}_1 = \begin{bmatrix} 1 \\ 2 \end{bmatrix}, \quad \mathbf{x}_2 = \begin{bmatrix} 2 \\ 1 \end{bmatrix}, \quad \text{and} \quad \mathbf{x}_3 = \begin{bmatrix} 1 \\ 3 \end{bmatrix}$$

are feasible solutions. For example,

$$\begin{bmatrix} 2 & 2 \\ 5 & 3 \end{bmatrix} \begin{bmatrix} 2 \\ 1 \end{bmatrix} = \begin{bmatrix} 6 \\ 13 \end{bmatrix} \leq \begin{bmatrix} 8 \\ 15 \end{bmatrix} \quad \text{and} \quad \begin{bmatrix} 2 \\ 1 \end{bmatrix} \geq \mathbf{0}$$

Therefore, $\mathbf{x}_2$ is a feasible solution. The vectors $\mathbf{x}_1$ and $\mathbf{x}_3$ can be checked in the same manner. Also, the same technique can be used to show that

$$\mathbf{x}_4 = \begin{bmatrix} 3 \\ 1 \end{bmatrix}, \quad \mathbf{x}_5 = \begin{bmatrix} 2 \\ 2 \end{bmatrix}, \quad \text{and} \quad \mathbf{x}_6 = \begin{bmatrix} -2 \\ 3 \end{bmatrix}$$

are not feasible solutions.

Later we will show that

$$\mathbf{x}_0 = \begin{bmatrix} \frac{3}{2} \\ \frac{5}{2} \end{bmatrix}$$

is an optimal solution to the problem.

Theorem 2.1. The set of all feasible solutions to a linear programming problem is a convex polyhedron.

Proof. The inequality constraints of the form

$$a_{i1}x_1 + a_{i2}x_2 + \cdots + a_{in}x_n \; (\leq) \; (\geq) \; b_i \quad \text{or} \quad x_j \geq 0$$

each define a closed half-space. Each equality constraint defines a hyperplane, which is the intersection of two closed half-spaces. Taking all the constraints at once defines a finite intersection of closed half-spaces which is a convex polyhedron.

The next theorem forms the basis for all the algorithms which are designed to solve linear programming problems.

Theorem 2.2. Suppose that the set S of all feasible solutions to a linear programming problem is bounded. Then one of the extreme points is an optimal solution.

Proof. We give the proof for a maximization problem.

Let $\mathbf{u}_1, \mathbf{u}_2, \ldots, \mathbf{u}_r$ be the extreme points of S. Among the values $z_1 = \mathbf{c}^T\mathbf{u}_1, z_2 = \mathbf{c}^T\mathbf{u}_2, \ldots, z_r = \mathbf{c}^T\mathbf{u}_r$, let $z_m = \mathbf{c}^T\mathbf{u}_m$ be the maximum value. (That is, pick the largest of the r numbers $z_1, z_2, \ldots, z_r$.) Let $\mathbf{x}$ be any feasible solution. By Theorem 1.10, the set of all feasible solutions is a convex polytope. Thus, using Theorem 1.12, we see that $\mathbf{x}$ can be written as a convex combination of $\mathbf{u}_1, \mathbf{u}_2, \ldots, \mathbf{u}_r$. Let

$$\mathbf{x} = \sum_{i=1}^{r} \lambda_i \mathbf{u}_i \tag{7}$$

where

$$\sum_{i=1}^{r} \lambda_i = 1 \quad \text{and} \quad \lambda_i \geq 0, \quad i = 1, 2, \ldots, r$$

Multiplying Equation (7) by $\mathbf{c}^T$, we obtain

$$z = \mathbf{c}^T \mathbf{x} = \mathbf{c}^T \left(\sum_{i=1}^{r} \lambda_i \mathbf{u}_i \right) = \sum_{i=1}^{r} \lambda_i (\mathbf{c}^T \mathbf{u}_i) = \sum_{i=1}^{r} \lambda_i z_i \leq \sum_{i=1}^{r} \lambda_i z_m$$

since $z_i \leq z_m$ and $\lambda_i \geq 0$ for $i = 1, 2, \ldots, r$. Therefore, $z \leq z_m$, and we see that the objective function takes on its maximum value at $\mathbf{u}_m$, an extreme point.

In solving a standard linear programming problem we are searching the convex set S of feasible solutions for an optimal solution. In general, S has infinitely many points so that such a search would be very lengthy. However, for many practical problems the convex set S is closed and bounded and has a finite number of extreme points. Thus, it is a convex polytope. Theorem 2.2 says that our search for an optimal solution may be limited to the extreme points of S. It will end in a finite number of steps.

EXAMPLE 3. Consider the linear programming problem of Example 1 of Section 2.1:

$$\text{Maximize} \quad z = 120x + 100y$$

subject to

$$2x + 2y \leq 8$$
$$5x + 3y \leq 15$$
$$x \geq 0, \quad y \geq 0$$

The convex set of all feasible solutions is shown as the shaded region in Figure 2.2.

The extreme points of the convex set S are $(0, 0)$, $(3, 0)$, $(0, 4)$ and $(\frac{3}{2}, \frac{5}{2})$. Since S is closed and bounded, the objective function attains its maximum at an extreme point of S (Theorem 2.2). We can find which extreme point is the optimal solution by evaluating the objective function at each extreme point. This evaluation is shown in Table 2.3. The maximum value of z occurs at the extreme point $(\frac{3}{2}, \frac{5}{2})$. Thus, an optimal

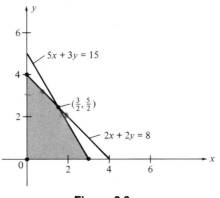

Figure 2.2

Table 2.3

Extreme point (x, y)	Value of objective function $z = 120x + 100y$
$(0, 0)$	0
$(3, 0)$	360
$(0, 4)$	400
$(\frac{3}{2}, \frac{5}{2})$	430

solution is $x = \frac{3}{2}$ and $y = \frac{5}{2}$. In terms of the model this means that the lumber mill should saw 1500 board feet of finish-grade lumber and 2500 board feet of construction-grade lumber per day. These amounts will yield the maximum profit of \$430 per day.

Some linear programming problems have no solution.

EXAMPLE 4. Consider the linear programming problem: *no solution*

$$\text{Maximize} \quad z = 2x + 5y$$

subject to

$$2x + 3y \geq 12$$

$$3x + 4y \leq 12$$

$$x \geq 0, \quad y \geq 0$$

The convex set of all feasible solutions consists of the points which lie in all four half-spaces defined by the constraints. The sketch of these half-spaces in Figure 2.3 shows that there are no such points. The set of

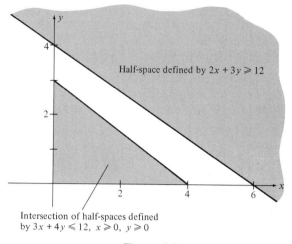

Half-space defined by $2x + 3y \geqslant 12$

Intersection of half-spaces defined
by $3x + 4y \leqslant 12$, $x \geqslant 0$, $y \geqslant 0$

Figure 2.3

feasible solutions is empty. This situation will arise when conflicting constraints are put on a problem. The assumptions for the model must be changed to yield a nonempty set of feasible solutions.

Other linear programming problems may have a large set of feasible solutions, but because the set is unbounded, it is possible that no optimal value will exist.

EXAMPLE 5. Consider the linear programming problem:

$$\text{Maximize} \quad z = 2x + 5y$$

$$\text{subject to}$$

$$-3x + 2y \leq 6$$

$$x + 2y \geq 2$$

$$x \geq 0, \quad y \geq 0$$

The convex set S of all feasible solutions is shown as the shaded area in Figure 2.4. Note that it is unbounded and that the line $y = x$ lies in S for all $x \geq \frac{2}{3}$. On this line, $z = 2x + 5x = 7x$, so that by choosing x large, z may also be made as large as one likes.

On the other hand, a linear programming problem with an unbounded convex set of feasible solutions may have an optimal solution.

EXAMPLE 6. Consider the same set of constraints as in Example 5 (Figure 2.4). Suppose the problem was instead to

$$\text{Minimize} \quad z = 2x + y$$

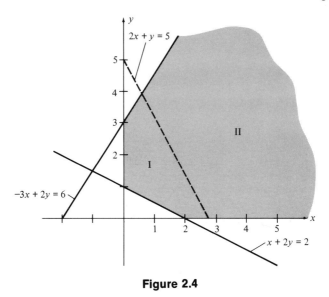

Figure 2.4

subject to these constraints. We divide the set of feasible solutions into two regions with the hyperplane $5 = 2x + y$ as shown in Figure 2.4. All the points in region II satisfy $5 < 2x + y$, and all the points in region I satisfy $5 > 2x + y$. Thus, we need only consider region I for the minimization problem. Region I is closed and bounded and has a finite number of extreme points. They are $(0, 3)$, $(0, 1)$, $(2, 0)$, $(\frac{4}{7}, \frac{27}{7})$, $(\frac{5}{2}, 0)$. Consequently, Theorem 2.2 applies. By evaluating the objective function at these five points, we find that the minimum value is $z = 1$ at $(0, 1)$. Note that other choices for the dividing hyperplane are possible.

A linear programming problem may have more than one optimal solution. In this case it will actually have an infinite number of optimal solutions. When this situation arises, the model may still be valid. There are real problems which physically have many optimal solutions. On the other hand, multiple optimal solutions may occur when not enough constraints have been built into the model.

EXAMPLE 7. Consider the linear programming problem:

$$\text{Maximize} \quad z = 2x + 3y$$

subject to

$$x + 3y \le 9$$
$$2x + 3y \le 12$$
$$x \ge 0, \quad y \ge 0$$

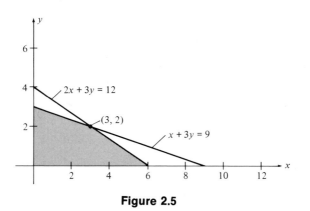

Figure 2.5

The convex set of all feasible solutions is shown in Figure 2.5. The extreme points and corresponding values of the objective function are given in Table 2.4. We see that both (6, 0) and (3, 2) are optimal solutions to the problem. The line segment joining these points is

$$(x, y) = \lambda(6, 0) + (1 - \lambda)(3, 2)$$
$$= (6\lambda, 0) + (3 - 3\lambda, 2 - 2\lambda)$$
$$= (3 + 3\lambda, 2 - 2\lambda) \quad \text{for} \quad 0 \le \lambda \le 1$$

For any point (x, y) on this line segment we have

$$z = 2x + 3y = 2(3 + 3\lambda) + 3(2 - 2\lambda)$$
$$= 6 + 6\lambda + 6 - 6\lambda$$
$$= 12$$

Any point on this line segment is an optimal solution.

We now describe the method for converting a standard linear programming problem into a problem in canonical form. To do this we must be able to change the inequality constraints into equality constraints. In canonical form the constraints form a system of linear equations, and we

Table 2.4

Extreme point	Value of $z = 2x + 3y$
(0, 0)	0
(0, 3)	9
(6, 0)	12
(3, 2)	12

can use the methods of linear algebra to solve such systems. In particular, we shall be able to employ the steps used in Gauss–Jordan reduction.

Changing an Inequality to an Equality

Consider the constraint

$$a_{i1}x_1 + a_{i2}x_2 + \cdots + a_{in}x_n \leq b_i \tag{8}$$

We may convert (8) into an equation by introducing a new variable u_i and writing

$$a_{i1}x_1 + a_{i2}x_2 + \cdots + a_{in}x_n + u_i = b_i \tag{9}$$

The variable u_i is nonnegative and is called a **slack variable** because it "takes up the slack" between the left side of constraint (8) and its right side.

We now convert the linear programming problem in standard form given by (1), (2), and (3) to a problem in canonical form by introducing a slack variable in each of the constraints. Note that each constraint will get a different slack variable. In the ith constraint

$$a_{i1}x_1 + a_{i2}x_2 + \cdots + a_{in}x_n \leq b_i$$

we introduce the slack variable x_{n+i} and write

$$a_{i1}x_1 + a_{i2}x_2 + \cdots + a_{in}x_n + x_{n+i} = b_i$$

Because of the direction of the inequality, we know that $x_{n+i} \geq 0$. Therefore, the canonical form of the problem is

$$\text{Maximize} \quad z = c_1x_1 + c_2x_2 + \cdots + c_nx_n \tag{10}$$

subject to

$$\left. \begin{array}{l} a_{11}x_1 + a_{12}x_2 + \cdots + a_{in}x_n + x_{n+1} \qquad\qquad\quad = b_1 \\ a_{21}x_1 + a_{22}x_2 + \cdots + a_{2n}x_n \qquad\quad + x_{n+2} \qquad\quad = b_2 \\ \vdots \qquad\quad \vdots \qquad\qquad \vdots \qquad\qquad\qquad\qquad\qquad \vdots \\ a_{m1}x_1 + a_{m2}x_2 + \cdots + a_{mn}x_n \qquad\qquad\quad + x_{n+m} = b_m \end{array} \right\} \tag{11}$$

$$x_1 \geq 0, \ x_2 \geq 0, \ \ldots, \ x_n \geq 0, \ x_{n+1} \geq 0, \ \ldots, \ x_{n+m} \geq 0 \tag{12}$$

The new problem has m equations in $m + n$ unknowns in addition to the nonnegativity restrictions on the variables $x_1, x_2, \ldots, x_n, x_{n+1}, \ldots, x_{n+m}$.

If $\mathbf{y} = [y_1 \ y_2 \ \cdots \ y_n]^T$ is a feasible solution to the problem given by (1), (2), and (3), then we define $y_{n+i}, \quad i = 1, 2, \ldots, m$, by

$$y_{n+i} = b_i - a_{i1}y_1 - a_{i2}y_2 - \cdots - a_{in}y_n$$

That is, y_{n+i} is the difference between the right side of the ith constraint in (2) and the value of the left side of this constraint at the feasible solution y. Since each constraint in (2) is of the $\le$ form, we conclude that

$$y_{n+i} \ge 0, \qquad i = 1, 2, \ldots, m$$

Thus, $[y_1 \quad y_2 \cdots y_n \quad y_{n+1} \cdots y_{n+m}]^T$ satisfies (11) and (12), and the point $\hat{y} = [y_1 \quad y_2 \cdots y_{n+m}]^T$ is a feasible solution to the problem in (10), (11), and (12).

Conversely, suppose $\hat{y} = [y_1 \quad y_2 \cdots y_{n+m}]^T$ is a feasible solution to the linear programming problem in canonical form given by (10), (11), and (12). Then clearly $y_1 \ge 0$, $y_2 \ge 0$, $\ldots$, $y_n \ge 0$. Since $y_{n+i} \ge 0$, $i = 1, 2, \ldots, m$, we see that

$$a_{i1}y_1 + a_{i1}y_2 + \cdots + a_{in}y_n \le b_i \quad \text{for} \quad i = 1, 2, \ldots, m$$

Hence, $y = [y_1 \quad y_2 \cdots y_n]^T$ is a feasible solution to the linear programming problem in standard form given by (1), (2), and (3).

The discussion above has shown that a feasible solution to a standard linear programming problem yields a feasible solution to a canonical linear programming problem by adjoining the values of the slack variables. Conversely, a feasible solution to a canonical linear programming problem yields a feasible solution to the corresponding standard linear programming problem by truncating the slack variables.

EXAMPLE 8. Consider Example 1 of Section 2.1. Introducing the slack variables u and v, our problem becomes

$$\text{Maximize} \quad z = 120x + 100y$$

$$\text{subject to}$$

$$2x + 2y + u \quad\quad = 8$$

$$5x + 3y \quad\quad + v = 15$$

$$x \ge 0, \quad y \ge 0, \quad u \ge 0, \quad v \ge 0$$

In terms of the model, the slack variable u is the difference between the total amount of time the saw is available, 8 hours, and the amount of time $2x + 2y$ (in hours) it is actually used. Similarly, the slack variable v is the difference between the total amount of time the plane is available, 15 hours, and the amount of time $5x + 3y$ (in hours) actually used.

We showed in Example 2 of this section that $x = 2$, $y = 1$ is a feasible solution to the problem in standard form. For this feasible solution we have

$$u = 8 - 2\cdot 2 - 2\cdot 1 = 2$$

$$v = 15 - 5\cdot 2 - 3\cdot 1 = 2$$

Thus,

$$x = 2, \quad y = 1, \quad u = 2, \quad v = 2$$

is a feasible solution to the new form of the problem.

Consider now the values

$$x = 1, \quad y = 1, \quad u = 4, \quad v = 7$$

These values are a feasible solution to the new problem, since

$$2 \cdot 1 + 2 \cdot 1 + 4 = 8$$

and

$$5 \cdot 1 + 3 \cdot 1 + 7 = 15$$

Consequently,

$$x = 1, \quad y = 1$$

is a feasible solution to the given problem.

In Example 3 of this section we showed that an optimal solution to this problem is

$$x = \tfrac{3}{2}, \quad y = \tfrac{5}{2}$$

In canonical form this feasible solution gives values of u and v as

$$u = \ 8 - 2 \cdot \tfrac{3}{2} - 2 \cdot \tfrac{5}{2} = 0$$

$$v = 15 - 5 \cdot \tfrac{3}{2} - 3 \cdot \tfrac{5}{2} = 0$$

That is, an optimal feasible solution to the canonical form of the problem is

$$x = \tfrac{3}{2}, \quad y = \tfrac{5}{2}, \quad u = 0, \quad v = 0$$

The linear programming problem given by (10), (11), and (12) can also be written in matrix form as follows. We now let

$$\mathbf{A} = \begin{bmatrix} a_{11} & a_{12} & \cdots & a_{1n} & 1 & 0 & \cdots & 0 \\ a_{21} & a_{22} & \cdots & a_{2n} & 0 & 1 & \cdots & 0 \\ \vdots & \vdots & & \vdots & \vdots & \vdots & & \vdots \\ a_{m1} & a_{m2} & \cdots & a_{mn} & 0 & 0 & \cdots & 1 \end{bmatrix},$$

$$\mathbf{b} = \begin{bmatrix} b_1 \\ b_2 \\ \vdots \\ b_m \end{bmatrix}, \quad \mathbf{x} = \begin{bmatrix} x_1 \\ x_2 \\ \vdots \\ x_n \\ x_{n+1} \\ \vdots \\ x_{n+m} \end{bmatrix}, \quad \mathbf{c} = \begin{bmatrix} c_1 \\ c_2 \\ \vdots \\ c_n \\ 0 \\ \vdots \\ 0 \end{bmatrix}$$

Then this problem can be written as

$$\text{Maximize} \quad z = \mathbf{c}^T\mathbf{x} \tag{13}$$

subject to

$$\mathbf{Ax} = \mathbf{b} \tag{14}$$

$$\mathbf{x} \geq \mathbf{0} \tag{15}$$

Note that this problem is in canonical form.

EXAMPLE 9. The linear programming problem which was formulated in Example 8 can be written in matrix form as

$$\text{Maximize} \quad z = \begin{bmatrix} 120 & 100 & 0 & 0 \end{bmatrix} \begin{bmatrix} x \\ y \\ u \\ v \end{bmatrix}$$

subject to

$$\begin{bmatrix} 2 & 2 & 1 & 0 \\ 5 & 3 & 0 & 1 \end{bmatrix} \begin{bmatrix} x \\ y \\ u \\ v \end{bmatrix} = \begin{bmatrix} 8 \\ 15 \end{bmatrix}$$

$$\begin{bmatrix} x \\ y \\ u \\ v \end{bmatrix} \geq \begin{bmatrix} 0 \\ 0 \\ 0 \\ 0 \end{bmatrix}$$

We shall now prove two very important general theorems about linear programming problems in canonical form. We shall then apply these theorems to the specific linear programming problem given by (13), (14), and (15). We already know from Theorem 2.2 that if the set S' of all feasible solutions to the new problem is closed, bounded, and has a finite number of extreme points, then the objective function (13) has a maximum at an extreme point of S'.

Consider the linear programming problem in canonical form given by

$$\text{Maximize} \quad z = \mathbf{c}^T\mathbf{x} \tag{16}$$

subject to

$$\mathbf{Ax} = \mathbf{b} \tag{17}$$

$$\mathbf{x} \geq \mathbf{0} \tag{18}$$

where $\mathbf{A}$ is an $m \times s$ matrix, $\mathbf{c} \in R^s$, $\mathbf{x} \in R^s$, and $\mathbf{b} \in R^m$. Let the columns of $\mathbf{A}$ be denoted by $\mathbf{A}_1$, $\mathbf{A}_2$, ..., $\mathbf{A}_s$. We can then write (17) as

$$x_1\mathbf{A}_1 + x_2\mathbf{A}_2 + \cdots + x_s\mathbf{A}_s = \mathbf{b} \tag{19}$$

We make two assumptions about the constraint matrix $\mathbf{A}$; both of these assumptions are true for the special form of the canonical linear programming problem given by (13), (14), and (15). We assume that $m \leq s$ and that there are m columns of $\mathbf{A}$ which are linearly independent. Any set of m linearly independent columns of $\mathbf{A}$ forms a basis for R^m. We can always renumber the columns of $\mathbf{A}$ (by reordering the components of $\mathbf{x}$) so that the last m columns of $\mathbf{A}$ are linearly independent. Let S be the convex set of all feasible solutions to the problem determined by (16), (17), and (18).

Theorem 2.3.   Suppose that the last m columns of $\mathbf{A}$, which we denote by $\mathbf{A}_1'$, $\mathbf{A}_2'$, ..., $\mathbf{A}_m'$, are linearly independent and suppose that

$$x_1'\mathbf{A}_1' + x_2'\mathbf{A}_2' + \cdots + x_m'\mathbf{A}_m' = \mathbf{b} \tag{20}$$

where $x_i' \geq 0$ for $i = 1, 2, \ldots, m$. Then the point

$$\mathbf{x} = (0, 0, \ldots, 0, x_1', x_2', \ldots, x_m')$$

is an extreme point of S.

Proof.   We assumed $\mathbf{x} \geq \mathbf{0}$ in the statement of the theorem. Equation (20) represents $\mathbf{Ax} = \mathbf{b}$, since the first $s - m$ components of $\mathbf{x}$ are zero. Thus, $\mathbf{x}$ is a feasible solution to the linear programming problem given by (16), (17), and (18). Assume that $\mathbf{x}$ is not an extreme point of S. Then, $\mathbf{x}$ lies in the interior of a line segment in S. That is, there are points $\mathbf{v}$, $\mathbf{w}$ in S both different from $\mathbf{x}$ and a number λ, $0 < \lambda < 1$, such that

$$\mathbf{x} = \lambda\mathbf{v} + (1 - \lambda)\mathbf{w} \tag{21}$$

Now

$$\mathbf{v} = (v_1, v_2, \ldots, v_{s-m}, v_1', v_2', \ldots, v_m')$$

and

$$\mathbf{w} = (w_1, w_2, \ldots, w_{s-m}, w_1', w_2', \ldots, w_m')$$

where all the components of $\mathbf{v}$ and $\mathbf{w}$ are nonnegative, since $\mathbf{v}$ and $\mathbf{w}$ are feasible solutions. Substituting the expressions for $\mathbf{x}$, $\mathbf{v}$, and $\mathbf{w}$ into (21), we have

$$0 = \lambda v_i + (1 - \lambda)w_i, \qquad 1 \leq i \leq s - m \tag{22}$$

$$x_j' = \lambda v_j' + (1 - \lambda)w_j', \qquad 1 \leq j \leq m$$

Since all the terms in (22) are nonnegative and λ and $1 - \lambda$ are positive, we conclude that $v_i = 0$ and $w_i = 0$ for $i = 1, 2, \ldots, s - m$. Since $\mathbf{v}$ is a feasible solution, we know that $\mathbf{Av} = \mathbf{b}$, and because the first $s - m$

components of **v** are zero, this equation can be written as

$$v_1' \mathbf{A}_1' + v_2' \mathbf{A}_2' + \cdots + v_m' \mathbf{A}_m' = \mathbf{b} \tag{23}$$

If we now subtract Equation (23) from Equation (20), we have

$$(x_1' - v_1') \mathbf{A}_1' + (x_2' - v_2') \mathbf{A}_2' + \cdots + (x_m' - v_m') \mathbf{A}_m' = \mathbf{0}$$

Since we assumed that $\mathbf{A}_1', \mathbf{A}_2', \ldots, \mathbf{A}_m'$ were linearly independent, we conclude that

$$x_i' = v_i' \quad \text{for} \quad 1 \le i \le m$$

and consequently that $\mathbf{x} = \mathbf{v}$. But we had assumed $\mathbf{x} \neq \mathbf{v}$. This contradiction implies that our assumption that $\mathbf{x}$ is not an extreme point of S is false. Thus, $\mathbf{x}$ is an extreme point of S, as we wanted to show.

Theorem 2.4.   If $\mathbf{x} = (x_1, x_2, \ldots, x_s)$ is an extreme point of S, then the columns of $\mathbf{A}$ which correspond to positive x_j form a linearly independent set of vectors in R^m.

Proof.   To simplify the notation, renumber the columns of $\mathbf{A}$ and the components of $\mathbf{x}$, if necessary, so that the last k components, denoted by $x_1', x_2', \ldots, x_k'$, are positive. Thus, Equation (17) can be written as

$$x_1' \mathbf{A}_1' + x_2' \mathbf{A}_2' + \cdots + x_k' \mathbf{A}_k' = \mathbf{b} \tag{24}$$

We must show that $\mathbf{A}_1', \mathbf{A}_2', \ldots, \mathbf{A}_k'$ are linearly independent. Suppose they are linearly dependent. This means that there are numbers c_j, $1 \le j \le k$, not all of which are zero, such that

$$c_1 \mathbf{A}_1' + c_2 \mathbf{A}_2' + \cdots + c_k \mathbf{A}_k' = \mathbf{0} \tag{25}$$

Say that $c_t \neq 0$. Multiply Equation (25) by a positive scalar d, and first add the resulting equation to Equation (24), getting Equation (26), and second subtract it from Equation (24), getting Equation (27). We now have

$$(x_1' + dc_1) \mathbf{A}_1' + (x_2' + dc_2) \mathbf{A}_2' + \cdots + (x_k' + dc_k) \mathbf{A}_k' = \mathbf{b} \tag{26}$$

$$(x_1' - dc_1) \mathbf{A}_1' + (x_2' - dc_2) \mathbf{A}_2' + \cdots + (x_k' - dc_k) \mathbf{A}_k' = \mathbf{b} \tag{27}$$

Now consider the points in R^s

$$\mathbf{v} = (0, 0, \ldots, 0, x_1' + dc_1, x_2' + dc_2, \ldots, x_k' + dc_k)$$

and

$$\mathbf{w} = (0, 0, \ldots, 0, x_1' - dc_1, x_2' - dc_2, \ldots, x_k' - dc_k)$$

Since d is any positive scalar, we may choose it so that

$$0 < d < \min_j \frac{x_j'}{|c_j|}, \qquad c_j \neq 0$$

With this choice of d, the last k coordinates of both v and w are positive. This fact together with Equations (26) and (27) imply that $\mathbf{v}$ and $\mathbf{w}$ are feasible solutions. But we also have

$$\mathbf{x} = \tfrac{1}{2}\mathbf{v} + \tfrac{1}{2}\mathbf{w}$$

contradicting the hypothesis that $\mathbf{x}$ is an extreme point of S. Thus our assumption that the last k columns of $\mathbf{A}$ are linearly dependent is false; they are linearly independent.

Corollary. If $\mathbf{x}$ is an extreme point and $x_{i_1}, \ldots, x_{i_r}$ are the r positive components of $\mathbf{x}$, then $r \leq m$, and the set of columns $\mathbf{A}_{i_1}, \ldots, \mathbf{A}_{i_r}$ can be extended to a set of linearly independent vectors in R^m by adjoining a suitably chosen set of $m - r$ columns of $\mathbf{A}$.

Theorem 2.5. At most m components of any extreme point of S can be positive. The rest must be zero.

Proof. Theorem 2.4 says that the columns of $\mathbf{A}$ corresponding to the positive components of an extreme point $\mathbf{x}$ of the set S of feasible solutions are linearly independent vectors in R^m. But there can be no more than m linearly independent vectors in R^m. Therefore, at most m of the components of $\mathbf{x}$ are nonzero.

An important feature of the canonical form of a linear programming problem is that the constraints $\mathbf{Ax} = \mathbf{b}$ form a system of m equations in s unknowns. Our assumption that there are m linearly independent column vectors means that redundant equations do not occur in the system of constraints. In Theorems 2.3 and 2.4 we showed the relationships between the extreme points of the set S of feasible solutions and the linearly independent columns of $\mathbf{A}$. We now use information about solutions to a system of equations to further describe the points of S. Note that S is just the set of solutions to $\mathbf{Ax} = \mathbf{b}$ with nonnegative components ($\mathbf{x} \geq \mathbf{0}$).

Consider a system of m equations in s unknowns ($m \leq s$) and write it in matrix form as $\mathbf{Ax} = \mathbf{b}$. Assume that at least one set of m columns of $\mathbf{A}$ is linearly independent. Choosing any set T of m linearly independent columns of $\mathbf{A}$ (which is choosing a basis for R^m), set the $s - m$ variables corresponding to the remaining columns equal to zero. The equation $\mathbf{Ax} = \mathbf{b}$ may be written as

$$x_1\mathbf{A}_1 + x_2\mathbf{A}_2 + \cdots + x_s\mathbf{A}_s = \mathbf{b} \tag{28}$$

where $\mathbf{A}_i$ is the ith column of $\mathbf{A}$. But we have set $s - m$ of the variables equal to zero. Let $i_1, i_2, \ldots, i_m$ be the indices of the variables which were not set to zero. They are also the indices of the columns of $\mathbf{A}$ in the set T. Consequently, (28) reduces to

$$x_{i_1}\mathbf{A}_{i_1} + x_{i_2}\mathbf{A}_{i_2} + \cdots + x_{i_m}\mathbf{A}_{i_m} = \mathbf{b}$$

which is a system of m equations in m unknowns and has a unique solution. (Why?) The values of the m variables obtained from solving this system along with the $s - m$ zeros form a vector $\mathbf{x}$ which is called a **basic solution** to $\mathbf{Ax} = \mathbf{b}$.

EXAMPLE 10.   Consider the system of three equations in six unknowns

$$\begin{bmatrix} 1 & 0 & 1 & 0 & 1 & 0 \\ 0 & -1 & -1 & 0 & -1 & -1 \\ 1 & 2 & 2 & 1 & 1 & 1 \end{bmatrix} \begin{bmatrix} x_1 \\ x_2 \\ x_3 \\ x_4 \\ x_5 \\ x_6 \end{bmatrix} = \begin{bmatrix} b_1 \\ b_2 \\ b_3 \end{bmatrix}$$

Setting $x_2 = x_3 = x_5 = 0$, we get the system of three equations in three unknowns given by

$$\begin{bmatrix} 1 & 0 & 0 \\ 0 & 0 & -1 \\ 1 & 1 & 1 \end{bmatrix} \begin{bmatrix} x_1 \\ x_4 \\ x_6 \end{bmatrix} = \begin{bmatrix} b_1 \\ b_2 \\ b_3 \end{bmatrix}$$

The columns of this coefficient matrix are linearly independent, and this system has the solution $x_1 = b_1$, $x_4 = b_2 + b_3 - b_1$, $x_6 = -b_2$. Consequently, a basic solution to the original system is

$$\mathbf{x} = (b_1, 0, 0, b_2 + b_3 - b_1, 0, -b_2)$$

On the other hand, if we set $x_1 = x_3 = x_5 = 0$, we obtain the system

$$\begin{bmatrix} 0 & 0 & 0 \\ -1 & 0 & -1 \\ 2 & 1 & 1 \end{bmatrix} \begin{bmatrix} x_2 \\ x_4 \\ x_6 \end{bmatrix} = \begin{bmatrix} b_1 \\ b_2 \\ b_3 \end{bmatrix}$$

Here the columns of the coefficient matrix are not linearly independent. In fact, column 1 is the sum of columns 2 and 3. This system cannot be solved if $b_1 \neq 0$. Consequently, this choice of variables does not lead to a basic solution.

In any basic solution, the $s - m$ variables which are set equal to zero are called **nonbasic variables**, and the m variables solved for are called **basic variables**. Although the term *basic solution* appears in all the literature describing linear programming, it can be misleading. A basic solution is a solution to the system $\mathbf{Ax} = \mathbf{b}$; it does not necessarily satisfy $\mathbf{x} \geq \mathbf{0}$, and therefore it is not necessarily a feasible solution to the linear programming problem given by (16), (17), and (18).

Definition.   A **basic feasible solution** to the linear programming problem given by (16), (17), and (18) is a basic solution which is also feasible.

Theorem 2.6.   For the linear programming problem determined by (16), (17), and (18), every basic feasible solution is an extreme point, and conversely, every extreme point is a basic feasible solution.

Proof.   Exercise.

Theorem 2.7.   The problem determined by (16), (17), and (18) has a finite number of basic feasible solutions.

Proof.   The number of basic solutions to the problem is not more than

$$\binom{s}{m} = \frac{s!}{m!(s-m)!} = \binom{s}{s-m}$$

because there are $s - m$ choices for which of the s variables will be set to zero. The number of basic feasible solutions may be smaller than the number of basic solutions, since not all basic solutions need to be feasible.

We now examine the relationship between the set S of feasible solutions to a standard linear programming problem (4), (5), (6) and the set S' of feasible solutions to the associated canonical linear programming problem (13), (14), (15). We have already discussed the method of adding slack variables to go from a point in S to a point in S'. Conversely, we truncated variables to move from S' to S.

Theorem 2.8.   Every extreme point of S yields an extreme point of S' when slack variables are added. Conversely, every extreme point of S', when truncated, yields an extreme point of S.

Theorem 2.9.   The convex set S of all feasible solutions to a standard linear programming problem (4), (5), (6) has a finite number of extreme points.

Proof.   Exercise.

By combining Theorems 2.2 and 2.9, we can give a procedure for solving a standard linear programming problem (4), (5), (6). First, set up the associated canonical form of the problem. Then find all basic solutions and eliminate those which are not feasible. Among the basic feasible solutions find those which are optimal. Since the objective function does not change between the standard and canonical forms of the problem, any optimal solution for the canonical form, found as described above, is an optimal solution to the standard problem.

From the situation under consideration, the number of basic solutions to be examined is no more than

$$\binom{m+n}{n}$$

Although this number is finite, it is still too large for actual problems.

For example, a moderate-size problem with $m = 200$ and $n = 300$ would have about 10^{144} basic solutions.

EXAMPLE 11. Consider the linear programming problem given in Example 9. In this example we can select two of the four variables x, y, u, v as nonbasic variables by setting them equal to zero and then solve for the basic variables. If in

$$\begin{bmatrix} 2 & 2 & 1 & 0 \\ 5 & 3 & 0 & 1 \end{bmatrix} \begin{bmatrix} x \\ y \\ u \\ v \end{bmatrix} = \begin{bmatrix} 8 \\ 15 \end{bmatrix}$$

we set

$$x = y = 0 \quad \text{(nonbasic variables)}$$

then

$$u = 8 \quad \text{and} \quad v = 15 \quad \text{(basic variables)}$$

The vector $[0 \ \ 0 \ \ 8 \ \ 15]^T$ is a basic feasible solution to the canonical form of the problem and hence an extreme point of S'. By truncating the slack variables we get $[0 \ \ 0]^T$, which is an extreme point of S and a feasible solution to the standard form of the problem.

If instead we set

$$x = v = 0 \quad \text{(nonbasic variables)}$$

then

$$y = 5 \quad \text{and} \quad u = -2 \quad \text{(basic variables)}$$

The vector $[0 \ \ 5 \ \ -2 \ \ 0]^T$ is a basic solution, but it is not feasible, since u is negative.

In Table 2.5 we list all the basic solutions, indicate whether they are feasible, and give the truncated vectors. The student should locate these truncated vectors on Figure 2.2. Once we discard the basic solutions which are not feasible, we select a basic feasible solution for which the

Table 2.5

x	y	u	v	Type of solution	Value of $z = 120x + 100y$	Truncated vector
0	0	8	15	Basic feasible	0	$(0, 0)$
0	4	0	3	Basic feasible	400	$(0, 4)$
0	5	-2	0	Basic, not feasible	—	$(0, 5)$
4	0	0	-5	Basic, not feasible	—	$(4, 0)$
3	0	2	0	Basic feasible	360	$(3, 0)$
$\frac{3}{2}$	$\frac{5}{2}$	0	0	Basic feasible	430	$(\frac{3}{2}, \frac{5}{2})$

objective function is a maximum. Thus, we again obtain the optimal solution

$$x = \tfrac{3}{2}, \quad y = \tfrac{5}{2}, \quad u = 0, \quad v = 0$$

2.2 Exercises

1. Write the following linear programming problems in Section 2.1 in matrix form:
   (a) Example 4 (page 63)
   (b) Example 8 (page 68)
   (c) Example 9 (page 69)

2. Convert the following linear programming problems in Section 2.1 to canonical form and express this form in matrix notation:
   (a) Example 7(a) (page 67)
   (b) Example 10 (page 70)
   (c) Example 11 (page 70)

3. (a) For the linear programming problem in Example 1 (page 59) show that

$$\mathbf{x}_2 = \begin{bmatrix} 2 \\ 1 \end{bmatrix} \quad \text{and} \quad \mathbf{x}_3 = \begin{bmatrix} 1 \\ 3 \end{bmatrix}$$

are feasible solutions. Also compute the values of the objective function for these feasible solutions.
   (b) Show that

$$\mathbf{x}_4 = \begin{bmatrix} 3 \\ 1 \end{bmatrix}, \quad \mathbf{x}_5 = \begin{bmatrix} 2 \\ 2 \end{bmatrix}, \quad \mathbf{x}_6 = \begin{bmatrix} -2 \\ 3 \end{bmatrix}$$

are not feasible solutions.

In Exercises 4–15: (a) sketch the set S of feasible solutions defined by the given constraints; (b) find the extreme points of S; (c) for the given objective function find the optimal solution(s).

4. Maximize $z = x + 2y$

subject to

$$3x + \quad y \le 6$$
$$3x + 4y \le 12$$
$$x \ge 0, \quad y \ge 0$$

6. Maximize $z = 3x + y$

subject to

$$-3x + \quad y \ge 6$$
$$3x + 5y \le 15$$
$$x \ge 0, \quad y \ge 0$$

5. Minimize $z = 5x - 3y$

subject to

$$x + 2y \le 4$$
$$x + 3y \ge 6$$
$$x \ge 0, \quad y \ge 0$$

7. Maximize $z = 2x + 3y$

subject to

$$3x + \quad y \le 6$$
$$x + \quad y \le 4$$
$$x + 2y \le 6$$
$$x \ge 0, \quad y \ge 0$$

8. Minimize $z = 3x + 5y$

subject to same constraints as in Exercise 7.

9. Maximize $z = \frac{1}{4}x + \frac{2}{3}y$

subject to

$$x + 3y \leq 6$$
$$x + y \geq 4$$
$$x \geq 0, \quad y \geq 0$$

10. Maximize $z = 2x + 5y$

subject to

$$2x + y \geq 2$$
$$x + y \leq 8$$
$$x + y \geq 3$$
$$2x + y \leq 12$$
$$x \geq 0, \quad y \geq 0$$

11. Maximize $z = 2x_1 + 4x_2$

subject to

$$5x_1 + 3x_2 + 5x_3 \leq 15$$
$$10x_1 + 8x_2 + 15x_3 \leq 40$$
$$x_1 \geq 0, \quad x_2 \geq 0, \quad x_3 \geq 0$$

12. Maximize $z = 2x_1 + 4x_2 + 3x_3$

subject to

$$x_1 + x_2 + x_3 \leq 12$$
$$x_1 + 3x_2 + 3x_3 \leq 24$$
$$3x_1 + 6x_2 + 4x_3 \leq 90$$
$$x_1 \geq 0, \quad x_2 \geq 0, \quad x_3 \geq 0$$

13. Minimize $z = 2x_1 + 3x_2 + x_3$

subject to same constraints as in Exercise 12.

14. Maximize $z = 5x_1 + 2x_2 + 3x_3$

subject to

$$x_1 + x_2 + x_3 = 1$$
$$2x_1 + 5x_2 + 3x_3 \leq 4$$
$$4x_1 + x_2 + 3x_3 \leq 2$$
$$x_1 \geq 0, \quad x_2 \geq 0, \quad x_3 \geq 0$$

15. Minimize $z = 2x_1 + x_3$

subject to

$$x_1 + x_2 + x_3 = 1$$
$$2x_1 + x_2 + 2x_3 \geq 3$$
$$x_1 \geq 0, \quad x_2 \geq 0, \quad x_3 \geq 0$$

16. Suppose the canonical form of a linear programming problem is given by the constraint matrix **A** and resource vector **b**, where

$$\mathbf{A} = \begin{bmatrix} 3 & 0 & 1 & 1 & 0 \\ 2 & 1 & 0 & 0 & 0 \\ 4 & 0 & 3 & 0 & 1 \end{bmatrix} \quad \text{and} \quad \mathbf{b} = \begin{bmatrix} 5 \\ 3 \\ 6 \end{bmatrix}$$

Determine which of the following points is
(i) a feasible solution to the linear programming problem
(ii) an extreme point of the set of feasible solutions
(iii) a basic solution

(iv) a basic feasible solution

For each basic feasible solution **x** given below, list the basic variables:

(a) $\begin{bmatrix} 0 \\ 3 \\ 0 \\ 5 \\ 6 \end{bmatrix}$
(b) $\begin{bmatrix} 0 \\ 3 \\ 5 \\ 0 \\ -9 \end{bmatrix}$
(c) $\begin{bmatrix} \frac{3}{2} \\ 0 \\ 0 \\ \frac{1}{2} \\ 0 \end{bmatrix}$
(d) $\begin{bmatrix} \frac{1}{2} \\ 1 \\ 1 \\ 0 \\ 2 \end{bmatrix}$
(e) $\begin{bmatrix} 1 \\ 1 \\ \frac{1}{2} \\ \frac{3}{2} \\ -\frac{1}{2} \end{bmatrix}$

In Exercises 17 and 18, set up a linear programming model for the situation described, sketch the set of feasible solutions, and find an optimal solution by examining the extreme points.

17. The Savory Potato Chip Company makes pizza-flavored and chili-flavored potato chips. These chips must go through three main processes: frying, flavoring, and packing. Each kilogram of pizza-flavored chips takes 3 minutes to fry, 5 minutes to flavor, and 2 minutes to pack. Each kilogram of chili-flavored chips takes 3 minutes to fry, 4 minutes to flavor, and 3 minutes to pack. The net profit on each kilogram of pizza chips is $.12, while the net profit on each kilogram of chili chips is $.10. The fryer is available 4 hours each day, the flavorer is available 8 hours each day, and the packer is available 6 hours each day. Maximize the net profit with your model.

18. Sugary Donut Bakers, Inc. is known for its excellent glazed doughnuts. The firm also bakes doughnuts which are then dipped in powdered sugar. It makes a profit of $.07 per glazed doughnut and $.05 per powdered sugar doughnut. The three main operations are: baking, dipping (for the powdered sugar doughnuts only), and glazing (for the glazed doughnuts only). Each day the plant has the capacity to bake 1400 doughnuts, dip 1200 doughnuts, and glaze 1000 doughnuts. The manager has instructed that at least 600 glazed doughnuts must be made each day. Maximize the total profit with your model.

19. For Exercise 17, write the linear programming problem in canonical form, compute the values of the slack variables for an optimal solution, and give a physical interpretation for these values.

20. For Exercise 18, write the linear programming problem in canonical form, compute the values of the slack variables for an optimal solution, and give a physical interpretation for these values.

21. Show that if the optimal value of the objective function of a linear programming problem is attained at several extreme points, then it is also attained at any convex combination of these extreme points.

22. Prove Theorem 2.6.

23. Prove Theorem 2.9.

24. Show that the set of solutions to $\mathbf{Ax} > \mathbf{b}$, if it is nonempty, forms a convex set.

25. Consider the system of equations $\mathbf{Ax} = \mathbf{b}$, where

$$\mathbf{A} = \begin{bmatrix} 2 & 3 & 4 & 0 & 4 \\ 1 & 0 & 0 & -2 & 1 \end{bmatrix} \quad \text{and} \quad \mathbf{b} = \begin{bmatrix} 2 \\ 0 \end{bmatrix}$$

Determine whether each of the following 5-tuples is a basic solution to the system:

(a) $(1, 0, 1, 0, 0)$ (b) $(0, 2, -1, 0, 0)$

(c) $(2, -2, 3, 0, -2)$ (d) $(0, 0, \frac{1}{2}, 0, 0)$

26. Consider the linear programming problem in standard form

$$\text{Maximize} \quad z = \mathbf{c}^T\mathbf{x}$$

$$\text{subject to}$$

$$\mathbf{Ax} \le \mathbf{b}$$

$$\mathbf{x} \ge \mathbf{0}$$

Show that the constraints $\mathbf{Ax} \le \mathbf{b}$ may be written as

(i) $[\mathbf{A} \mid \mathbf{I}] \begin{bmatrix} \mathbf{x} \\ \mathbf{x}' \end{bmatrix} = \mathbf{b}$

or as

(ii) $\mathbf{Ax} + \mathbf{Ix}' = \mathbf{b}$

where $\mathbf{x}'$ is a vector of slack variables.

2.3 THE SIMPLEX METHOD

We already know from Section 2.2 that a linear programming problem in canonical form can be solved by finding all the basic solutions, discarding those which are not feasible, and finding an optimal solution among the remaining. Since this procedure can still be a lengthy one, we seek a more efficient method for solving linear programming problems. The simplex algorithm is such a method; in this section we shall describe and carefully illustrate it. Even though the method is an algebraic one, it is helpful to examine it geometrically.

Consider a linear programming problem in standard form

$$\text{Maximize} \quad z = \mathbf{c}^T\mathbf{x} \tag{1}$$

$$\text{subject to}$$

$$\mathbf{Ax} \le \mathbf{b} \tag{2}$$

$$\mathbf{x} \ge \mathbf{0} \tag{3}$$

where $\mathbf{A} = [a_{ij}]$ is an $m \times n$ matrix and

$$\mathbf{b} = \begin{bmatrix} b_1 \\ b_2 \\ \vdots \\ b_m \end{bmatrix}, \qquad \mathbf{c} = \begin{bmatrix} c_1 \\ c_2 \\ \vdots \\ c_n \end{bmatrix}, \qquad \mathbf{x} = \begin{bmatrix} x_1 \\ x_2 \\ \vdots \\ x_n \end{bmatrix}$$

In this section we shall make the additional assumption that $\mathbf{b} \geq \mathbf{0}$. In Section 2.5 we will describe a procedure for handling problems in which $\mathbf{b}$ is not nonnegative.

We now transform each of the constraints in (2) into an equation by introducing a slack variable. We obtain the canonical form of the problem

$$\text{Maximize} \quad z = \mathbf{c}^T \mathbf{x} \tag{4}$$

subject to

$$\mathbf{A}\mathbf{x} = \mathbf{b} \tag{5}$$

$$\mathbf{x} \geq \mathbf{0} \tag{6}$$

where in this case $\mathbf{A}$ is the $m \times (n + m)$ matrix

$$\mathbf{A} = \begin{bmatrix} a_{11} & a_{12} & \cdots & a_{1n} & 1 & 0 & \cdots & 0 \\ a_{21} & a_{22} & \cdots & a_{2n} & 0 & 1 & \cdots & 0 \\ \vdots & \vdots & & \vdots & \vdots & \vdots & & \vdots \\ a_{m1} & a_{m2} & \cdots & a_{mn} & 0 & 0 & \cdots & 1 \end{bmatrix}$$

$$\mathbf{c} = \begin{bmatrix} c_1 \\ c_2 \\ \vdots \\ c_n \\ 0 \\ 0 \\ \vdots \\ 0 \end{bmatrix}, \qquad \mathbf{x} = \begin{bmatrix} x_1 \\ x_2 \\ \vdots \\ x_n \\ x_{n+1} \\ \vdots \\ x_{n+m} \end{bmatrix}$$

and $\mathbf{b}$ is as before.

Recall from Section 2.2 that a basic feasible solution to the canonical form of the problem (4), (5), (6) is an extreme point of the convex set S' of all feasible solutions to the problem.

Definition. Two distinct extreme points in S' are said to be **adjacent** if as basic feasible solutions they have all but one basic variable in common.

EXAMPLE 1. Consider Example 11 of Section 2.2 and especially Table 2.5 in that example. The extreme points (0, 0, 8, 15) and (0, 4, 0, 3) are adjacent, since the basic variables in the first extreme point are u and v

and the basic variables in the second extreme point are y and v. In fact, the only extreme point which is not adjacent to $(0, 0, 8, 15)$ is $(\frac{3}{2}, \frac{5}{2}, 0, 0)$.

The simplex method developed by George B. Dantzig in 1947 is a method which proceeds from a given extreme point (basic feasible solution) to an adjacent extreme point in such a way that the value of the objective function increases. The method proceeds until we either obtain an optimal solution or find that the given problem has no finite optimal solution. The simplex algorithm consists of two steps: (1) a way of finding out whether a given basic feasible solution is an optimal solution, and (2) a way of obtaining an adjacent basic feasible solution with a larger value for the objective function. In actual use, the simplex method does not examine every basic feasible solution; it checks only a small number of them. However, examples have been given where a large number of basic feasible solutions have been examined by the simplex method.

We shall demonstrate parts of our description of the simplex method on the linear programming problem in Example 1 of Section 2.1. The associated canonical form of the problem was described in Example 7 of Section 2.2. In this form it is:

$$\text{Maximize} \quad z = 120x + 100y \tag{7}$$

subject to

$$\left. \begin{array}{l} 2x + 2y + u \quad = \quad 8 \\ 5x + 3y \quad + v = 15 \end{array} \right\} \tag{8}$$

$$x \geq 0, \quad y \geq 0, \quad u \geq 0, \quad v \geq 0 \tag{9}$$

The Initial Basic Feasible Solution

To start the simplex method, we must find a basic feasible solution. The assumption that $\mathbf{b} \geq \mathbf{0}$ allows the following procedure to work. If it is not true that $\mathbf{b} \geq \mathbf{0}$, another procedure (discussed in Section 2.5) must be used. We take all the nonslack variables as nonbasic variables; that is, we set all the nonslack variables in the system $\mathbf{Ax} = \mathbf{b}$ equal to zero. The basic variables are then just the slack variables. We have

$$x_1 = x_2 = \cdots = x_n = 0 \quad \text{and} \quad x_{n+1} = b_1, x_{n+2} = b_2, \ldots, x_{n+m} = b_m$$

This is a feasible solution, since $\mathbf{b} \geq \mathbf{0}$; and it is a basic solution, since $(n + m) - m = n$ of the variables are zero and the columns

$$\begin{bmatrix} 1 \\ 0 \\ \vdots \\ 0 \end{bmatrix} \quad \begin{bmatrix} 0 \\ 1 \\ \vdots \\ 0 \end{bmatrix} \quad \cdots \quad \begin{bmatrix} 0 \\ 0 \\ \vdots \\ 1 \end{bmatrix}$$

corresponding to the nonzero variables are linearly independent.

In our example, we let

$$x = y = 0$$

Solving for u and v, we obtain

$$u = 8, \quad v = 15$$

The initial basic feasible solution constructed by this method is $(0, 0, 8, 15)$. The basic feasible solution yields the extreme point $(0, 0)$ in Figure 2.2 (Section 2.2).

It is useful to set up our example problem and its initial basic feasible solution in tabular form. To do this, we write (7) as

$$-120x - 100y + z = 0 \qquad (10)$$

where z is now viewed as another variable. The **initial tableau** is now formed (Tableau 2.1). At the top we list the variables x, y, u, v, z as labels on the corresponding columns. The last row, called the **objective row**, is Equation (10). The constraints (8) are on the first two rows. Along the left side of each row we indicate which variable is basic in the corresponding equation. Thus, in the first equation u is the basic variable, and v is the basic variable in the second equation.

Tableau 2.1

	x	y	u	v	z	
u	2	2	1	0	0	8
v	5	3	0	1	0	15
	-120	-100	0	0	1	0

In the tableau, a basic variable has the following properties:

1. It appears in exactly one equation and in that equation it has a coefficient of $+1$.
2. The column which it labels has all zeros (including the objective row entry) except for the $+1$ in the row which is labeled by the basic variable.
3. The value of a basic variable is the entry in the same row in the rightmost column.

The initial tableau for the general problem (4), (5), (6) is shown in Tableau 2.2. The value of the objective function

$$z = c_1 x_1 + c_2 x_2 + \cdots + c_n x_n + 0 \cdot x_{n+1} + \cdots + 0 \cdot x_{n+m}$$

for the initial basic feasible solution is

$$z = c_1 \cdot 0 + c_2 \cdot 0 + \cdots + c_n \cdot 0 + 0 \cdot b_1 + 0 \cdot b_2 + \cdots + 0 \cdot b_m = 0$$

Tableau 2.2

	x_1	x_2	$\cdots$	x_n	x_{n+1}	x_{n+2}	$\cdots$	x_{n+m}	z	
x_{n+1}	a_{11}	a_{12}	$\cdots$	a_{1n}	1	0	$\cdots$	0	0	b_1
x_{n+2}	a_{21}	a_{22}	$\cdots$	a_{2n}	0	1	$\cdots$	0	0	b_2
$\vdots$	$\vdots$	$\vdots$		$\vdots$	$\vdots$	$\vdots$		$\vdots$	$\vdots$	$\vdots$
x_{n+m}	a_{m1}	a_{m2}	$\cdots$	a_{mn}	0	0	$\cdots$	1	0	b_m
	$-c_1$	$-c_2$	$\cdots$	$-c_n$	0	0	$\cdots$	0	1	0

In our example we have

$$z = 120 \cdot 0 + 100 \cdot 0 + 0 \cdot 8 + 0 \cdot 15 = 0$$

We can increase the value of z from its initial value of 0 by increasing any one of the nonbasic variables having a positive coefficient from its current value of 0 to some positive value. For our example,

$$z = 120x + 100y + 0 \cdot u + 0 \cdot v$$

so that z can be increased by increasing either x or y.

If in the general situation we write the objective function so that the coefficients of the basic variables are zero, then we have

$$z = \sum_{\text{nonbasic}} d_j x_j + \sum_{\text{basic}} 0 \cdot x_i \tag{11}$$

where the d_j's are the negatives of the entries in the objective row of the tableau. We see that (11) has some terms with positive coefficients if and only if the objective row has negative entries under some of the labeled columns. Now the value of z can be increased by increasing the value of any nonbasic variable with a negative entry in the objective row from its current value of 0. For setting a basic variable to zero will not change the value of the objective function, since the coefficient of the basic variable was 0, and changing a nonbasic variable with positive coefficient in (11) to a positive value will increase the value of z. We summarize this discussion by stating the following optimality criterion for testing whether a feasible solution shown in a tableau is an optimal solution.

Optimality Criterion. If the objective row of a tableau has zero entries in the columns labeled by basic variables and no negative entries in the columns labeled by nonbasic variables, then the solution represented by the tableau is optimal.

As soon as the optimality criterion has been met we can stop our computations, for we have found an optimal solution.

Selecting the Entering Variable

Suppose now that the objective row of a tableau has negative entries in the labeled columns. Then the solution shown in the tableau is not optimal, and some adjustment of the values of the variables must be made.

The simplex method proceeds from a given extreme point (basic feasible solution) to an *adjacent* extreme point in such a way that the objective function increases in value. From the definition of adjacent extreme point, it is clear that we reach such a point by increasing a single variable from zero to a positive value. The largest increase in z per unit increase in a variable occurs for the most negative entry in the objective row. We shall see below that if the feasible set is bounded, there is a limit on the amount by which we can increase a variable. Because of this limit, it may turn out that a larger increase in z may be achieved by *not* increasing the variable with the most negative entry in the objective row. However, this rule is most commonly followed because of its computational simplicity. Some computer implementations of the simplex algorithm provide other strategies for choosing the variable to be increased, including one as simple as choosing the first negative entry. Another compares increases in the objective function for several likely candidates for the entering variable. In Tableau 2.1, the most negative entry, -120, in the objective row occurs under the x column, so that x is chosen to be the variable to be increased from zero to a positive value. The variable to be increased is called the **entering variable**, since in the next iteration it will become a basic variable; that is, it will *enter* the set of basic variables. If there are several possible entering variables, choose one. (This situation will occur when the most negative entry in the objective row occurs in more than one column.) Now an increase in one variable must be accompanied by a decrease in some of the other variables to maintain a solution to $\mathbf{A}\mathbf{x} = \mathbf{b}$.

Solving (8) for the basic variables u and v, we have

$$u = 8 - 2x - 2y$$
$$v = 15 - 5x - 3y$$

We increase only x and keep y at zero. We have

$$\left.\begin{array}{l} u = 8 - 2x \\ v = 15 - 5x \end{array}\right\} \tag{12}$$

which shows that as x increases both u and v decrease. By how much can we increase x? It can be increased until either u or v becomes negative. That is, from (9) we have

$$0 \le u = 8 - 2x$$
$$0 \le v = 15 - 5x$$

Solving these inequalities for x, we find

$$2x \leq 8 \quad \text{or} \quad x \leq 8/2 = 4$$

and

$$5x \leq 15 \quad \text{or} \quad x \leq 15/5 = 3$$

We see that we cannot increase x by more than the smaller of the two ratios 8/2 and 15/5. Letting $x = 3$, we obtain a new feasible solution

$$x = 3, \quad y = 0, \quad u = 2, \quad v = 0$$

In fact, this is a basic feasible solution, and it was constructed to be adjacent to the previous basic feasible solution, since we changed only one basic variable. The new basic variables are x and u; the nonbasic variables are y and v. The objective function now has the value

$$z = 120 \cdot 3 + 100 \cdot 0 + 0 \cdot 2 + 0 \cdot 0 = 360$$

which is a considerable improvement over the previous value of zero.

The new basic feasible solution yields the extreme point (3, 0) in Figure 2.2, and it is adjacent to (0, 0).

Choosing the Departing Variable

In the new basic feasible solution to our example, we have the variable $v = 0$. It is no longer a basic variable because it is zero, and it is called a **departing variable** since it has *departed* from the set of basic variables. The column of the entering variable is called the **pivotal column**; the row which is labeled with the departing variable is called the **pivotal row**.

We now examine more carefully the selection of the departing variable. Recall that the ratios of the rightmost column entries to the corresponding entries in the pivotal column determined by how much we could increase the entering variable (x in our example). These ratios are called θ-**ratios**. The smallest nonnegative θ-ratio is the largest possible value for the entering variable. The basic variable labeling the row where the smallest nonnegative θ-ratio occurs is the departing variable, and the row is the pivotal row. In our example,

$$\min \{8/2, \ 15/5\} = 3$$

and the second row in Tableau 2.1 is the pivotal row.

If the smallest nonnegative θ-ratio is not chosen, then the next basic solution is not feasible. Suppose we had chosen u as the departing variable by choosing the θ-ratio as 4. Then $x = 4$, and from (12) we have

$$u = 8 - 2 \cdot 4 = 0$$
$$v = 15 - 5 \cdot 4 = -5$$

and the next basic solution is

$$x = 4, \quad y = 0, \quad u = 0, \quad v = -5$$

which is not feasible.

In the general case, we have assumed that the rightmost column will contain only nonnegative entries. However the entries in the pivotal column may be positive, negative, or zero. Positive entries lead to non-negative θ-ratios, which are fine. Negative entries lead to nonpositive θ-ratios. In this case, there is no restriction imposed on how far the entering variable can be increased. For example, suppose the pivotal column in our example were

$$\begin{bmatrix} -2 \\ 5 \end{bmatrix} \text{ instead of } \begin{bmatrix} 2 \\ 5 \end{bmatrix}$$

Then we would have, instead of (12),

$$u = 8 + 2x$$
$$v = 15 - 5x$$

Since u must be nonnegative, we find that

$$8 + 2x \geq 0 \quad \text{or} \quad x \geq -4$$

which puts no restriction on how far we can increase x. Thus, in calculating θ-ratios we can ignore any negative entries in the pivotal column.

If an entry in the pivotal column is zero, the corresponding θ-ratio is undefined. However, checking the equations corresponding to (12) but with one of the entries in the pivotal column equal to zero will show that no restriction is placed on the size of x by the zero entry. Consequently, in forming the θ-ratios we only use the positive entries in the pivotal column which are above the objective row.

If all the entries in the pivotal column above the objective row are either zero or negative, then the entering variable can be made as large as we wish. Hence, the given problem has no finite optimal solution, and we can stop.

Forming a New Tableau

Having determined the entering and departing variables, we must obtain a new tableau showing the new basic variables and the new basic feasible solution.

We illustrate the procedure with our continuing example. Solving the second equation of (8) (it corresponds to the departing variable) for x, the entering variable, we have

$$x = 3 - \tfrac{3}{5}y - \tfrac{1}{5}v \tag{13}$$

Substituting (13) into the first equation of (8), we get

$$2(3 - \tfrac{3}{5}y - \tfrac{1}{5}v) + 2y + u = 8$$

or

$$\tfrac{4}{5}y + u - \tfrac{2}{5}v = 2 \qquad (14)$$

We also rewrite (13) as

$$x + \tfrac{3}{5}y + \tfrac{1}{5}v = 3 \qquad (15)$$

Substituting (13) into (7), we have

$$(-120)(3 - \tfrac{3}{5}y - \tfrac{1}{5}v) - 100y + z = 0$$

or

$$-28y + 24v + z = 360 \qquad (16)$$

Since in the new basic feasible solution we have $y = v = 0$, the value of z for this solution is 360.

Equations (14), (15), and (16) yield the new tableau (Tableau 2.3).

Tableau 2.3

	x	y	u	v	z	
u	0	$\tfrac{4}{5}$	1	$-\tfrac{2}{5}$	0	2
x	1	$\tfrac{3}{5}$	0	$\tfrac{1}{5}$	0	3
	0	-28	0	24	1	360

Observe that the basic variables in Tableau 2.3 are x and u. By comparing Tableaus 2.1 and 2.3, we see that the steps which were used to obtain Tableau 2.3 from Tableau 2.1 are:

Step a. Locate and circle the entry at the intersection of the pivotal row and pivotal column. This entry is called the **pivot**. Mark the pivotal column by placing an arrow $\downarrow$ above the entering variable, and mark the pivotal row by placing an arrow $\leftarrow$ to the left of the departing variable.

Step b. If the pivot is k, multiply the pivotal row by $1/k$, making the entry in the pivot position in the new tableau equal to 1.

Step c. Add suitable multiples of the new pivotal row to all other rows (including the objective row) so that all other elements in the pivotal column become zero.

Step d. In the new tableau replace the label on the pivotal row by the entering variable.

These four steps constitute a process called **pivoting**. Steps b and c use what are called elementary row operations and form one iteration of the procedure used to transform a given matrix to reduced row echelon form.

We now repeat Tableau 2.1 with the arrows placed next to the entering and departing variables and with the pivot circled (Tableau 2.1a).

Tableau 2.1a

		x	y	u	v	z	
	u	2	2	1	0	0	8
←	v	⑤	3	0	1	0	15
		-120	-100	0	0	1	0

Tableau 2.3 was obtained from Tableau 2.1 by pivoting. We now repeat the process with Tableau 2.3. Since the most negative entry in the objective row of Tableau 2.3, -28, occurs in the second column, y is the entering variable of this tableau and the second column is the pivotal column. To find the departing variable we form the θ-ratios, that is, the ratios of the entries in the rightmost column (except for the objective row) to the corresponding entries of the pivotal column for those entries in the pivotal column which are positive. The θ-ratios are

$$\frac{2}{\frac{4}{5}} = \frac{5}{2} \quad \text{and} \quad \frac{3}{\frac{3}{5}} = 5$$

The minimum of these is $\frac{5}{2}$, which occurs for the first row. Therefore, the pivotal row is the first row, the pivot is $\frac{4}{5}$, and the departing variable is u. We now show Tableau 2.3 with the pivot, entering, and departing variables marked (Tableau 2.3a).

Tableau 2.3a

		x	y	u	v	z	
←	u	0	④⁄₅	1	$-\frac{2}{5}$	0	2
	x	1	$\frac{3}{5}$	0	$\frac{1}{5}$	0	3
		0	-28	0	24	1	360

We obtain Tableau 2.4 from Tableau 2.3 by pivoting.

Since the objective row in Tableau 2.4 has no negative entries, we are finished, by the optimality criterion. That is, the indicated solution

$$x = \tfrac{3}{2}, \quad y = \tfrac{5}{2}, \quad u = 0, \quad v = 0$$

Tableau 2.4

	x	y	u	v	z	
y	0	1	$\frac{5}{4}$	$-\frac{1}{2}$	0	$\frac{5}{2}$
x	1	0	$-\frac{3}{4}$	$\frac{1}{2}$	0	$\frac{3}{2}$
	0	0	35	10	1	430

is optimal, and the maximum value of z is 430. Notice from Figure 2.2 that we moved from the extreme point (0, 0) to the adjacent extreme point (3, 0) and then to the adjacent extreme point $(\frac{3}{2}, \frac{5}{2})$. The value of the objective function started at 0, increased to 360, and then to 430.

Summary of the Simplex Method

We assume that the linear programming problem is in standard form and that $\mathbf{b} \geq \mathbf{0}$. In this case the initial basic feasible solution is

$$\mathbf{x} = \begin{bmatrix} \mathbf{0} \\ \mathbf{b} \end{bmatrix}$$

In subsequent sections we will show how to extend the simplex method to other linear programming problems.

Step 1. Set up the initial tableau.

Step 2. Apply the optimality test: If the objective row has no negative entries in the labeled columns, then the indicated solution is optimal. Stop our computation.

Step 3. Find the pivotal column by determining the column with the most negative entry in the objective row. If there are several possible pivotal columns, choose any one.

Step 4. Find the pivotal row. This is done by forming the **θ-ratios**—the ratios formed by dividing the entries of the rightmost column (except for the objective row) by the corresponding entries of the pivotal columns using only those entries in the pivotal column which are positive. The **pivotal row** is the row for which the minimum ratio occurs. If two or more θ-ratios are the same, choose one of the possible rows. If none of the entries in the pivotal column above the objective row is positive, the problem has no finite optimum. We stop our computation in this case.

Step 5. Obtain a new tableau by pivoting. Then return to Step 2.

In Figure 2.6 we give a flowchart and in Figure 2.7 a structure diagram for the simplex algorithm.

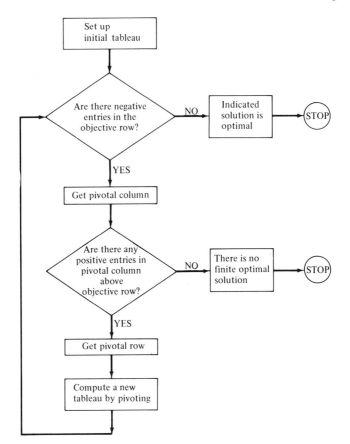

Figure 2.6 Flowchart for simplex algorithm (standard form, **b** ≥ **0**).

The reader should note that the z column always appears in the form

z
0
0
⋮
0
1

in any simplex tableau. We included it initially to remind the reader that each row of a tableau including the objective row represents an equation in the variables x_1, x_2, . . ., x_s, z. From this point on we will not include the z column in tableaux. The student should remember to read the

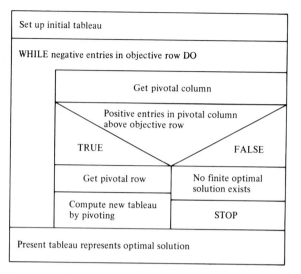

Figure 2.7 Structure diagram of simplex algorithm (standard form, $\mathbf{b} \geq \mathbf{0}$).

objective row of a tableau as an equation which involves z with coefficient $+1$.

EXAMPLE 2. We solve the following linear programming problem in standard form by using the simplex method:

$$\text{Maximize} \quad z = 8x_1 + 9x_2 + 5x_3$$

subject to

$$x_1 + x_2 + 2x_3 \leq 2$$
$$2x_1 + 3x_2 + 4x_3 \leq 3$$
$$6x_1 + 6x_2 + 2x_3 \leq 8$$
$$x_1 \geq 0, \quad x_2 \geq 0, \quad x_3 \geq 0$$

We first convert the problem to canonical form by adding slack variables, obtaining:

$$\text{Maximize} \quad z = 8x_1 + 9x_2 + 5x_3$$

subject to

$$x_1 + x_2 + 2x_3 + x_4 \qquad\qquad = 2$$
$$2x_1 + 3x_2 + 4x_3 \qquad + x_5 \qquad = 3$$
$$6x_1 + 6x_2 + 2x_3 \qquad\qquad + x_6 = 8$$
$$x_j \geq 0, \quad j = 1, 2, \ldots, 6$$

Tableau 2.5

	x_1	x_2	x_3	x_4	x_5	x_6	
x_4	1	1	2	1	0	0	2
← x_5	2	③	4	0	1	0	3
x_6	6	6	2	0	0	1	8
	-8	-9	-5	0	0	0	0

Tableau 2.6

	x_1	x_2	x_3	x_4	x_5	x_6	
x_4	$\frac{1}{3}$	0	$\frac{2}{3}$	1	$-\frac{1}{3}$	0	1
x_2	$\frac{2}{3}$	1	$\frac{4}{3}$	0	$\frac{1}{3}$	0	1
← x_6	②	0	-6	0	-2	1	2
	-2	0	7	0	3	0	9

Tableau 2.7

	x_1	x_2	x_3	x_4	x_5	x_6	
x_4	0	0	$\frac{5}{3}$	1	0	$-\frac{1}{6}$	$\frac{2}{3}$
x_2	0	1	$\frac{10}{3}$	0	1	$-\frac{1}{3}$	$\frac{1}{3}$
x_1	1	0	-3	0	-1	$\frac{1}{2}$	1
	0	0	1	0	1	1	11

The initial tableau is Tableau 2.5; the succeeding tableaux are Tableaux 2.6 and 2.7.

Hence, an optimal solution to the standard form of the problem is

$$x_1 = 1, \quad x_2 = \tfrac{1}{3}, \quad x_3 = 0$$

The values of the slack variables are

$$x_4 = \tfrac{2}{3}, \quad x_5 = 0, \quad x_6 = 0$$

The optimal value of z is 11.

2.3 Exercises

In Exercises 1 and 2, set up the initial simplex tableau.

1. Maximize $z = 2x + 5y$

subject to

$$3x + 5y \le 8$$

$$2x + 7y \le 12$$

$$x \ge 0, \quad y \ge 0$$

2. Maximize $z = x_1 + 3x_2 + 5x_3$

subject to

$$2x_1 - 5x_2 + x_3 \leq 3$$
$$x_1 + 4x_2 \qquad \leq 5$$
$$x_1 \geq 0, \quad x_2 \geq 0, \quad x_3 \geq 0$$

3. Consider the simplex tableau:

	x_1	x_2	x_3	x_4	x_5	x_6	x_7	
x_4	0	0	2	1	$\frac{5}{2}$	0	0	$\frac{6}{7}$
x_1	1	0	5	0	-3	0	-2	$\frac{2}{7}$
x_6	0	0	3	0	4	1	-4	$\frac{5}{7}$
x_2	0	1	0	0	$\frac{3}{2}$	0	0	$\frac{1}{7}$

Determine the departing variable if the entering variable is (a) x_5; (b) x_3; (c) x_7.

In Exercises 4–7 use one iteration of the simplex algorithm to obtain the next tableau from the given tableau.

4.

	x_1	x_2	x_3	x_4	
x_4	$\frac{3}{2}$	0	$\frac{5}{3}$	1	6
x_2	$\frac{2}{3}$	1	2	0	8
	-4	0	-2	0	12

5.

	x_1	x_2	x_3	x_4	
x_1	1	2	0	1	3
x_3	0	$\frac{1}{2}$	1	-1	$\frac{3}{2}$
	0	-4	0	-4	$\frac{11}{2}$

6.

	x_1	x_2	x_3	x_4	x_5	
x_3	$\frac{2}{3}$	0	1	$\frac{3}{5}$	0	$\frac{3}{2}$
x_2	$\frac{3}{2}$	1	0	1	0	$\frac{5}{3}$
x_5	5	0	0	$\frac{2}{9}$	1	$\frac{2}{3}$
	4	0	0	-5	0	$\frac{7}{3}$

7.

	x_1	x_2	x_3	x_4	
x_2	1	1	5	0	4
x_4	−1	0	2	1	6
	−3	0	−2	0	7

8. The following tableau arose in the course of using the simplex algorithm to solve a linear programming problem. What basic feasible solution does this tableau represent?

x_1	x_2	x_3	x_4	x_5	x_6	x_7	
0	$\frac{4}{3}$	$\frac{2}{3}$	0	1	0	$-\frac{1}{3}$	4
0	$\frac{1}{3}$	$\frac{2}{3}$	1	0	1	$-\frac{1}{3}$	10
1	$\frac{1}{3}$	$\frac{1}{6}$	$\frac{1}{2}$	0	0	$\frac{1}{6}$	4
0	$-\frac{5}{3}$	$-\frac{4}{3}$	−1	0	0	$\frac{5}{3}$	12

9. Perform one operation of the simplex algorithm on the tableau in Exercise 8. What basic feasible solution does this new tableau represent?

In Exercises 10–20 solve the indicated linear programming problem using the simplex method.

10. Example 4, Section 2.1 (page 63).

11. Example 7(a), Section 2.1 (page 67).

12. Example 7(b), Section 2.1 (page 67).

13. Example 10, Section 2.1 (page 70).

14. Exercise 4, Section 2.1 (page 72).

15. Exercise 5, Section 2.1 (page 72).

16. Exercise 7, Section 2.1 (page 73).

17. Exercise 9, Section 2.1 (page 74).

18. Exercise 17, Section 2.2 (page 98).

19. Maximize $z = x_1 + 2x_2 + x_3 + x_4$

subject to

$$2x_1 + x_2 + 3x_3 + x_4 \le 8$$
$$2x_1 + 3x_2 \qquad + 4x_4 \le 12$$
$$3x_1 + x_2 + 2x_3 \qquad \le 18$$
$$x_j \ge 0, \qquad j = 1, 2, 3, 4$$

20. Maximize $z = 5x_1 + 2x_2 + x_3 + x_4$

subject to

$$2x_1 + x_2 + x_3 + 2x_4 \leq 6$$

$$3x_1 + x_3 \leq 15$$

$$5x_1 + 4x_2 + x_4 \leq 24$$

$$x_j \geq 0, \quad j = 1, 2, 3, 4$$

21. Suppose a linear programming problem has a constraint of the form

$$3x_1 + 2x_2 + 5x_3 - 2x_4 \geq 12$$

Why can we not solve this problem using the simplex method as described up to this point? (In Section 2.5 we develop techniques for handling this situation.)

2.4 DEGENERACY AND CYCLING (Optional)

In choosing the departing variable we computed the minimum θ-ratio. If the minimum θ-ratio occurs, say, in the rth row of a tableau, we drop the variable which labels that row. Now suppose that there is a tie for minimum θ-ratio so that several variables are candidates for departing variable. We choose one of the candidates by using an arbitrary rule such as dropping the variable with the smallest subscript. However, there are potential difficulties any time such an arbitrary choice must be made. We now examine these difficulties.

Suppose that the θ-ratios for the rth and sth rows of a tableau are the same and their value is the minimum value of all the θ-ratios. These two rows of the tableau are shown in Tableau 2.8 with the label on the rth row marked as the departing variable. The θ-ratios of these two rows are

$$b_r/a_{rj} = b_s/a_{sj}$$

When we do pivoting to Tableau 2.8 we obtain Tableau 2.9, where *

Tableau 2.8

		x_1	x_2	$\cdots$	x_j	$\cdots$	x_{n+m}	
		$\vdots$	$\vdots$		$\vdots$		$\vdots$	$\vdots$
$\leftarrow$	x_{i_r}	a_{r1}	a_{r2}	$\cdots$	$\boxed{a_{rj}}$	$\cdots$	$a_{r,n+m}$	b_r
		$\vdots$	$\vdots$		$\vdots$		$\vdots$	$\vdots$
	x_{i_s}	a_{s1}	a_{s2}	$\cdots$	a_{sj}	$\cdots$	$a_{s,n+m}$	b_s

Tableau 2.9

	x_1	x_2	$\cdots$	x_j	$\cdots$	x_{n+m}	
$\vdots$	$\vdots$	$\vdots$		$\vdots$		$\vdots$	$\vdots$
x_j	a_{r1}/a_{rj}	a_{r2}/a_{rj}	$\cdots$	1	$\cdots$	$a_{r,n+m}/a_{rj}$	b_r/a_{rj}
$\vdots$	$\vdots$	$\vdots$		$\vdots$		$\vdots$	$\vdots$
x_{i_s}	$*$	$*$	$\cdots$	0	$\cdots$	$*$	$b_s - a_{sj} \cdot b_r / a_{rj}$

indicates an entry whose value we are not concerned about. Setting the nonbasic variables in Tableau 2.9 equal to zero, we find that

$$x_j = b_r/a_{rj}$$

and

$$x_{i_s} = b_s - a_{sj} \cdot b_r/a_{rj} = a_{sj}(b_s/a_{sj} - b_r/a_{rj}) = 0$$

Consequently, the tie among the θ-ratios has produced a basic variable whose value is 0.

Definition. A basic feasible solution in which some basic variables are zero is called **degenerate**.

If no degenerate solution occurs in the course of the simplex method, then the value of z increases as we go from one basic feasible solution to an adjacent basic feasible solution. Since the number of basic feasible solutions is finite, the simplex method eventually stops. However, if we have a degenerate basic feasible solution and if a basic variable whose value is zero departs, then the value of z does not change. To see this, remember that z increases by a multiple of the value in the rightmost column of the pivotal row. But this value is zero, so that z does not increase. Therefore, we may return to a basic feasible solution which we already encountered. Now the simplex method is in a cycle and will never terminate. Thus, we cannot prove that the version of the simplex method presented earlier will terminate in all cases.

Degeneracy occurs often in problems arising from real situations. However, cycling is encountered only occasionally in practical problems. Kotiah and Steinberg (see the Further Readings at the end of this chapter) have discovered a linear programming problem arising in the solution of a practical queuing model which cycles. Workers in linear programming who deal extensively with computing problems have encountered other examples of cycling in practice.

Computer programs designed for large linear programming problems provide several options for dealing with degeneracy. One option is to ignore it. Another allows the objective function to remain constant for a specified number of iterations and then invokes some degeneracy-breaking strategy. One such strategy is called **perturbation**, which slightly alters the right side of the constraint equations, giving a new problem where ties among the θ-ratios no longer occur. The infrequent occurrence of cycling has been attributed to the fact that roundoff perturbs the problem.

EXAMPLE 1. **(Degeneracy)** Consider the linear programming problem in standard form:

$$\text{Maximize} \quad z = 5x_1 + 3x_2$$

subject to

$$x_1 - x_2 \le 2$$
$$2x_1 + x_2 \le 4$$
$$-3x_1 + 2x_2 \le 6$$
$$x_1 \ge 0, \quad x_2 \ge 0$$

The region of all feasible solutions is shown in Figure 2.8. The extreme points and corresponding values of the objective function are given in Table 2.6. The simplex method leads to the following tableaux. In Tableau 2.10 we have two candidates for the departing variable: x_3 and x_4. Choosing x_3 gives Tableaux 2.10, 2.11, 2.12, and 2.13. Choosing x_4 gives Tableaux 2.10a, 2.11a, and 2.12a. Note that Tableaux 2.12a and 2.13 are the same except for the order of the constraint rows.

The optimal solution is

$$x_1 = \tfrac{2}{7} \quad \text{and} \quad x_2 = \tfrac{24}{7}$$

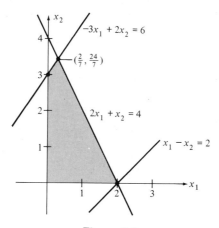

Figure 2.8

Table 2.6

Extreme point	Value of $z = 5x_1 + 3x_2$
$(0, 0)$	0
$(2, 0)$	10
$(0, 3)$	9
$(\frac{2}{7}, \frac{24}{7})$	$\frac{82}{7}$

Tableau 2.10

$\downarrow$

	x_1	x_2	x_3	x_4	x_5	
$\leftarrow$ x_3	①	-1	1	0	0	2
x_4	2	1	0	1	0	4
x_5	-3	2	0	0	1	6
	-5	-3	0	0	0	0

Tableau 2.11

$\downarrow$

	x_1	x_2	x_3	x_4	x_5	
x_1	1	-1	1	0	0	2
$\leftarrow$ x_4	0	③	-2	1	0	0
x_5	0	-1	3	0	1	12
	0	-8	5	0	0	10

Tableau 2.12

$\downarrow$

	x_1	x_2	x_3	x_4	x_5	
x_1	1	0	$\frac{1}{3}$	$\frac{1}{3}$	0	2
x_2	0	1	$-\frac{2}{3}$	$\frac{1}{3}$	0	0
$\leftarrow$ x_5	0	0	$\boxed{\frac{7}{3}}$	$\frac{1}{3}$	1	12
	0	0	$-\frac{1}{3}$	$\frac{8}{3}$	0	10

Tableau 2.13

	x_1	x_2	x_3	x_4	x_5	
x_1	1	0	0	$\frac{2}{7}$	$-\frac{1}{7}$	$\frac{2}{7}$
x_2	0	1	0	$\frac{3}{7}$	$\frac{2}{7}$	$\frac{24}{7}$
x_3	0	0	1	$\frac{1}{7}$	$\frac{3}{7}$	$\frac{36}{7}$
	0	0	0	$\frac{19}{7}$	$\frac{1}{7}$	$\frac{82}{7}$

Tableau 2.10a

	x_1	x_2	x_3	x_4	x_5	
x_3	1	-1	1	0	0	2
x_4	②	1	0	1	0	4
x_5	-3	2	0	0	1	6
	-5	-3	0	0	0	0

(← at x_4 row; ↓ at x_1 column)

Tableau 2.11a

	x_1	x_2	x_3	x_4	x_5	
x_3	0	$-\frac{3}{2}$	1	$-\frac{1}{2}$	0	0
x_1	1	$\frac{1}{2}$	0	$\frac{1}{2}$	0	2
x_5	0	⑦⁄₂	0	$\frac{3}{2}$	1	12
	0	$-\frac{1}{2}$	0	$\frac{5}{2}$	0	10

(← at x_5 row; ↓ at x_2 column)

Tableau 2.12a

	x_1	x_2	x_3	x_4	x_5	
x_3	0	0	1	$\frac{1}{4}$	$\frac{3}{7}$	$\frac{36}{7}$
x_1	1	0	0	$\frac{2}{7}$	$-\frac{1}{7}$	$\frac{2}{7}$
x_2	0	1	0	$\frac{3}{7}$	$\frac{2}{7}$	$\frac{24}{7}$
	0	0	0	$\frac{19}{7}$	$\frac{1}{7}$	$\frac{82}{7}$

with the optimal value of the objective function as

$$z = \tfrac{82}{7}$$

The slack variables have values

$$x_3 = \tfrac{36}{7}, \quad x_4 = 0, \quad x_5 = 0$$

What is happening geometrically? We start with the initial basic feasible solution as the origin $(0, 0)$, where $z = 0$. If we choose to replace x_3 with x_1, we move to the adjacent extreme point $(2, 0)$ where $z = 10$ (Tableau 2.11). Now we replace x_4 by x_2 and remain at $(2, 0)$ (Tableau 2.12). Finally we replace x_5 with x_3 and move to $(\tfrac{2}{7}, \tfrac{24}{7})$, where $z = \tfrac{82}{7}$. This is our optimal solution (Tableau 2.13).

If we choose to replace x_4 with x_1 at $(0,0)$ where the Θ-ratios are equal we again move to $(2,0)$, where $z = 10$ (Tableau 211a). However, at the next stage, x_3, which has value 0 and is the degenerate variable, is not a departing variable. Instead, x_5 is the departing variable, and we move immediately to the optimal solution (Tableau 2.12a) at $(\tfrac{2}{7}, \tfrac{24}{7})$.

2.4 Exercises

In Exercises 1–5 solve the indicated linear programming problem, noting where degeneracies occur. Sketch the set of feasible solutions, indicating the order in which the extreme points are examined by the simplex algorithm.

1. Maximize $z = 6x_1 + 5x_2$

   subject to

$$3x_1 - 2x_2 \leq 0$$
$$3x_1 + 2x_2 \leq 15$$
$$x_1 \geq 0, \quad x_2 \geq 0$$

2. Maximize $z = 5x_1 + 4x_2$

   subject to

$$x_1 + 2x_2 \leq 8$$
$$x_1 - 2x_2 \leq 4$$
$$3x_1 + 2x_2 \leq 12$$
$$x_1 \geq 0, \quad x_2 \geq 0$$

3. Maximize $z = 3x_1 + 2x_2 + 5x_3$

   subject to

$$2x_1 - x_2 + 4x_3 \leq 12$$
$$4x_1 + 3x_2 + 6x_3 \leq 18$$
$$x_1 \geq 0, \quad x_2 \geq 0, \quad x_3 \geq 0$$

4. Maximize $z = 5x_1 + 8x_2 + x_3$

   subject to

$$x_1 + x_2 + x_3 \leq 7$$
$$2x_1 + 3x_2 + 3x_3 \leq 12$$
$$3x_1 + 6x_2 + 5x_3 \leq 24$$
$$x_1 \geq 0, \quad x_2 \geq 0, \quad x_3 \geq 0$$

5. Maximize $z = 6x_1 + 5x_2$

   subject to

$$4x_1 + 3x_2 \leq 19$$

$$x_1 - x_2 \leq 3$$

$$x_1 - 2x_2 \leq 2$$

$$3x_1 + 4x_2 \leq 18$$

$$x_1 \geq 0, \quad x_2 \geq 0$$

6. If a degeneracy arose in any of the exercises above, use all choices of departing variable.

2.5   ARTIFICIAL VARIABLES

In the previous two sections we discussed the simplex algorithm as a method of solving certain types of linear programming problems. Recall that we restricted our discussion to those linear programming problems which could be put in standard form, and in that form we considered only those problems which had a *nonnegative right-hand side*. That is, we have assumed that all our constraints were of the form

$$\sum_{j=1}^{n} a_{ij}x_j \leq b_i, \qquad i = 1, 2, \ldots, m \tag{1}$$

where

$$b_i \geq 0$$

The assumption that $\mathbf{b} \geq \mathbf{0}$ enabled us to easily find an initial basic feasible solution. For when we introduced slack variables, we found that we could set the nonslack variables equal to zero and obtain the basic solution

$$\mathbf{x} = (0, 0, \ldots, 0, b_1, b_2, \ldots, b_m)$$

Furthermore, this solution was feasible since $b_i \geq 0$ for $i = 1, 2, \ldots, m$.

Unfortunately, there are many linear programming problems where not all the constraints can be transformed to the form of (1). For example, the constraint

$$2x + 3y \geq 4 \tag{2}$$

can be changed to

$$-2x - 3y \leq -4$$

but then the right-hand side is negative. By adding a slack variable, we obtain

$$-2x - 3y + u = -4 \tag{3}$$

Setting the nonslack variables equal to zero gives us $u = -4$, which will not yield a feasible solution.

When we examine the procedure described above for finding an initial basic feasible solution, we see that it was *not* important that the procedure was applied to a problem in canonical form coming from a problem in standard form. It was important that the problem was in canonical form, that the right-hand side of $\mathbf{Ax} = \mathbf{b}$ was nonnegative, and that in each equation of the system of constraints there was a variable with coefficient $+1$ which appeared in no other equation. Then setting all but these "special" variables equal to zero, we again would have an initial basic feasible solution.

However, there may be equations in the canonical form of a problem where no such "special" variable exists. In Equation (3) we can make the right-hand side positive so that the equation reads

$$2x + 3y - u = 4$$

But now no coefficient is $+1$, and presumably x and y appear in other equations of the system of constraints, so that they cannot be chosen as "special" variables.

We now develop a procedure for introducing a variable which can serve as a basic variable for an initial basic feasible solution. We start with the general linear programming problem

$$\text{Maximize} \quad z = \mathbf{c}^T\mathbf{x} \tag{4}$$

subject to

$$\left.\begin{array}{l} a_{11}x_1 + a_{12}x_2 + \cdots + a_{1n}x_n \ (\leq)\,(=)\,(\geq)\ b_1 \\ a_{21}x_1 + a_{22}x_2 + \cdots + a_{2n}x_n \ (\leq)\,(=)\,(\geq)\ b_2 \\ \quad\vdots \qquad\quad \vdots \qquad\qquad\quad \vdots \qquad\qquad\quad\ \vdots \\ a_{m1}x_1 + a_{m2}x_2 + \cdots + a_{mn}x_n \ (\leq)\,(=)\,(\geq)\ b_m \end{array}\right\} \tag{5}$$

$$x_j \geq 0, \qquad j = 1, 2, \ldots, n \tag{6}$$

Note that we have assumed that the problem is a maximization problem. Any minimization problem can be converted to a maximization problem by multiplying the objective function by -1.

We now write each constraint in (5) so that its right-hand side is nonnegative. This can be done by multiplying those constraints with negative right-hand sides by -1. The constraints in (5) can now be renumbered so that the $\leq$ constraints come first, then the $\geq$ constraints, and finally the $=$ constraints. Assume there are $r_1 \leq$ constraints, $r_2 \geq$ constraints and $r_3 =$ constraints. Then we can assume (5) is in the following form:

$$
\left.\begin{array}{rl}
a_{11}x_1 + a_{12}x_2 + \cdots + a_{1n}x_n \leq b_1, & \quad b_1 \geq 0 \\
a_{21}x_1 + a_{22}x_2 + \cdots + a_{2n}x_n \leq b_2, & \quad b_2 \geq 0 \\
\vdots \quad \vdots \qquad\qquad \vdots \qquad\quad \vdots & \quad \vdots \\
a_{r_1 1}x_1 + a_{r_1 2}x_2 + \cdots + a_{r_1 n}x_n \leq b_{r_1}, & \quad b_{r_1} \geq 0
\end{array}\right\} \quad (7a)
$$

$$
\left.\begin{array}{rl}
a'_{11}x_1 + a'_{12}x_2 + \cdots + a'_{1n}x_n \geq b'_1, & \quad b'_1 \geq 0 \\
a'_{21}x_1 + a'_{22}x_2 + \cdots + a'_{2n}x_n \geq b'_2, & \quad b'_2 \geq 0 \\
\vdots \quad \vdots \qquad\qquad \vdots \qquad\quad \vdots & \quad \vdots \\
a'_{r_2 1}x_1 + a'_{r_2 2}x_2 + \cdots + a'_{r_2 n}x_n \geq b'_{r_2}, & \quad b'_{r_2} \geq 0
\end{array}\right\} \quad (7b)
$$

$$
\left.\begin{array}{rl}
a''_{11}x_1 + a''_{12}x_2 + \cdots + a''_{1n}x_n = b''_1, & \quad b''_1 \geq 0 \\
a''_{21}x_1 + a''_{22}x_2 + \cdots + a''_{2n}x_n = b''_2, & \quad b''_2 \geq 0 \\
\vdots \quad \vdots \qquad\qquad \vdots \qquad\quad \vdots & \quad \vdots \\
a''_{r_3 1}x_1 + a''_{r_3 2}x_2 + \cdots + a''_{r_3 n}x_n = b''_{r_3}, & \quad b''_{r_3} \geq 0
\end{array}\right\} \quad (7c)
$$

Next we make each inequality in (7a) and (7b) into an equality by introducing a slack variable. Each slack variable must be nonnegative so that in (7a) it is introduced with a $+$ sign and in (7b) it is introduced with a $-$ sign. We now write (7a) and (7b) as:

$$
\sum_{j=1}^{n} a_{ij}x_j + x_{n+i} = b_i, \qquad b_i \geq 0, \qquad i = 1, 2, \ldots, r_1 \qquad (8a)
$$

$$
\sum_{j=1}^{n} a'_{ij}x_j - x_{n+r_1+i} = b'_i, \qquad b'_i \geq 0, \qquad i = 1, 2, \ldots, r_2 \qquad (8b)
$$

EXAMPLE 1. Consider the linear programming problem

$$
\text{Maximize} \quad z = 2x_1 + 5x_2
$$

subject to

$$
2x_1 + 3x_2 \leq 6
$$
$$
-2x_1 + x_2 \leq -2
$$
$$
x_1 - 6x_2 = -2
$$
$$
x_1 \geq 0, \quad x_2 \geq 0
$$

We first rewrite the constraints so that the right-hand sides are nonnegative. The problem becomes

$$\text{Maximize} \quad z = 2x_1 + 5x_2$$

subject to

$$2x_1 + 3x_2 \le 6$$
$$2x_1 - x_2 \ge 2$$
$$-x_1 + 6x_2 = 2$$
$$x_1 \ge 0, \quad x_2 \ge 0$$

We now insert slack variables in each of the inequalities, obtaining an associated canonical form problem:

$$\text{Maximize} \quad z = 2x_1 + 5x_2$$

subject to

$$\left. \begin{array}{l} 2x_1 + 3x_2 + x_3 \quad\quad = 6 \\ 2x_1 - x_2 \quad\quad - x_4 = 2 \\ -x_1 + 6x_2 \quad\quad = 2 \end{array} \right\} \qquad (9)$$

$$x_j \ge 0, \quad j = 1, 2, 3, 4$$

However, we do not have a variable in each equation which can act as a basic variable for an initial basic feasible solution. In fact, in (9) both the second and third equations do not have such a variable.

As Example 1 shows, we can convert the general linear programming problem in (4), (5), and (6) to an associated problem in canonical form with $\mathbf{b} \ge \mathbf{0}$. Thus, we may now assume that we have a problem in the following form:

$$\text{Maximize} \quad z = \sum_{j=1}^{s} c_j x_j \qquad (10)$$

subject to

$$\sum_{j=1}^{s} a_{ij} x_j = b_i, \qquad i = 1, 2, \ldots, m \qquad (11)$$

$$x_j \ge 0, \qquad j = 1, 2, \ldots, s \qquad (12)$$

with $b_i \ge 0, \quad i = 1, 2, \ldots, m$.

The following method of finding an initial basic feasible solution is widely used in modern computer codes.

Two-Phase Method

To enable us to obtain an initial basic feasible solution, we introduce another variable into each equation in (11). We denote the variable for the ith equation by y_i. The variables y_i, $i = 1, 2, \ldots, m$ are called **artificial variables** and have no physical significance. Assuming the profit associated with each y_i to be zero, we obtain the problem

$$\text{Maximize} \quad z = \sum_{j=1}^{s} c_j x_j \tag{13}$$

subject to

$$\sum_{j=1}^{s} a_{ij} x_j + y_i = b_i, \qquad i = 1, 2, \ldots, m \tag{14}$$

$$x_j \geq 0, \quad j = 1, 2, \ldots, s; \qquad y_i \geq 0, \quad i = 1, 2, \ldots, m \tag{15}$$

with $b_i \geq 0$, $i = 1, 2, \ldots, m$.

It is easy to see (Exercise 22) that the vector $\mathbf{x}$ in R^s is a feasible solution to the problem given by (10), (11), and (12) if and only if the vector

$$\begin{bmatrix} \mathbf{x} \\ \mathbf{0} \end{bmatrix}$$

in R^{s+m} is a feasible solution to the problem given by (13), (14), and (15). Note that it is easy to find an initial basic feasible solution to the latter problem, namely, $\mathbf{x} = \mathbf{0}$, $\mathbf{y} = \mathbf{b}$. We now develop a way to use the simplex algorithm to change this initial basic feasible solution into a basic feasible solution to the same problem in which $\mathbf{y} = \mathbf{0}$. Thus, we will have found a basic feasible solution to the problem given by (10), (11), and (12). This procedure is Phase 1 of the **two-phase method**.

Phase 1

Since each y_i is constrained to be nonnegative, one way of guaranteeing that each y_i is zero is to make the sum of the y_i's zero. Thus, we set up an auxiliary problem in which we minimize the sum of the y_i's subject to the constraints (14) and (15) and hope that this minimum value is zero. If this minimum value is not zero, then it must be positive, and at least one of the y_i's must be positive. Furthermore, the y_i's will never all be zero, since we have found the minimum value of their sum. Thus, in this case the original problem has no feasible solution.

We convert the canonical linear programming problem given by (10), (11), and (12) to the form involving artificial variables and introduce the new objective function. The resulting problem is

$$\text{Minimize} \quad z = \sum_{i=1}^{m} y_i \tag{16}$$

subject to

$$\sum_{j=1}^{s} a_{ij}x_j + y_i = b_i, \qquad i = 1, 2, \ldots, m \tag{17}$$

$$x_j \geq 0, \quad j = 1, 2, \ldots, s; \qquad y_i \geq 0, \quad i = 1, 2, \ldots, m \tag{18}$$

with $b_i \geq 0$, $i = 1, 2, \ldots, m$.

This problem has the initial basic feasible solution

$$[0 \quad 0 \cdots 0 \quad b_1 \quad b_2 \cdots b_m]^{\mathsf{T}}$$

obtained by setting

$$x_1 = 0, \, x_2 = 0, \, \ldots, \, x_s = 0$$

as nonbasic variables and solving (17) for

$$y_1 = b_1, \, y_2 = b_2, \, \ldots, \, y_m = b_m$$

This procedure gives a basic solution because the columns corresponding to $y_1, y_2, \ldots, y_m$ are linearly independent. (They are the columns of an $m \times m$ identity matrix.)

To use the simplex method as it was developed earlier, we must multiply (16) by -1 to convert to a maximization problem and then write the result as

$$z + \sum_{i=1}^{m} y_i = 0 \tag{19}$$

Recall that when the simplex method was first described, the initial basic variables were the slack variables, and these had zero objective function coefficients. Consequently, the entries in the objective row in the columns labeled by the basic variables were zero initially and remained so after each pivoting step. This was necessary for the use of the optimality criterion. Therefore, we must eliminate y_i, $i = 1, 2, \ldots, m$ from (19). We can do this by solving (17) for y_i.

Now

$$y_i = b_i - \sum_{j=1}^{s} a_{ij}x_j$$

and substituting into (19), we obtain

$$z + \sum_{i=1}^{m} \left(b_i - \sum_{j=1}^{s} a_{ij}x_j \right) = 0$$

Rearranging, we can write the objective equation as

$$z - \sum_{j=1}^{s} \left(\sum_{i=1}^{m} a_{ij} \right) x_j = -\sum_{i=1}^{m} b_i \qquad (20)$$

We can now solve the problem as given by (20), (17), and (18) using the simplex method.

EXAMPLE 2. Consider the linear programming problem in canonical form

Maximize $z = x_1 - 2x_2 - 3x_3 - x_4 - x_5 + 2x_6$

subject to

$$x_1 + 2x_2 + 2x_3 + x_4 + x_5 = 12$$
$$x_1 + 2x_2 + x_3 + x_4 + 2x_5 + x_6 = 18$$
$$3x_1 + 6x_2 + 2x_3 + x_4 + 3x_5 = 24$$
$$x_j \geq 0, \qquad j = 1, 2, \ldots, 6$$

Introducing the artificial variables y_1, y_2, y_3, we can write the auxiliary problem as

$$\text{or}\ Max\ w = -(y_1 + y_2 + y_3)$$

Minimize $z = y_1 + y_2 + y_3 \qquad (21)$

subject to

$$\left.\begin{array}{l} x_1 + 2x_2 + 2x_3 + x_4 + x_5 + y_1 = 12 \\ x_1 + 2x_2 + x_3 + x_4 + 2x_5 + x_6 + y_2 = 18 \\ 3x_1 + 6x_2 + 2x_3 + x_4 + 3x_5 + y_3 = 24 \end{array}\right\} \qquad (22)$$

$$x_j \geq 0, \quad j = 1, 2, \ldots, 6; \qquad y_i \geq 0, \quad i = 1, 2, 3 \qquad (23)$$

After converting to a maximization problem, the objective function (21) can be written as

$$z + y_1 + y_2 + y_3 = 0 \qquad (24)$$

To eliminate y_1, y_2, and y_3 from (24), we add the constraints in (22) and subtract the result from (24). We obtain

$$z - 5x_1 - 10x_2 - 5x_3 - 3x_4 - 6x_5 - x_6 = -54 \qquad (25)$$

The initial basic feasible solution is

$$x_1 = x_2 = x_3 = x_4 = x_5 = x_6 = 0 \qquad \text{(nonbasic variables)}$$

$$y_1 = 12, \quad y_2 = 18, \quad y_3 = 24 \qquad \text{(basic variables)}$$

Using (25), (22), and (23), we can write the initial tableau (Tableau 2.14).

Tableau 2.14

	x_1	x_2	x_3	x_4	x_5	x_6	y_1	y_2	y_3	
y_1	1	2	2	1	1	0	1	0	0	12
y_2	1	2	1	1	2	1	0	1	0	18
y_3	3	⑥	2	1	3	0	0	0	1	24
	−5	−10	−5	−3	−6	−1	0	0	0	−54

$w + (y_1 + y_2 + y_3) = w + 54 - (5x_1 + 10x_2 + 5 \ldots + \ldots)$

The most negative entry in the objective row is -10, so that x_2 is the entering variable and the second column is the pivotal column. The departing variable is determined by computing the minimum of the θ-ratios. We have

$$\min \left\{ \tfrac{12}{2}, \tfrac{18}{2}, \tfrac{24}{6} \right\} = \tfrac{24}{6} = 4$$

so that the row labeled by y_3 is the pivotal row and y_3 is the departing variable. The pivot is 6.

We obtain Tableau 2.15 from Tableau 2.14 by pivoting.

In Tableau 2.15 we choose x_3 as the entering variable and y_1 as the departing variable. We form Tableau 2.16 by pivoting.

There is a tie in determining the entering variable in Tableau 2.16. We choose x_6 as the entering variable and obtain Tableau 2.17.

At this point all the artificial variables are nonbasic variables and have

Tableau 2.15

	x_1	x_2	x_3	x_4	x_5	x_6	y_1	y_2	y_3	
y_1	0	0	④⁄₃	$\frac{2}{3}$	0	0	1	0	$-\frac{1}{3}$	4
y_2	0	0	$\frac{1}{3}$	$\frac{2}{3}$	1	1	0	1	$-\frac{1}{3}$	10
x_2	$\frac{1}{2}$	1	$\frac{1}{3}$	$\frac{1}{6}$	$\frac{1}{2}$	0	0	0	$\frac{1}{6}$	4
	0	0	$-\frac{5}{3}$	$-\frac{4}{3}$	−1	−1	0	0	$\frac{5}{3}$	−14

Tableau 2.16

	x_1	x_2	x_3	x_4	x_5	x_6	y_1	y_2	y_3	
x_3	0	0	1	$\frac{1}{2}$	0	0	$\frac{3}{4}$	0	$-\frac{1}{4}$	3
y_2	0	0	0	$\frac{1}{2}$	1	①	$-\frac{1}{4}$	1	$-\frac{1}{4}$	9
x_2	$\frac{1}{2}$	1	0	0	$\frac{1}{2}$	0	$-\frac{1}{4}$	0	$\frac{1}{4}$	3
	0	0	0	$-\frac{1}{2}$	−1	−1	$\frac{5}{4}$	0	$\frac{5}{4}$	−9

Tableau 2.17

	x_1	x_2	x_3	x_4	x_5	x_6	y_1	y_2	y_3	
x_3	0	0	1	$\frac{1}{2}$	0	0	$\frac{3}{4}$	0	$-\frac{1}{4}$	3
x_6	0	0	0	$\frac{1}{2}$	1	1	$-\frac{1}{4}$	1	$-\frac{1}{4}$	9
x_2	$\frac{1}{2}$	1	0	0	$\frac{1}{2}$	0	$-\frac{1}{4}$	0	$\frac{1}{4}$	3
	0	0	0	0	0	0	1	1	1	0

value zero. Tableau 2.17 gives the basic feasible solution

$$x_2 = 3, \quad x_3 = 3, \quad x_6 = 9$$

$$x_1 = x_4 = x_5 = y_1 = y_2 = y_3 = 0$$

with objective function value $z = 0$.

The reader can verify that $\mathbf{x} = [0 \ \ 3 \ \ 3 \ \ 0 \ \ 0 \ \ 9]^T$ is a basic feasible solution to the original problem without artificial variables. The introduction of the artificial variables gave a systematic way of finding such a basic feasible solution.

If the solution to Phase 1 is a set of values for the variables which makes the objective function of the auxiliary problem equal to zero, then we may start Phase 2. There are two possibilities: (1) Every artificial variable is a nonbasic variable in the final tableau of Phase 1, or (2) some artificial variables are still basic variables, with value 0, in the final tableau (see Example 5). At this time we shall only discuss the first case. The second case is discussed in Section 3.2. For Phase 2, then, we assume that no artificial variable is a basic variable at the end of Phase 1 and that the value of the objective function of the auxiliary problem is zero.

Phase 2

The optimal solution obtained in Phase 1 is used to obtain an initial basic feasible solution for the original problem (10), (11), (12). By deleting the y_i's from the optimal solution, we obtain a basic feasible solution to the original problem because we have assumed that no artificial variables appear in the optimal solution. The initial tableau for Phase 2 is the final tableau of Phase 1, with the following modifications:

(a) Delete the columns labeled with artificial variables.
(b) Calculate a new objective row as follows: Start with the objective row of the original problem [Equation (10)]. For each of the basic variables in the final tableau of Phase 1, make the entry in the objective row in the column labeled by that basic variable equal to

zero by adding a suitable multiple of the row labeled by that basic variable.

These two steps construct an initial tableau for Phase 2. We can now apply the simplex algorithm to this tableau.

EXAMPLE 2 (continued).   We now form the initial tableau for Phase 2 from Tableau 2.17. We delete the columns labeled with y_1, y_2, and y_3 and now use the original objective function,

$$z = x_1 - 2x_2 - 3x_3 - x_4 - x_5 + 2x_6$$

The objective row would be

$$\boxed{\begin{array}{ccccccc} -1 & 2 & 3 & 1 & 1 & -2 & 0 \end{array}} \qquad (26)$$

But the entries in the second, third, and sixth columns must be zeroed for the optimality criterion to hold, since x_2, x_3, and x_6 are basic variables. We do this by adding -2 times the x_2 row to (26); also, we add -3 times the x_3 row and 2 times the x_6 row to (26). This calculation yields

	x_1	x_2	x_3	x_4	x_5	x_6		
	-1	2	3	1	1	-2	0	[Equation (26)]
x_2	-1	-2	0	0	-1	0	-6	(-2 times x_2 row)
x_3	0	0	-3	$-\frac{3}{2}$	0	0	-9	(-3 times x_3 row)
x_6	0	0	0	1	2	2	18	(2 times x_6 row)
	-2	0	0	$\frac{1}{2}$	2	0	3	(objective row for Phase 2)

We then have the initial tableau for Phase 2 (Tableau 2.18), in which, as usual, we have not included the z column.

We now continue with the simplex algorithm to find an optimal solution. The next tableau is Tableau 2.19.

Since the objective row in Tableau 2.19 has no negative entries, we have found an optimal solution:

$$x_1 = 6, \quad x_2 = 0, \quad x_3 = 3, \quad x_4 = 0, \quad x_5 = 0, \quad x_6 = 9$$

which gives $z = 15$ as the value of the objective function.

Tableau 2.18

		x_1	x_2	x_3	x_4	x_5	x_6	
	x_3	0	0	1	$\frac{1}{2}$	0	0	3
	x_6	0	0	0	$\frac{1}{2}$	1	1	9
$\leftarrow$	x_2	$\textcircled{\frac{1}{2}}$	1	0	0	$\frac{1}{2}$	0	3
		-2	0	0	$\frac{1}{2}$	2	0	3

Tableau 2.19

	x_1	x_2	x_3	x_4	x_5	x_6	
x_3	0	0	1	$\frac{1}{2}$	0	0	3
x_6	0	0	0	$\frac{1}{2}$	1	1	9
x_1	1	2	0	0	1	0	6
	0	4	0	$\frac{1}{2}$	4	0	15

We had originally put an artificial variable in every equation. This method requires no decision steps but may cause more tableaux to be computed. If some of the equations have a variable which can be used as a basic variable, then it is not necessary to introduce an artificial variable into each of those equations.

EXAMPLE 3. In the problem discussed in Example 2, we see that x_6 can be used as a basic variable. That is, it appears in only one equation and in that equation has a coefficient of $+1$. Consequently, we need only introduce artificial variables into the first and third equations. Doing this, we obtain the auxiliary problem

$$\text{Minimize} \quad z = y_1 + y_2$$

subject to

$$x_1 + 2x_2 + 2x_3 + x_4 + x_5 \qquad\qquad + y_1 \qquad = 12$$
$$x_1 + 2x_2 + x_3 + x_4 + 2x_5 + x_6 \qquad\qquad = 18$$
$$3x_1 + 6x_2 + 2x_3 + x_4 + 3x_5 \qquad\qquad + y_2 = 24$$
$$x_j \geq 0, \quad j = 1, 2, \ldots, 6; \qquad y_1 \geq 0, \qquad y_2 \geq 0$$

As before, we must eliminate the basic variables from the objective function rewritten as in (24) by adding the constraints which include artificial variables and subtracting the result from the rewritten objective function. We obtain

$$z - 4x_1 - 8x_2 - 4x_3 - 2x_4 - 4x_5 = -36$$

This now leads to the sequence of Tableaux 2.20, 2.21, and 2.22.

In Tableau 2.22 we have an optimal solution to the auxiliary problem in which all the artificial variables have value zero and are nonbasic variables. Thus, a basic feasible solution for the original problem is $[0 \ 3 \ 3 \ 0 \ 0 \ 9]^T$, as we obtained in Example 2.

The system of equations (11) which gives the constraints for the canonical form of a linear programming problem always must have a solution

Tableau 2.20

	x_1	x_2	x_3	x_4	x_5	x_6	y_1	y_2	
y_1	1	2	2	1	1	0	1	0	12
x_6	1	2	1	1	2	1	0	0	18
← y_2	3	⑥	2	1	3	0	0	1	24
	−4	−8	−4	−2	−4	0	0	0	−36

Tableau 2.21

	x_1	x_2	x_3	x_4	x_5	x_6	y_1	y_2	
← y_1	0	0	$\tfrac{4}{3}$	$\tfrac{2}{3}$	0	0	1	$-\tfrac{1}{3}$	4
x_6	0	0	$\tfrac{1}{3}$	$\tfrac{2}{3}$	1	1	0	$-\tfrac{1}{3}$	10
x_2	$\tfrac{1}{2}$	1	$\tfrac{1}{3}$	$\tfrac{1}{6}$	$\tfrac{1}{2}$	0	0	$\tfrac{1}{6}$	4
	0	0	$-\tfrac{4}{3}$	$-\tfrac{2}{3}$	0	0	0	$\tfrac{4}{3}$	−4

Tableau 2.22

	x_1	x_2	x_3	x_4	x_5	x_6	y_1	y_2	
x_3	0	0	1	$\tfrac{1}{2}$	0	0	$\tfrac{3}{4}$	$-\tfrac{1}{4}$	3
x_6	0	0	0	$\tfrac{1}{2}$	1	1	$-\tfrac{1}{4}$	$-\tfrac{1}{4}$	9
x_2	$\tfrac{1}{2}$	1	0	0	$\tfrac{1}{2}$	0	$-\tfrac{1}{4}$	$\tfrac{1}{4}$	3
	0	0	0	0	0	0	1	1	0

if $m \leq s$ and if there are m linearly independent columns in the coefficient matrix. But it is possible that none of the solutions satisfies the nonnegativity criterion (12). When the artificial variables are added to the system of constraints (14), there is always a solution

$$[0 \quad 0 \cdots 0 \quad b_1 \quad b_2 \cdots b_m]^{\mathrm{T}}$$

which satisfies the nonnegativity conditions (15). A solution to the system in (14) is a solution to the system in (11) if all the artificial variables have values equal to zero. We use the simplex algorithm to try to find a solution to (14) in which all the artificial variables are zero. The simplex algorithm is designed to find only solutions which satisfy the nonnegativity requirements. Thus, if one of the y_i is positive when the simplex algorithm reaches an optimal solution to the auxiliary problem, this indicates that there are no solutions to (11) which satisfy the nonnegativity constraints (12). In this case, the original problem (10), (11), (12) has no feasible solutions. Consider the following example of this situation.

EXAMPLE 4. Consider the general linear programming problem

$$\text{Maximize} \quad z = 2x_1 + 5x_2$$

subject to

$$2x_1 + 3x_2 \le 6$$

$$x_1 + x_2 \ge 4$$

$$x_1 \ge 0, \quad x_2 \ge 0$$

By inserting slack variables x_3 and x_4, we can write the problem in canonical form as

$$\text{Maximize} \quad z = 2x_1 + 5x_2$$

subject to

$$2x_1 + 3x_2 + x_3 \qquad = 6$$

$$x_1 + x_2 \qquad - x_4 = 4$$

$$x_j \ge 0, \quad j = 1, 2, 3, 4$$

We insert an artificial variable y into the second equation; x_3 can serve as the basic variable in the first equation. The auxiliary problem then has the form

$$\text{Minimize} \quad z = y$$

subject to

$$2x_1 + 3x_2 + x_3 \qquad = 6$$

$$x_1 + x_2 \qquad - x_4 + y = 4$$

$$x_j \ge 0, \quad j = 1, 2, 3, 4; \qquad y \ge 0$$

After subtracting the second constraint from the rewritten objective function, we obtain

$$z - x_1 - x_2 + x_4 = -4$$

We then calculate the sequence of Tableaux 2.23, 2.24, and 2.25.

Tableau 2.23

	x_1	x_2	x_3	x_4	y	
x_3	2	③	1	0	0	6
y	1	1	0	-1	1	4
	-1	-1	0	1	0	-4

Tableau 2.24

↓

	x_1	x_2	x_3	x_4	y	
← x_2	$\boxed{\tfrac{2}{3}}$	1	$\tfrac{1}{3}$	0	0	2
y	$\tfrac{1}{3}$	0	$-\tfrac{1}{3}$	-1	1	2
	$-\tfrac{1}{3}$	0	$\tfrac{1}{3}$	1	0	-2

Tableau 2.25

	x_1	x_2	x_3	x_4	y	
x_1	1	$\tfrac{3}{2}$	$\tfrac{1}{2}$	0	0	3
y	0	$-\tfrac{1}{2}$	$-\tfrac{1}{2}$	-1	1	1
	0	$\tfrac{1}{2}$	$\tfrac{1}{2}$	1	0	-1

Tableau 2.25 represents an optimal solution to Phase 1 in which the artificial variable y has the value 1. This means that the given problem has no feasible solution. When we look at the graphs of the constraints (Figure 2.9), we see that there is no intersection between the two half-spaces. The set of feasible solutions is empty.

We have already pointed out that an artificial variable can appear in an optimal solution to the auxiliary problem with a value of zero. In this case the given problem has a feasible solution. The following example illustrates these ideas. We can complete Phase 1 in this section; in Section 3.2 we will develop tools for handling Phase 2.

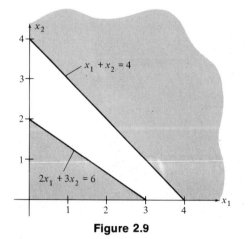

Figure 2.9

EXAMPLE 5. Consider the linear programming problem in canonical form

$$\text{Maximize}\quad z = x_1 + 2x_2 + x_3$$

subject to

$$3x_1 + x_2 - x_3 = 15$$
$$8x_1 + 4x_2 - x_3 = 50$$
$$2x_1 + 2x_2 + x_3 = 20$$
$$x_1 \geq 0, \quad x_2 \geq 0, \quad x_3 \geq 0$$

In this problem we must introduce an artificial variable in each of the constraint equations. We obtain the auxiliary problem

$$\text{Minimize}\quad z = y_1 + y_2 + y_3$$

subject to

$$3x_1 + x_2 - x_3 + y_1 \qquad\qquad = 15$$
$$8x_1 + 4x_2 - x_3 \qquad + y_2 \qquad = 50$$
$$2x_1 + 2x_2 + x_3 \qquad\qquad + y_3 = 20$$
$$x_j \geq 0, \quad j = 1, 2, 3; \qquad y_i \geq 0, \quad i = 1, 2, 3$$

Rewriting the objective function in the same manner as before, we obtain the initial tableau (Tableau 2.26) for the auxiliary problem. At the end of Phase 1 we obtain Tableau 2.27, which represents an optimal solution to the auxiliary problem with $y_2 = 0$ as a basic variable.

Tableau 2.26 ↓

	x_1	x_2	x_3	y_1	y_2	y_3	
← y_1	③	1	-1	1	0	0	15
y_2	8	4	-1	0	1	0	50
y_3	2	2	1	0	0	1	20
	-13	-7	1	0	0	0	-85

Tableau 2.27

	x_1	x_2	x_3	y_1	y_2	y_3	
x_1	1	$\frac{3}{5}$	0	$\frac{1}{5}$	0	$\frac{1}{5}$	7
y_2	0	0	0	-2	1	-1	0
x_3	0	$\frac{4}{5}$	1	$-\frac{2}{5}$	0	$\frac{3}{5}$	6
	0	0	0	3	0	2	0

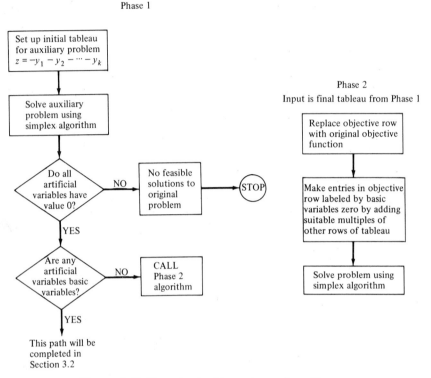

Figure 2.10 Flowchart of two-phase algorithm.

Figure 2.10 gives a flowchart and Figure 2.11 a structure diagram which summarize the two-phase method.

Big *M* Method (Optional)

The following method of solving linear programming problems which require artificial variables is attributed to Charnes. Historically it precedes the two-phase method; it has been replaced by the latter in computer codes due to the greater efficiency of the two-phase method. It is still of interest for theoretical computations.

Instead of introducing an auxiliary problem, the **big *M* method** insures that the artificial variables are zero in the final solution by assigning to each y_i a **penalty cost** $-M$, where M is a very large positive number. This means that we use an objective function of the form

$$z = \sum_{j=1}^{s} c_j x_j - \sum_{i=1}^{m} M y_i$$

If any y_i is positive, the $-M$ serves to decrease z drastically.

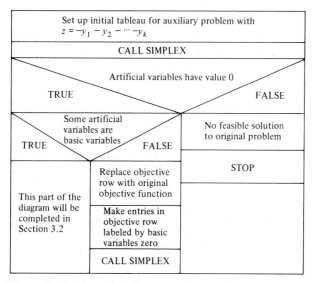

Figure 2.11 Structure diagram of two-phase algorithm.

We convert the canonical linear programming problem given by (10), (11), and (12) to the form involving artificial variables. We obtain

$$\text{Maximize} \quad z = \sum_{j=1}^{s} c_j x_j - \sum_{i=1}^{m} M y_i \qquad (27)$$

subject to

$$\sum_{j=1}^{s} a_{ij} x_j + y_i = b_i, \qquad i = 1, 2, \ldots, m \qquad (28)$$

$$x_j \geq 0, \quad j = 1, 2, \ldots, s; \qquad y_i \geq 0, \quad i = 1, 2, \ldots, m \qquad (29)$$

with $b_i \geq 0$, $i = 1, 2, \ldots, m$.

This problem has an initial basic feasible solution

$$[0 \quad 0 \cdots 0 \quad b_1 \quad b_2 \cdots b_m]^{\mathrm{T}}$$

obtained by setting

$$x_1 = 0, \quad x_2 = 0, \ldots, x_s = 0$$

as nonbasic variables and solving (28) for

$$y_1 = b_1, \quad y_2 = b_2, \ldots, y_m = b_m$$

To use the simplex method as it was developed earlier, we must write (27) as

$$z - \sum_{j=1}^{s} c_j x_j + M \sum_{i=1}^{m} y_i = 0 \qquad (30)$$

and eliminate the y_i from this equation. This is done by summing the constraints in (28), multiplying by $-M$, and subtracting the result from (30).

We obtain

$$z - \sum_{j=1}^{s} c_j x_j + M \sum_{i=1}^{m} \left(b_i - \sum_{j=1}^{s} a_{ij} x_j \right) = 0$$

Rearranging, we can write the objective equation as

$$z - \sum_{j=1}^{s} \left(c_j + M \sum_{i=1}^{m} a_{ij} \right) x_j = -M \sum_{i=1}^{m} b_i \tag{31}$$

We can now solve the problem as given by (31), (28), and (29) using the simplex method.

EXAMPLE 6. Consider the linear programming problem in Example 2. Introducing artificial variables into the first and third equations, we have the problem

$$\text{Maximize} \quad z = x_1 - 2x_2 - 3x_3 - x_4 - x_5 + 2x_6 - My_1 - My_2$$

subject to

$$x_1 + 2x_2 + 2x_3 + x_4 + x_5 \qquad + y_1 \qquad = 12$$
$$x_1 + 2x_2 + x_3 + x_4 + 2x_5 + x_6 \qquad = 18$$
$$3x_1 + 6x_2 + 2x_3 + x_4 + 3x_5 \qquad + y_2 = 24$$
$$x_j \geq 0, \quad j = 1, 2, \ldots, 6; \qquad y_1 \geq 0, \quad y_2 \geq 0$$

Adding the first and third constraints, multiplying by $-M$, and subtracting from the given objective function, we may write the objective function as

$$z + (1 - 4M)x_1 + (6 - 8M)x_2 + (5 - 4M)x_3$$
$$+ (3 - 2M)x_4 + (5 - 4M)x_5 = 36 - 36M$$

We obtain the initial tableau (Tableau 2.28).

Tableau 2.28

	x_1	x_2	x_3	x_4	x_5	x_6	y_1	y_2	
y_1	1	2	2	1	1	0	1	0	12
x_6	1	2	1	1	2	1	0	0	18
y_2	3	⑥	2	1	3	0	0	1	24
	$1 - 4M$	$6 - 8M$	$5 - 4M$	$3 - 2M$	$5 - 4M$	0	0	0	$36 - 36M$

Since M is a large positive number, the most negative entry in the objective row is $6 - 8M$, so that x_2 is the entering variable. The departing variable, obtained as usual, is y_2. Pivoting, we obtain Tableau 2.29. Using the same reasoning, we obtain Tableaux 2.30 and 2.31.

Since the objective row has no negative entries, we have found an optimal solution:

$$x_1 = 6, \quad x_2 = 0, \quad x_3 = 3, \quad x_4 = 0, \quad x_5 = 0, \quad x_6 = 9$$

$$y_1 = 0, \quad y_2 = 0$$

which gives $z = 15$ as the value of the objective function. This solution coincides with the one obtained in Example 2.

Tableau 2.29

	x_1	x_2	x_3	x_4	x_5	x_6	y_1	y_2	
y_1	0	0	$\frac{4}{3}$	$\frac{2}{3}$	0	0	1	$-\frac{1}{3}$	4
x_6	0	0	$\frac{1}{3}$	$\frac{2}{3}$	1	1	0	$-\frac{1}{3}$	10
x_2	$\frac{1}{2}$	1	$\frac{1}{3}$	$\frac{1}{6}$	$\frac{1}{2}$	0	0	$\frac{1}{6}$	4
	-2	0	$3 - \frac{4}{3}M$	$2 - \frac{2}{3}M$	2	0	0	$-1 + \frac{4}{3}M$	$12 - 4M$

Tableau 2.30

	x_1	x_2	x_3	x_4	x_5	x_6	y_1	y_2	
x_3	0	0	1	$\frac{1}{2}$	0	0	$\frac{3}{4}$	$-\frac{1}{4}$	3
x_6	0	0	0	$\frac{1}{2}$	1	1	$-\frac{1}{4}$	$-\frac{1}{4}$	9
x_2	$\frac{1}{2}$	1	0	0	$\frac{1}{2}$	0	$-\frac{1}{4}$	$\frac{1}{4}$	3
	-2	0	0	$\frac{1}{2}$	2	0	$-\frac{9}{4} + M$	$-\frac{1}{4} + M$	3

Tableau 2.31

	x_1	x_2	x_3	x_4	x_5	x_6	y_1	y_2	
x_3	0	0	1	$\frac{1}{2}$	0	0	$\frac{3}{4}$	$-\frac{1}{4}$	3
x_6	0	0	0	$\frac{1}{2}$	1	1	$-\frac{1}{4}$	$-\frac{1}{4}$	9
x_1	1	2	0	0	1	0	$-\frac{1}{2}$	$\frac{1}{2}$	6
	0	4	0	$\frac{1}{2}$	4	0	$-\frac{13}{4} + M$	$\frac{3}{4} + M$	15

2.5 Exercises

In Exercises 1–4 set up the initial simplex tableau (a) for solving the problem using the two-phase method, and (b) for solving the problem using the big M method.

1. Maximize $z = x_1 + 3x_3$

subject to

$$x_1 + 2x_2 + 7x_3 = 4$$
$$x_1 + 3x_2 + \ x_3 = 5$$
$$x_1 \geq 0, \quad x_2 \geq 0, \quad x_3 \geq 0$$

2. Maximize $z = x_1 + 2x_2 + x_4$

subject to

$$x_1 + 3x_2 - x_3 + x_4 \leq 5$$
$$x_1 + 7x_2 + x_3 \qquad \geq 4$$
$$4x_1 + 2x_2 \qquad + x_4 = 3$$
$$x_j \geq 0, \quad j = 1, 2, 3, 4$$

3. Minimize $z = 3x_1 - 2x_2$

subject to

$$x_1 + 3x_2 + 2x_3 \geq 7$$
$$2x_1 + \ x_2 + \ x_3 \geq 4$$
$$x_1 \geq 0, \quad x_2 \geq 0, \quad x_3 \geq 0$$

4. Minimize $z = x_1 + 2x_2 + 7x_3 - x_4$

subject to

$$3x_1 + \ x_2 - 2x_3 - x_4 = 2$$
$$2x_1 + 4x_2 + 7x_3 \qquad \geq 3$$
$$x_j \geq 0, \quad j = 1, 2, 3, 4$$

In Exercises 5 and 6 carry out Phase 1 for the given problems.

5. Maximize $z = 3x_1 - 4x_3$

subject to

$$2x_1 + x_2 + 3x_3 \geq 5$$
$$x_1 - x_2 + \ x_3 \geq 1$$
$$x_1 \geq 0, \quad x_2 \geq 0, \quad x_3 \geq 0$$

6. Maximize $z = x_1 + x_2 + 2x_4$

subject to

$$3x_1 + x_2 + 3x_3 + 2x_4 = 10$$
$$x_1 - 3x_2 + 2x_3 \le 7$$
$$x_1 + 2x_2 + 3x_3 + x_4 \ge 4$$
$$x_j \ge 0, \quad j = 1, 2, 3, 4$$

In Exercises 7–9 we give the final tableaux for Phase 1 of the two-phase method along with the original objective function. (a) Form the initial tableau for Phase 2 using the given information; (b) apply the simplex method to the tableaux in (a) to find optimal solutions to the given problems.

7. Maximize $z = 2x_1 + x_2 + x_3$

	x_1	x_2	x_3	x_4	x_5	
x_2	-1	1	$\frac{3}{10}$	0	$-\frac{1}{2}$	$\frac{3}{5}$
x_4	$\frac{1}{2}$	0	$\frac{1}{5}$	1	$-\frac{1}{10}$	$\frac{7}{10}$
	0	0	0	0	0	0

8. Maximize $z = x_2 + 3x_3 + x_4$

	x_1	x_2	x_3	x_4	x_5	x_6	x_7	
x_7	$-\frac{3}{2}$	0	-2	-1	0	$-\frac{3}{4}$	1	0
x_2	0	1	1	3	0	$\frac{1}{2}$	0	2
x_5	1	0	3	0	1	$-\frac{1}{2}$	0	4
	0	0	0	0	0	0	0	0

9. Maximize $z = 2x_1 + x_2 + x_4$

	x_1	x_2	x_3	x_4	x_5	x_6	y_1	y_2	
x_2	$\frac{3}{8}$	1	1	0	0	$\frac{3}{4}$	0	0	1
x_5	$-\frac{5}{8}$	0	1	1	1	$\frac{7}{4}$	0	0	2
y_1	-1	0	0	-2	0	$-\frac{5}{4}$	1	0	$\frac{3}{2}$
y_2	0	0	$-\frac{1}{2}$	1	0	$-\frac{1}{4}$	0	1	0
	1	0	$\frac{1}{2}$	1	0	$\frac{3}{2}$	0	0	$-\frac{3}{2}$

In Exercises 10–21 solve the indicated linear programming problem using either the two-phase method or the big M method.

10. Example 2, Section 2.1 (page 60).

11. Example 5, Section 2.1 (page 64).

12. Example 6, Section 2.1 (page 66).

13. Example 7(c), Section 2.1 (page 67).

14. Example 7(d), Section 2.1 (page 67).

15. Exercise 1, Section 2.1 (page 71).

16. Exercise 11, Section 2.1 (page 75).

17. Exercise 15, Section 2.2 (page 97).

18. Exercise 18, Section 2.2 (page 98).

19. Exercise 8, Section 2.1 (page 73).

20. Maximize $z = 2x_1 + 5x_2 - x_3$

    subject to

    $$-4x_1 + 2x_2 + 6x_3 = 4$$
    $$6x_1 + 9x_2 + 12x_3 = 3$$
    $$x_1 \geq 0, \quad x_2 \geq 0, \quad x_3 \geq 0$$

21. Maximize $z = 3x_1 - x_2 + 2x_3 + 4x_4$

    subject to

    $$x_2 + 7x_3 + 2x_4 \geq 3$$
    $$x_1 + 2x_2 + x_3 \qquad = 9$$
    $$2x_1 + 3x_2 + x_3 - 4x_4 \leq 7$$
    $$x_j \geq 0, \quad j = 1, 2, 3, 4$$

22. Show that the vector x in R^s is a feasible solution to the problem in canonical form given by (10), (11), and (12) if and only if the vector

$$\begin{bmatrix} x \\ 0 \end{bmatrix}$$

in R^{s+m} is a feasible solution to the auxiliary problem given by (13), (14), and (15).

Further Readings

Kotiah, Thoddi C. T., and David I. Steinberg. "Occurrences in Cycling and Other Phenomena Arising in a Class of Linear Programming Models." *Communications of the ACM,* 20 (1977), 107–112.

Kotiah, Thoddi C. T., and David I. Steinberg. "On the Possibility of Cycling with the Simplex Method." *Operations Research,* 26 (1978), 374–375.

CHAPTER
012345

FURTHER TOPICS IN
LINEAR PROGRAMMING

This chapter covers several topics in linear programming which have important computational consequences. The idea of duality, which is introduced in the first section, is particularly useful in modeling, since it provides economic interpretations of the solution to a linear programming problem. We present a brief discussion of sensitivity analysis, another tool which is useful in interpreting the solution to a linear programming problem. We discuss several variants of the simplex algorithm, including the one which is used in most computer codes. Finally, we deal with computational considerations from the viewpoint of the user of a packaged linear programming system.

3.1 DUALITY

In this section we shall show how to associate a minimization problem with each linear programming problem in standard form. There are some very interesting interpretations of the associated problem which we will discuss. Generally, a problem in standard form can be thought of as a manufacturing problem, one in which scarce resources are allocated in a way that maximizes profit. The associated minimization problem is one that seeks to minimize cost.

Consider the pair of linear programming problems

$$\left. \begin{array}{l} \text{Maximize} \quad z = \mathbf{c}^T \mathbf{x} \\[6pt] \text{subject to} \\[6pt] \mathbf{Ax} \leq \mathbf{b} \\ \mathbf{x} \geq \mathbf{0} \end{array} \right\} \tag{1}$$

143

and

$$\left.\begin{aligned}\text{Minimize} \quad & z' = \mathbf{b}^T\mathbf{w} \\ \text{subject to} \quad & \\ & \mathbf{A}^T\mathbf{w} \geq \mathbf{c} \\ & \mathbf{w} \geq \mathbf{0} \end{aligned}\right\} \tag{2}$$

where $\mathbf{A}$ is an $m \times n$ matrix, $\mathbf{c}$ and $\mathbf{x}$ are $n \times 1$ column vectors, and $\mathbf{b}$ and $\mathbf{w}$ are $m \times 1$ column vectors.

These problems are called **dual problems**. The problem given by (1) is called the **primal problem**; the problem given by (2) is called the **dual problem**.

EXAMPLE 1. If the primal problem is

$$\text{Maximize} \quad z = [2 \quad 3]\begin{bmatrix} x_1 \\ x_2 \end{bmatrix}$$

subject to

$$\begin{bmatrix} 3 & 2 \\ -1 & 2 \\ 4 & 1 \end{bmatrix}\begin{bmatrix} x_1 \\ x_2 \end{bmatrix} \leq \begin{bmatrix} 2 \\ 5 \\ 1 \end{bmatrix}$$

$$x_1 \geq 0, \quad x_2 \geq 0$$

then the dual problem is

$$\text{Minimize} \quad z' = [2 \quad 5 \quad 1]\begin{bmatrix} w_1 \\ w_2 \\ w_3 \end{bmatrix}$$

subject to

$$\begin{bmatrix} 3 & -1 & 4 \\ 2 & 2 & 1 \end{bmatrix}\begin{bmatrix} w_1 \\ w_2 \\ w_3 \end{bmatrix} \geq \begin{bmatrix} 2 \\ 3 \end{bmatrix}$$

$$w_1 \geq 0, \quad w_2 \geq 0, \quad w_3 \geq 0$$

Observe that in forming the dual problem, the coefficients of the ith constraint of the primal problem became the coefficients of the variable w_i in the constraints of the dual problem. Conversely, the coefficients of x_j became the coefficients of the jth constraint in the dual problem. Also, the coefficients of the objective function of the primal problem became the right-hand sides of the constraints of the dual problem, and conversely.

Theorem 3.1. Given a primal problem as in (1), the dual of its dual problem is again the primal problem.

Proof. The dual problem as given by (2) is

$$
\left.
\begin{aligned}
&\text{Minimize} \quad z' = \mathbf{b}^T\mathbf{w} \\
&\text{subject to} \\
&\qquad \mathbf{A}^T\mathbf{w} \geq \mathbf{c} \\
&\qquad \mathbf{w} \geq \mathbf{0}
\end{aligned}
\right\} \tag{3}
$$

We can rewrite (3) as

$$
\left.
\begin{aligned}
&\text{Maximize} \quad z' = -\mathbf{b}^T\mathbf{w} \\
&\text{subject to} \\
&\qquad -\mathbf{A}^T\mathbf{w} \leq -\mathbf{c} \\
&\qquad \mathbf{w} \geq \mathbf{0}
\end{aligned}
\right\} \tag{4}
$$

Now the dual problem to (4) is

$$
\begin{aligned}
&\text{Minimize} \quad z = -\mathbf{c}^T\mathbf{x} \\
&\text{subject to} \\
&\qquad (-\mathbf{A}^T)^T\mathbf{x} \geq -\mathbf{b} \\
&\qquad \mathbf{x} \geq \mathbf{0}
\end{aligned}
$$

This problem can be rewritten as

$$
\begin{aligned}
&\text{Maximize} \quad z = \mathbf{c}^T\mathbf{x} \\
&\text{subject to} \\
&\qquad \mathbf{A}\mathbf{x} \leq \mathbf{b} \\
&\qquad \mathbf{x} \geq \mathbf{0}
\end{aligned}
$$

which is the primal problem.

Theorem 3.2. The linear programming problem in canonical form given by

$$
\begin{aligned}
&\text{Maximize} \quad z = \mathbf{c}^T\mathbf{x} \\
&\text{subject to} \\
&\qquad \mathbf{A}\mathbf{x} = \mathbf{b} \\
&\qquad \mathbf{x} \geq \mathbf{0}
\end{aligned}
$$

has for its dual the linear programming problem

$$\text{Minimize} \quad z' = \mathbf{b}^T\mathbf{w}$$

subject to

$$\mathbf{A}^T\mathbf{w} \geq \mathbf{c}$$

where $\mathbf{w}$ is unrestricted.

Proof. The primal problem can be written as

$$\text{Maximize} \quad z = \mathbf{c}^T\mathbf{x}$$

subject to

$$\mathbf{A}\mathbf{x} \leq \mathbf{b}$$

$$-\mathbf{A}\mathbf{x} \leq -\mathbf{b}$$

$$\mathbf{x} \geq \mathbf{0}$$

by using the method of converting equalities which we described in Section 2.1. In matrix form the primal problem is

$$\text{Maximize} \quad z = \mathbf{c}^T\mathbf{x}$$

subject to

$$\begin{bmatrix} \mathbf{A} \\ -\mathbf{A} \end{bmatrix} \mathbf{x} \leq \begin{bmatrix} \mathbf{b} \\ -\mathbf{b} \end{bmatrix}$$

$$\mathbf{x} \geq \mathbf{0}$$

The dual problem is then

$$\text{Minimize} \quad z' = [\mathbf{b}^T \quad -\mathbf{b}^T] \begin{bmatrix} \mathbf{u} \\ \mathbf{v} \end{bmatrix}$$

subject to

$$[\mathbf{A}^T \quad -\mathbf{A}^T] \begin{bmatrix} \mathbf{u} \\ \mathbf{v} \end{bmatrix} \geq \mathbf{c}$$

$$\mathbf{u} \geq \mathbf{0}, \quad \mathbf{v} \geq \mathbf{0}$$

When we multiply out the matrices, we have

$$\text{Minimize} \quad z' = \mathbf{b}^T\mathbf{u} - \mathbf{b}^T\mathbf{v} = \mathbf{b}^T(\mathbf{u} - \mathbf{v})$$

subject to

$$\mathbf{A}^T\mathbf{u} - \mathbf{A}^T\mathbf{v} = \mathbf{A}^T(\mathbf{u} - \mathbf{v}) \geq \mathbf{c}$$

$$\mathbf{u} \geq \mathbf{0}, \quad \mathbf{v} \geq \mathbf{0}$$

If we let $\mathbf{w} = \mathbf{u} - \mathbf{v}$, then the dual problem has the form

$$\left.\begin{aligned} \text{Minimize} \quad & z' = \mathbf{b}^T\mathbf{w} \\ \text{subject to} \\ & \mathbf{A}^T\mathbf{w} \geq \mathbf{c} \\ & \mathbf{w} \text{ unrestricted} \end{aligned}\right\} \tag{5}$$

since any vector may be written as the difference of two nonnegative vectors.

Theorem 3.3. The linear programming problem

$$\left.\begin{aligned} \text{Maximize} \quad & z = \mathbf{c}^T\mathbf{x} \\ \text{subject to} \\ & \mathbf{A}\mathbf{x} \leq \mathbf{b} \\ & \mathbf{x} \text{ unrestricted} \end{aligned}\right\} \tag{6}$$

has as its dual problem

$$\begin{aligned} \text{Minimize} \quad & z' = \mathbf{b}^T\mathbf{w} \\ \text{subject to} \\ & \mathbf{A}^T\mathbf{w} = \mathbf{c} \\ & \mathbf{w} \geq \mathbf{0} \end{aligned}$$

Proof. We can rewrite the given problem as

$$\begin{aligned} \text{Minimize} \quad & z = -\mathbf{c}^T\mathbf{x} \\ \text{subject to} \\ & -\mathbf{A}\mathbf{x} \geq -\mathbf{b} \\ & \mathbf{x} \text{ unrestricted} \end{aligned}$$

Comparing this statement of the problem with (5), we see that it is the dual of

$$\begin{aligned} \text{Maximize} \quad & z' = -\mathbf{b}^T\mathbf{w} \\ \text{subject to} \\ & -\mathbf{A}^T\mathbf{w} = -\mathbf{c} \\ & \mathbf{w} \geq \mathbf{0} \end{aligned}$$

This last problem statement can be written as

$$\left.\begin{array}{l} \text{Minimize} \quad z' = \mathbf{b}^{\mathrm{T}}\mathbf{w} \\[6pt] \text{subject to} \\[6pt] \qquad \mathbf{A}^{\mathrm{T}}\mathbf{w} = \mathbf{c} \\[6pt] \qquad \mathbf{w} \geq \mathbf{0} \end{array}\right\} \tag{7}$$

We have shown that the dual of problem (7) is problem (6). Therefore, the dual of the dual of problem (7) is the dual of problem (6). Applying Theorem 2.1, it follows that problem (7) is the dual of problem (6), as we were to show.

We summarize the relationships between the primal and dual problems in Table 3.1.

Remember that Theorem 3.1 allows us to also read the table from right to left. That is, if the headings "Primal problem" and "Dual problem" are interchanged, the resulting table remains valid. For example, Table 3.1 shows that if the jth constraint in a minimization problem is a $\geq$ inequality, then the jth variable of the dual problem is ≥ 0. Note that the table shows how to find the dual of a maximization problem with $\leq$ and $=$ constraints and of a minimization problem with $\geq$ and $=$ constraints. If we have a maximization problem with a $\geq$ constraint, this constraint must be converted to $\leq$ (by multiplying by -1) before the dual problem can be constructed. The same procedure must be used on a minimization problem with a $\leq$ constraint. A diagrammatic illustration of duality is given in Figure 3.1.

There are also connections between the values of the primal variables and the values of the dual variables. These will be discussed later. However, it should be noted here that it is *not* possible to determine a feasible solution to the dual problem from values of the primal variables unless one has an optimal solution to the primal problem.

Table 3.1

Primal problem	Dual problem
Maximization	Minimization
Coefficients of objective function	Right-hand sides of constraints
Coefficients of ith constraint	Coefficients of ith variable, one in each constraint
ith constraint is an inequality $\leq$	ith variable is ≥ 0
ith constraint is an equality	ith variable is unrestricted
jth variable is unrestricted	jth constraint is an equality
jth variable is ≥ 0	jth constraint is an inequality $\geq$
Number of variables	Number of constraints

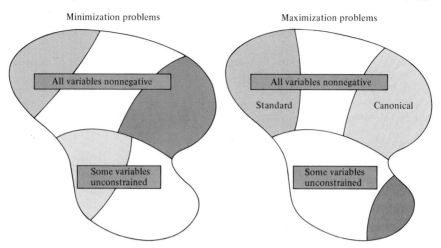

Figure 3.1

Minimization problems

Maximization problems

All variables nonnegative

Standard

All variables nonnegative

Canonical

Some variables
unconstrained

Some variables
unconstrained

Figure 3.1 Duality. Dual problems have the same shading.

EXAMPLE 2. If the primal problem is

$$\text{Maximize} \quad z = 3x_1 + 2x_2 + x_3$$

subject to

$$x_1 + 2x_2 - x_3 \le 4$$
$$2x_1 - x_2 + x_3 = 8$$
$$x_1 - x_2 \qquad \le 6$$
$$x_1 \ge 0, \quad x_2 \ge 0$$

then the dual problem is

$$\text{Minimize} \quad z' = 4w_1 + 8w_2 + 6w_3$$

subject to

$$w_1 + 2w_2 + w_3 \ge 3$$
$$2w_1 - w_2 - w_3 \ge 2$$
$$-w_1 + w_2 \qquad = 1$$
$$w_1 \ge 0, \quad w_3 \ge 0$$

EXAMPLE 3. If the primal problem is

$$\text{Minimize} \quad z = 2x_1 - 3x_2 + x_4$$

subject to

$$x_1 + 2x_2 + x_3 \qquad\quad \le 7 \qquad |-x_1 - 2v_2 - x_3 \ge -7$$

$$x_1 + 4x_2 \qquad\quad - x_4 = 5 \qquad w_2 \text{ unres.}$$

$$x_2 + x_3 + 5x_4 \ge 3$$

$$x_1 \ge 0, \quad x_2 \ge 0, \quad x_3 \ge 0, \quad x_4 \ge 0$$

then the dual problem is

$$\text{Maximize} \quad z' = -7w_1 + 5w_2 + 3w_3$$

subject to

$$-w_1 + w_2 \qquad\qquad \le 2$$

$$-2w_1 + 4w_2 + w_3 \le -3$$

$$-w_1 \qquad\quad + w_3 \le 0$$

$$-w_2 + 5w_3 \le 1$$

$$w_1 \ge 0, \quad w_3 \ge 0$$

To find this dual problem, we wrote the first constraint of the primal problem as

$$-x_1 - 2x_2 - x_3 \ge -7$$

An alternate method would be to change the primal problem to a maximization problem and multiply the third constraint by -1.

Properties of Solutions to the Primal and Dual Problems

We now present several theorems which will allow us to give conditions for a feasible solution to be an optimal solution.

Theorem 3.4. If x_0 is a feasible solution to the primal problem

$$\text{Maximize} \quad z = c^T x$$

subject to

$$Ax \le b \tag{8}$$

$$x \ge 0$$

and w_0 is a feasible solution to the dual problem

$$\text{Minimize} \quad z' = b^T w$$

subject to

$$A^T w \ge c \tag{9}$$

$$w \ge 0$$

then

$$c^Tx_0 \le b^Tw_0 \qquad (10)$$

That is, the value of the objective function of the dual problem is always greater than or equal to the value of the objective function of the primal problem.

Proof. Since x_0 is a feasible solution to (8), we have

$$Ax_0 \le b \qquad (11)$$

It follows from (11) that

$$w_0^TAx_0 \le w_0^Tb = b^Tw_0 \qquad (12)$$

since $w_0 \ge 0$. The equality in (12) comes from the fact that w_0^Tb is a 1×1 matrix and consequently is equal to its transpose.

Since w_0 is a feasible solution to (9), we have

$$A^Tw_0 \ge c$$

or, taking transposes,

$$w_0^TA \ge c^T$$

Again we can multiply by x_0, which is nonnegative, without changing the inequality. We get

$$w_0^TAx_0 \ge c^Tx_0 \qquad (13)$$

Combining the inequalities (12) and (13) gives the desired result.

Another interpretation of Theorem 3.4 comes in the case where the primal problem has an unbounded optimal solution. This means that given any number N, we can find a feasible solution to the primal problem for which the value of the objective function is greater than N. Consequently the objective function of the dual problem (using Theorem 3.4) is greater than N for any feasible solution w_0 to the dual problem. This means that there are no feasible solutions to the dual problem. For if w were such a solution, the value of the dual objective function would be b^Tw. But N can be chosen greater than this value, and we have

$$b^Tw < N < b^Tw$$

the second inequality coming from the argument above. We have shown:

If the primal problem has an unbounded optimal solution, then the dual problem has no feasible solutions.

Theorem 3.1 can be applied to this statement to conclude:

If the dual problem has an unbounded optimal solution, then the primal problem has no feasible solutions.

We now give a condition that a feasible solution to the primal problem be an optimal solution.

Theorem 3.5. If x_0 and w_0 are feasible solutions to the primal and dual problems (8) and (9), respectively, and $c^Tx_0 = b^Tw_0$, then both x_0 and w_0 are optimal solutions to their respective problems.

Proof. Suppose x_1 is any feasible solution to the primal problem. Then, from (10),

$$c^Tx_1 \le b^Tw_0 = c^Tx_0$$

Hence, x_0 is an optimal solution. Similarly, if w_1 is any feasible solution to the dual problem, then, from (10),

$$b^Tw_0 = c^Tx_0 \le b^Tw_1$$

and we see that w_0 is an optimal solution to the dual problem.

Theorem 3.6. (Duality Theorem)
   (a) If the primal (8) and dual (9) problems have feasible solutions, then (i) the primal problem has an optimal solution—say, x_0; (ii) the dual problem has an optimal solution—say, w_0; and (iii) $c^Tx_0 = b^Tw_0$.
   (b) If either the primal or dual problem has a feasible solution with a finite optimal objective value, then the other problem has a feasible solution with the same objective value.

Note that our discussion following Theorem 3.4 shows that for part b of the duality theorem (Theorem 3.6) we must assume that the problem has a finite optimal objective value.

Using the duality theorem and the other preceding theorems, we can state the following results:

The feasible solution x_0 to the primal problem is optimal if and only if the dual problem has a feasible solution w_0 which satisfies $c^Tx_0 = b^Tw_0$.

The feasible solution w_0 to the dual problem is optimal if and only if the primal problem has a feasible solution x_0 which satisfies $b^Tw_0 = c^Tx_0$.

If the primal problem has a feasible solution but the dual problem has no feasible solutions, then the primal problem has an unbounded optimal solution.

If the dual problem has a feasible solution but the primal problem has no feasible solutions, then the dual problem has an unbounded optimal solution.

EXAMPLE 4.   Consider the linear programming problem

$$\text{Maximize} \quad z = 2x_1 + x_2$$

subject to

$$3x_1 - 2x_2 \le 6$$

$$x_1 - 2x_2 \le 1 \qquad\qquad (14)$$

$$x_1 \ge 0, \quad x_2 \ge 0$$

The set of feasible solutions is shown in Figure 3.2a. It is evident that this problem has an unbounded optimal solution. For setting $x_1 = 0$ allows x_2 to be as large as we please. In this case $z = x_2$ is also as large as we please.

The dual problem to (14) is

$$\text{Minimize} \quad z' = 6w_1 + w_2 \qquad ,$$

subject to

$$3w_1 + w_2 \ge 2$$

$$-2w_1 - 2w_2 \ge 1$$

$$w_1 \ge 0, \quad w_2 \ge 0$$

The constraints are shown in Figure 3.2b. There are no feasible solutions to the problem, since the second constraint can never hold for nonnegative values of w_1 and w_2.

It is also possible, as the next example shows, that neither the primal problem nor its dual will have a feasible solution.

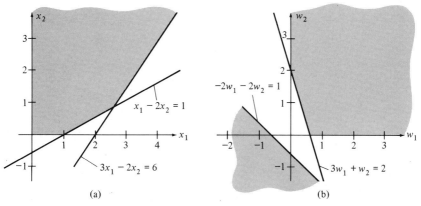

(a)   (b)

Figure 3.2

EXAMPLE 5. Consider the linear programming problem

$$\text{Maximize} \quad z = 3x_1 + 2x_2$$

subject to

$$2x_1 - 2x_2 \le -1$$
$$-2x_1 + 2x_2 \le -4$$
$$x_1 \ge 0, \quad x_2 \ge 0$$

Its dual problem is

$$\text{Minimize} \quad z' = -w_1 - 4w_2$$

subject to

$$2w_1 - 2w_2 \ge 3$$
$$-2w_1 + 2w_2 \ge 2$$
$$w_1 \ge 0, \quad w_2 \ge 0$$

The graphs of the constraints of the primal problem are shown in Figure 3.3a, and the graphs for the dual problem are shown in Figure 3.3b. Neither of these problems has a feasible solution.

Economic Interpretation of the Dual Problem

The dual problem can have various economic interpretations depending on the point of view one takes. We shall describe two possible interpretations here. As we discover more relationships between the primal and dual problems, we will be able to present some additional interpretations.

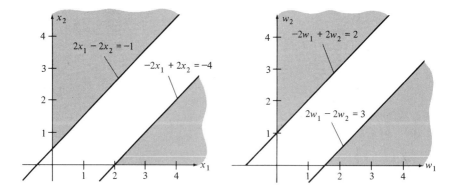

Figure 3.3

In Example 1 of Section 2.1 we gave a model of a sawmill. It is

$$\text{Maximize} \quad z = 120x_1 + 100x_2$$

subject to

$$2x_1 + 2x_2 \le 8$$

$$5x_1 + 3x_2 \le 15$$

$$x_1 \ge 0, \quad x_2 \ge 0$$

This is a simple example of a typical manufacturing problem—a problem that is in standard form (8). The first constraint in the sawmill problem deals with the number of hours the saw is available. The second constraint deals with the number of hours the plane is available. In general, in the ith constraint of (8)

$$\sum_{j=1}^{n} a_{ij}x_j \le b_i$$

we may think of b_i as the total supply of the ith resource, or raw material, or input. For the first constraint of the sawmill problem it was sawing time. The coefficient a_{ij} in the general problem represents the amount of the ith input required per unit of the jth product, or output. For example, $a_{21} = 5$ in the sawmill example represents the 5 hours planing time required for each 1000 board feet of finish-grade lumber. The variable x_j is the unknown amount of the jth output which is to be produced. The coefficient c_j in the objective function represents the profit, or value, derived from one unit of the jth output. The optimal solution maximizes the total value of all outputs

$$z = \sum_{j=1}^{n} c_j x_j$$

The dual problem to the sawmill example is

$$\text{Minimize} \quad z' = 8w_1 + 15w_2$$

subject to

$$2w_1 + 5w_2 \ge 120$$

$$2w_1 + 3w_2 \ge 100$$

$$w_1 \ge 0, \quad w_2 \ge 0$$

The coefficients of the first constraint are the amounts of each input which are needed to make one unit of the first output. That is, to make 1000 board feet of finish-grade lumber we need 2 hours of sawing and 5 hours of planing. The right-hand side of the first constraint is the profit, or value of one unit of the first output. Likewise, the second constraint of the dual problem of the sawmill example says that to make 1000 board feet of construction-grade lumber we need 2 hours of sawing and 3 hours of planing and the value of this amount of lumber is \$100.

The dual problem of the general linear programming problem in standard form is

$$\text{Minimize} \quad z' = \mathbf{b}^T\mathbf{w}$$

$$\text{subject to}$$

$$\mathbf{A}^T\mathbf{w} \geq \mathbf{c}$$

$$\mathbf{w} \geq \mathbf{0}$$

The jth constraint of this problem is

$$\sum_{i=1}^{m} a_{ij}w_i \geq c_j$$

As above, the coefficient a_{ij} represents the amount of input i per unit of output j, and the right-hand side is the value per unit of output j. This means that the units of the dual variable w_i are "value per unit of input i"; the dual variables act as prices, or costs, or values of one unit of each of the inputs. They are called by several names, including **accounting prices, fictitious prices, shadow prices,** and **imputed values**. The values of the dual variables are not directly related to the actual costs of the inputs. Just because the optimal solution to the dual of the sawmill problem is $w_1 = 35$, $w_2 = 10$ does not mean that the cost of the saw is \$35 per hour and the cost of the plane is \$10 per hour. The actual costs are hidden in whatever computations were done to figure out that the profits were \$120 per 1000 board feet of finish-grade and \$100 per board feet of construction-grade lumber. In the sense that the dual variables do not represent actual costs, they are fictitious prices. At an optimal solution to the dual problem, w_i represents the contribution to profit of one unit of the ith input, because in this case

$$z = \mathbf{c}^T\mathbf{x} = \text{Profit} = \mathbf{b}^T\mathbf{w}$$

The left-hand side of the jth constraint of the dual problem gives the total value of the inputs used in making one unit of the jth output. The constraint says that this value must be at least as much as the profit of one unit of the jth output. But at an optimal solution the value of the left-hand side represents the total contribution to profit of one unit of the jth output, and it is reasonable to expect to operate where this contribution to profit is at least as much as the actual profit. If this were not the case, the manufacturer would be well advised to use the available inputs in a better way.

We see then that the dual problem seeks shadow prices for each of the inputs which minimize their total price, subject to the restriction that these prices, or values, yield a corresponding value for each unit of output which is at least the profit for a unit of output.

Another interpretation of the dual variables comes from the fact that at the optimum

$$\text{Profit} = \mathbf{b}^T\mathbf{w}$$

To increase this profit the manufacturer must increase the availability of at least one of the resources. If b_i is increased by one unit, the profit will increase by w_i. Thus, w_i represents the **marginal value** of the ith input. In the same way, w_i is the loss incurred if one unit of the ith resource is not used. Thus it can be considered as a **replacement value** of the ith resource for insurance purposes. In fact, the insurance company would want to use the dual problem in case of a claim for lost resources; it wants to pay out as little as possible to settle the claim.

It is also interesting to look at an interpretation of the dual of the diet problem which was given in Example 2 of Section 2.1. The model of the diet problem is

$$\text{Minimize} \quad z = 20x_1 + 25x_2$$

subject to

$$2x_1 + 3x_2 \geq 18$$
$$x_1 + 3x_2 \geq 12$$
$$4x_1 + 3x_2 \geq 24$$
$$x_1 \geq 0, \quad x_2 \geq 0$$

The dual problem is

$$\text{Maximize} \quad z' = 18w_1 + 12w_2 + 24w_3$$

subject to

$$2w_1 + \ w_2 + 4w_3 \leq 20$$
$$3w_1 + 3w_2 + 3w_3 \leq 25$$
$$w_1 \geq 0, \quad w_2 \geq 0, \quad w_3 \geq 0$$

To discuss the dual problem we introduce some notation. Let N_1, N_2, and N_3 denote the nutrients fat, carbohydrate, and protein, respectively. It is also convenient to denote foods A and B by F_1 and F_2, respectively. Now introduce a manufacturer who makes artificial foods P_1, P_2, and P_3 with the property that for each $i = 1, 2, 3$, one unit of P_i provides one unit of nutrient N_i. Assume that the manufacturer sets w_i as the price of P_i $(i = 1, 2, 3)$. Recall that in the original statement of the problem, a_{ij} is the number of units of nutrient N_i in 1 ounce of food F_j. For example, $a_{12} = 3$ is the number of units of fat (N_1) in 1 ounce of food F_2, and $a_{31} = 4$ is the number of units of protein (N_3) in 1 ounce of food F_1. The artificial food manufacturer will set its prices so that

$$2w_1 + w_2 + 4w_3 \leq 20$$

and

$$3w_1 + 3w_2 + 3w_3 \leq 25$$

That is, it will set the prices on the foods P_i so that when these foods are taken in the proportions necessary to give the same nutrition as foods F_1 and F_2, the cost of the substitute for F_1 is no greater than the cost of F_1 itself and the cost of the substitute for F_2 is no greater than the cost of F_2 itself. Thus, the nutritionist will always find it at least as economical to buy the three artificial foods. Since we require 18 units of fat (from P_1), 12 units of carbohydrate (from P_2), and 24 units of protein (from P_3), the manufacturer's revenue is

$$z' = 18w_1 + 12w_2 + 24w_3$$

It seeks to maximize this revenue.

Thus, the fictitious prices w_1, w_2, and w_3 of the nutrients are those prices which the artificial food manufacturer should charge to maximize the revenue, yet still be able to compete with the producer of foods F_1 and F_2. The fictitious prices represent competitive prices.

Complementary Slackness

In addition to the relations between an optimal solution to the primal problem and an optimal solution to the corresponding dual problem which we discussed before, we can determine information about which constraints may be "acting" on the solution to the primal problem. Specifically, we can show that if an optimal solution to the primal problem makes a constraint into a strict inequality, then the corresponding dual variable must be zero.

We consider a linear programming problem in standard form (1) as the primal problem. Its dual is given in (2). For these problems we state and prove the theorem on complementary slackness:

Theorem 3.7.   For any pair of optimal solutions to the primal problem and its dual, we have:

(a) For $i = 1, 2, \ldots, m$, the product of the ith slack variable for the primal problem and the ith dual variable is zero. That is,

$x_{n+i}w_i = 0, \quad i = 1, 2, \ldots, m$, where x_{n+i} is the ith slack variable for the primal problem.

(b) For $j = 1, 2, \ldots, n$, the product of the jth slack variable for the dual problem and the jth variable for the primal problem is zero.

Another way of stating the theorem is to say that if the ith slack variable of the primal problem is not zero, then the ith dual variable must be zero. Likewise, if the jth slack variable for the dual problem is not zero, then the jth primal variable must be zero. Note that it is possible for both the slack variable and corresponding dual variable to be zero.

Proof. We add slack variables to the primal problem and write

$$\mathbf{Ax} + \mathbf{Ix'} = \mathbf{b}$$

where

$$\mathbf{x'} = \begin{bmatrix} x_{n+1} \\ x_{n+2} \\ \vdots \\ x_{n+m} \end{bmatrix}$$

is the vector of slack variables. Then for any vector of dual variables

$$\mathbf{w} = \begin{bmatrix} w_1 \\ w_2 \\ \vdots \\ w_n \end{bmatrix}$$

we have

$$\mathbf{w}^T\mathbf{Ax} + \mathbf{w}^T\mathbf{Ix'} = \mathbf{w}^T\mathbf{b}$$

This equality is an equality between numbers so that it is preserved when the transpose of each side is taken. We obtain

$$\mathbf{x}^T\mathbf{A}^T\mathbf{w} + \mathbf{x'}^T\mathbf{w} = \mathbf{b}^T\mathbf{w} \tag{15}$$

At the optimal solutions $\begin{bmatrix} \mathbf{x}_0 \\ \mathbf{x}_0' \end{bmatrix}$ and $\mathbf{w}_0$, we have, from the duality theorem,

$$\mathbf{c}^T\mathbf{x}_0 = \mathbf{b}^T\mathbf{w}_0 \tag{16}$$

Using (15), we may rewrite (16) as

$$\mathbf{c}^T\mathbf{x}_0 = \mathbf{x}_0^T\mathbf{A}^T\mathbf{w}_0 + \mathbf{x}_0'^T\mathbf{w}_0 \tag{17}$$

In the dual problem, $\mathbf{A}^T\mathbf{w}_0 \geq \mathbf{c}$; hence, we may write (17) as

$$\mathbf{c}^T\mathbf{x}_0 \geq \mathbf{x}_0^T\mathbf{c} + \mathbf{x}_0'^T\mathbf{w}_0 \tag{18}$$

Now $\mathbf{c}^T\mathbf{x}_0$ is a number so that $\mathbf{c}^T\mathbf{x}_0 = (\mathbf{c}^T\mathbf{x}_0)^T = \mathbf{x}_0^T\mathbf{c}$. Since $\mathbf{x}_0' \geq \mathbf{0}$ and

$\mathbf{w}_0 \geq \mathbf{0}$, we have equality in (18) and $\mathbf{x}_0'^T \mathbf{w}_0 = 0$. That is,

$$x_{n+1}w_1 + x_{n+2}w_2 + \cdots + x_{n+m}w_m = 0$$

and each term is nonnegative. Therefore, for each $i = 1, 2, \ldots, m$, we have $x_{n+i}w_i = 0$. The proof of part b is similar.

The economic interpretation of complementary slackness is related to the understanding of the dual variables as marginal costs. Suppose that in an optimal solution to the primal problem the ith slack variable is nonzero. This means that there is more of the ith input available than is needed for this optimal solution. The value of the ith slack variable is exactly the excess of the ith input. But there is no need for any of this excess of the ith input; its marginal value is zero. The theorem on complementary slackness tells us that if the ith primal slack variable is positive, then the ith dual variable, which can be thought of as the marginal value of the ith input, is zero. On the other hand, if in an optimal solution to the dual problem the ith dual variable is nonzero, its value can be thought of as the marginal value of the ith input. For in this case the ith primal slack variable is zero, indicating that all the ith input has been used and that it is desirable to have more. Its marginal value is positive.

EXAMPLE 6. Consider the linear programming problem

$$\text{Maximize} \quad z = 2x_1 + x_2$$

subject to

$$x_1 + 2x_2 \leq 8$$
$$3x_1 + 4x_2 \leq 18$$
$$x_1 \geq 0, \quad x_2 \geq 0$$

The dual of this problem is

$$\text{Minimize} \quad z' = 8w_1 + 18w_2$$

subject to

$$w_1 + 3w_2 \geq 2$$
$$2w_1 + 4w_2 \geq 1$$
$$w_1 \geq 0, \quad w_2 \geq 0$$

The feasible regions for the primal and dual problems are shown in Figures 3.4a and 3.4b, respectively.

An optimal solution to the primal problem is (verify)

$$x_1 = 6, \qquad x_2 = 0, \quad \text{and} \quad z = 12$$

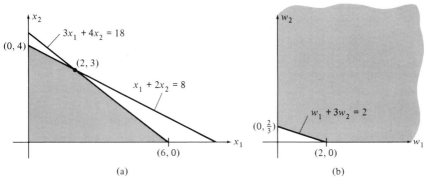

Figure 3.4

Since there is slack in the first constraint, the principle of complementary slackness says that the first dual variable must be zero in an optimal solution to the dual problem. Thus, without evaluating the objective function at the extreme points, we see that

$$w_1 = 0, \quad w_2 = \tfrac{2}{3}$$

must be an optimal solution to the dual problem. Furthermore, the value of the dual objective function at this extreme point must be 12. If there were several points for which $w_1 = 0$, a point at which the dual objective function has value 12 would be an optimal solution to the dual problem by the duality theorem (Theorem 3.6).

The dual variables do provide some measure of the contribution of each resource to the maximum profit. Recall that at optimal solutions, $z = z'$ or

$$\mathbf{b}^T\mathbf{w} = \mathbf{c}^T\mathbf{x} = \text{Maximum profit} \tag{19}$$

The constraints of the primal problem in which there is slack at an optimal solution correspond to dual variables which have value zero by complementary slackness. Thus, the corresponding resources do not contribute to the maximum profit. Considering the constraints of the primal problem in which there is no slack at an optimal solution, we see from equation (19) that the values of the corresponding dual variables divide the profit per unit of output proportionately among the corresponding resources.

For example, in the sawmill problem, each $120 of profit from finish-grade lumber comes from the value of the saw at $35 per hour ($w_1 = 35$) and from the plane at $10 per hour ($w_2 = 10$). Furthermore, it says that the maximum profit of $430 is attributable to the saw and the plane at the respective rates of $35 per hour and $10 per hour. That is, to make this profit, the saw is $3\frac{1}{2}$ times more valuable than the plane. In this sense the dual variables are accounting costs and would be useful for cost-accounting procedures.

3.1 Exercises

In Exercises 1–6 find the dual of the given linear programming problem.

1. Minimize $z = 3x_1 + 4x_2$

subject to

$$x_1 + 4x_2 \geq 8$$
$$2x_1 + 3x_2 \geq 12$$
$$2x_1 + x_2 \geq 6$$
$$x_1 \geq 0, \quad x_2 \geq 0$$

2. Minimize $z = 6x_1 + 6x_2 + 8x_3 + 9x_4$

subject to

$$x_1 + 2x_2 + x_3 + x_4 \geq 3$$
$$2x_1 + x_2 + 4x_3 + 9x_4 \geq 8$$
$$x_1 \geq 0, \quad x_2 \geq 0, \quad x_3 \geq 0, \quad x_4 \geq 0$$

3. Maximize $z = 3x_1 + 2x_2 + 5x_3 + 7x_4$

subject to

$$3x_1 + 2x_2 + x_3 \leq 8$$
$$5x_1 + x_2 + 2x_3 + 4x_4 = 7$$
$$4x_1 + x_3 - 2x_4 \leq 12$$
$$x_1 \geq 0, \quad x_2 \geq 0, \quad x_3 \geq 0, \quad x_4 \geq 0$$

4. Maximize $z = 2x_1 + x_2 + 3x_3 + 4x_4$

subject to

$$4x_1 + 2x_2 + 5x_3 + 5x_4 \leq 10$$
$$4x_1 + 2x_2 + 5x_3 + 5x_4 \geq 5$$
$$3x_1 + 5x_2 + 4x_3 + x_4 \geq 8$$
$$3x_1 + 5x_2 + 4x_3 + x_4 \leq 15$$
$$x_1 + x_2 + x_3 + x_4 = 20$$
$$x_1 \geq 0, \quad x_2 \geq 0, \quad x_3 \geq 0, \quad x_4 \geq 0$$

5. Maximize $z = 3x_1 + x_2 + 4x_3$

   subject to

   $$3x_1 + 3x_2 + x_3 \leq 18$$
   $$2x_1 + 2x_2 + 4x_3 = 12$$
   $$x_1 \geq 0, \quad x_3 \geq 0$$

6. Minimize $z = 5x_1 + 2x_2 + 6x_3$

   subject to

   $$4x_1 + 2x_2 + x_3 \geq 12$$
   $$3x_1 + 2x_2 + 3x_3 \leq 6$$
   $$x_1 \geq 0, \quad x_2 \geq 0$$

In Exercises 7 and 8, formulate the dual problem for the given linear programming problem.

7. Exercise 2, Section 2.1 (page 72).

8. Exercise 4, Section 2.1 (page 72). Give an economic interpretation of the dual problem.

9. A health food store packages a nut sampler consisting of walnuts, pecans, and almonds. Suppose each ounce of walnuts contains 12 units of protein and 3 units of iron and costs 12 cents, each ounce of pecans contains 1 unit of protein and 3 units of iron and costs 9 cents, and each ounce of almonds contains 2 units of protein and 1 unit of iron and costs 6 cents. If each package of the nut sampler is to contain at least 24 units of protein and at least 18 units of iron, how many ounces of each type of nut should be used to minimize the cost of the sampler? (*Hint:* Set up and solve the dual problem. Then use the principle of complementary slackness to solve the given problem.)

10. Show that $x_1 = \frac{5}{26}$, $x_2 = \frac{3}{2}$, $x_3 = \frac{27}{26}$ is an optimal solution to the following linear programming problem:

    Maximize $z = 9x_1 + 14x_2 + 7x_3$

    subject to

    $$2x_1 + x_2 + 3x_3 \leq 6$$
    $$5x_1 + 4x_2 + x_3 \leq 12$$
    $$2x_2 \leq 5$$
    $$x_1, x_2, x_3 \text{ unrestricted}$$

(*Hint:* Formulate the dual of this problem.)

11. Consider the linear programming problem

$X_1 = 7$
$X_2 = 1$

$$\text{Maximize} \quad z = 3x_1 + 4x_2$$

subject to

$$x_1 + 2x_2 \le 10$$
$$x_1 + x_2 \le 8$$
$$3x_1 + 5x_2 \le 26$$
$$x_1 \ge 0, \quad x_2 \ge 0$$

By using the principle of complementary slackness, show that $w_1 = 0$ in any optimal solution to the dual problem.

12. Suppose that $x_1 = 2$, $x_2 = 0$, $x_3 = 4$ is an optimal solution to the linear programming problem

$$\text{Maximize} \quad z = 4x_1 + 2x_2 + 3x_3$$

subject to

$$2x_1 + 3x_2 + x_3 \le 12$$
$$x_1 + 4x_2 + 2x_3 \le 10$$
$$3x_1 + x_2 + x_3 \le 10$$
$$x_1 \ge 0, \quad x_2 \ge 0, \quad x_3 \ge 0$$

Using the principle of complementary slackness and the duality theorem (Theorem 3.6), find an optimal solution to the dual problem. What value will the objective function of the dual problem have at this optimal solution?

13. Using the definition of the dual of a problem in standard form, find the dual of the linear programming problem

$$\text{Maximize} \quad z = c^T x + d^T x'$$

subject to

$$Ax + Bx' \le b$$
$$x \ge 0, \quad x' \text{ unrestricted}$$

(*Hint:* Write $x' = u - v$, $u \ge 0$, $v \ge 0$, and express the given problem as a standard linear programming problem in matrix notation.)

3.2 COMPUTATIONAL RELATIONS BETWEEN THE PRIMAL AND DUAL PROBLEMS

We have seen how to use the simplex method to solve a linear programming problem. We shall show in this section that if the problem is considered to be our primal problem, the simplex method when applied to it also yields a solution to the dual problem. This result comes from a more careful examination of the information available in the final tableau of the simplex algorithm.

Suppose that we are given a general linear programming problem, and suppose that we have introduced slack and artificial variables to convert the problem to canonical form. That is, we take our problem as

$$\left.\begin{array}{c} \text{Maximize}\quad z = \mathbf{c}^T\mathbf{x} \\[6pt] \text{subject to} \\[6pt] \mathbf{A}\mathbf{x} = \mathbf{b} \\[6pt] \mathbf{x} \geq \mathbf{0} \end{array}\right\} \tag{1}$$

where $\mathbf{A}$ is an $m \times s$ matrix which contains an $m \times m$ identity submatrix; $\mathbf{b} \geq \mathbf{0}$ is an $m \times 1$ matrix; and $\mathbf{c}$ is an $s \times 1$ matrix. Remember that if x_j is a slack variable, then $c_j = 0$. Also, we use the two-phase method for the artificial variables so that if x_j is an artificial variable, then $c_j = 0$.

We now examine a tableau constructed by the simplex algorithm during the solution of our problem. This tableau represents a basic feasible solution. Let i_1 be the subscript of the first basic variable in this solution; let i_2 be the subscript of the second basic variable. Continuing in this manner, we end with i_m being the subscript of the mth basic variable. Let N denote the set of indices of the nonbasic variables. We also let $\mathbf{A}_j$ denote the jth column of $\mathbf{A}$. Using this notation, we may write (1) as

$$\sum_{r=1}^{m} x_{i_r}\mathbf{A}_{i_r} + \sum_{j \in N} x_j\mathbf{A}_j = \mathbf{b} \tag{2}$$

Recall that the nonbasic variables are set equal to zero, so that (2) may be simplified to

$$x_{i_1}\mathbf{A}_{i_1} + x_{i_2}\mathbf{A}_{i_2} + \cdots + x_{i_m}\mathbf{A}_{i_m} = \mathbf{b} \tag{3}$$

We make an $m \times m$ matrix out of the m columns $\mathbf{A}_{i_1}, \mathbf{A}_{i_2}, \ldots, \mathbf{A}_{i_m}$ of $\mathbf{A}$ corresponding to basic variables and denote this matrix by $\mathbf{B}$. Let

$$\mathbf{x}_B = \begin{bmatrix} x_{i_1} \\ x_{i_2} \\ \vdots \\ x_{i_m} \end{bmatrix} \quad \text{and} \quad \mathbf{c}_B = \begin{bmatrix} c_{i_1} \\ c_{i_2} \\ \vdots \\ c_{i_m} \end{bmatrix}$$

Then (3) may be written as

$$\mathbf{B}\mathbf{x}_B = \mathbf{b} \tag{4}$$

Since the columns of $\mathbf{B}$ are linearly independent, $\mathbf{B}$ must be an invertible matrix. We may write (4) as

$$\mathbf{x}_B = \mathbf{B}^{-1}\mathbf{b} \tag{5}$$

That is, the m-tuple of basic variables has the value $\mathbf{B}^{-1}\mathbf{b}$ in any tableau.

We may also note that the m columns of $\mathbf{B}$ considered as m-tuples form a basis for R^m. Thus, the jth column of our initial tableau, $\mathbf{A}_j$, can be written as a linear combination of the columns $\mathbf{A}_{i_1}, \mathbf{A}_{i_2}, \ldots, \mathbf{A}_{i_m}$, where the indices $i_1, i_2, \ldots, i_m$ are the indices of the basic variables of our current tableau in the order of the labels on the left-hand side.

We have

$$\mathbf{A}_j = t_{1j}\mathbf{A}_{i_1} + t_{2j}\mathbf{A}_{i_2} + \cdots + t_{mj}\mathbf{A}_{i_m} \tag{6}$$

It can be shown that the vector of coefficients in (6)

$$\mathbf{t}_j = \begin{bmatrix} t_{1j} \\ t_{2j} \\ \vdots \\ t_{mj} \end{bmatrix}$$

which is the coordinate vector of $\mathbf{A}_j$ with respect to the basis

$$\{\mathbf{A}_{i_1}, \mathbf{A}_{i_2}, \ldots, \mathbf{A}_{i_m}\}$$

of R^m is also the jth column of our current tableau. In the same way as in (4), we may write (6) as

$$\mathbf{B}\mathbf{t}_j = \mathbf{A}_j$$

or

$$\mathbf{t}_j = \mathbf{B}^{-1}\mathbf{A}_j \tag{7}$$

Equation (7) says that any column of a tableau can be found from the corresponding column of the initial tableau by multiplying on the left by $\mathbf{B}^{-1}$.

Using the notation that has been developed, we define

$$z_j = \mathbf{c}_B^T\mathbf{t}_j \tag{8}$$

or, using (7),

$$z_j = \mathbf{c}_B^T\mathbf{B}^{-1}\mathbf{A}_j \tag{9}$$

From this definition we see that for the rth basic variable x_{i_r} we have

$$z_{i_r} = [c_{i_1} \quad c_{i_2} \ldots c_{i_m}] \begin{bmatrix} 0 \\ \vdots \\ 0 \\ 1 \\ 0 \\ \vdots \\ 0 \end{bmatrix} \quad \leftarrow r\text{th entry}$$

$$= c_{i_r}$$

The objective function for this problem takes on a particularly simple form when we use the definition of z_j. Recall that

$$z = \mathbf{c}^T\mathbf{x} = \sum_{j=1}^{s} c_j x_j \tag{10}$$

We divide the sum on the right-hand side of (10) into two parts: one part contains the basic variables and one part contains the nonbasic variables. We write

$$z = \sum_{r=1}^{m} c_{i_r} x_{i_r} + \sum_{j \in N} c_j x_j \tag{11}$$

In order to apply the optimality criterion to our tableau, we must modify (11) so that the coefficients of the basic variables are zero. To make this modification we add $(-c_{i_1}) \times$ first row of the current tableau $+ (-c_{i_2}) \times$ second row $+ \cdots + (-c_{i_m}) \times$ mth row to (11). In symbols, the rth row of the current tableau can be expressed as

$$\sum_{j=1}^{s} t_{rj} x_j = \mathbf{x}_{Br}$$

where $\mathbf{x}_{Br}$ is the rth component of the vector $\mathbf{x}_B$.

Adding the appropriate multiples of the rows of the tableau to (11), we obtain

$$z - \sum_{r=1}^{m} c_{i_r} \mathbf{x}_{Br} = \sum_{j \in N} c_j x_j - \sum_{r=1}^{m} c_{i_r} \sum_{j \in N} t_{rj} x_j$$

Simplifying, we have

$$z - \mathbf{c}_B^T \mathbf{x}_B = \sum_{j \in N} (c_j - \mathbf{c}_B^T \mathbf{t}_j) x_j$$

or

$$\sum_{j \in N} (z_j - c_j) x_j + z = \mathbf{c}_B^T \mathbf{x}_B$$

Since $z_{i_r} - c_{i_r} = 0$, we obtain

$$\sum_{j=1}^{s} (z_j - c_j)x_j + z = \mathbf{c}_B^T \mathbf{x}_B$$

This equation shows that the entries in the objective row of a tableau are simply

$$z_j - c_j = \mathbf{c}_B^T \mathbf{t}_j - c_j \tag{12}$$

Consider now the dual problems

$$\left.\begin{array}{l} \text{Maximize} \quad z = \mathbf{c}^T \mathbf{x} \\[4pt] \text{subject to} \\[4pt] \qquad \mathbf{Ax} = \mathbf{b} \\[4pt] \qquad \mathbf{x} \geq \mathbf{0} \end{array}\right\} \tag{13}$$

and

$$\left.\begin{array}{l} \text{Minimize} \quad z' = \mathbf{b}^T \mathbf{w} \\[4pt] \text{subject to} \\[4pt] \qquad \mathbf{A}^T \mathbf{w} \geq \mathbf{c} \\[4pt] \quad \mathbf{w} \text{ unrestricted} \end{array}\right\} \tag{14}$$

For any basic feasible solution $\mathbf{x}$ to (13) with corresponding matrix $\mathbf{B}$, we set

$$\mathbf{w} = (\mathbf{B}^{-1})^T \mathbf{c}_B \tag{15}$$

The objective function for (13) can be written as

$$z = \mathbf{c}_B^T \mathbf{x}_B = \mathbf{c}_B^T \mathbf{B}^{-1} \mathbf{b} = \mathbf{b}^T (\mathbf{B}^{-1})^T \mathbf{c}_B = \mathbf{b}^T \mathbf{w}$$

Thus our choice for $\mathbf{w}$ is a vector which gives the value of the primal objective function when substituted into the dual objective function. This property is necessary if $\mathbf{x}$ is an optimal solution in light of Theorem 3.6.

If we substitute the value for $\mathbf{w}$ given by (15) into the left-hand side of the jth constraint of the dual problem

$$\sum_{i=1}^{m} a_{ij} w_i \geq c_j$$

we obtain

$$\sum_{i=1}^{m} a_{ij}w_i = \mathbf{A}_j^\mathsf{T}\mathbf{w}$$

$$= \mathbf{A}_j^\mathsf{T}(\mathbf{B}^{-1})^\mathsf{T}\mathbf{c}_B$$

$$= (\mathbf{B}^{-1}\mathbf{A}_j)^\mathsf{T}\mathbf{c}_B$$

$$= \mathbf{t}_j^\mathsf{T}\mathbf{c}_B$$

$$= z_j$$

Thus, this value for $\mathbf{w}$ satisfies the jth constraint of the dual problem if and only if $z_j \geq c_j$ or $z_j - c_j \geq 0$. Of course, at an optimal solution we have $z_j - c_j \geq 0$ for all j.

The optimality criterion for the simplex method tells us that the jth variable of the primal problem is a candidate for an entering variable if $z_j - c_j < 0$. We have already noted that w_i is a value, or cost, attached to each unit of resource i. Then z_j represents the total value assigned to one unit of the jth product. In economics this value is called an **imputed value** because it is not directly determined by the marketplace; it is a value assigned using some other basis than an exchange of money. The jth variable is a candidate for an entering variable if $z_j < c_j$; that is, x_j may enter if the imputed value z_j of the jth product is less than the profit c_j per unit of the jth product.

We now modify the form of a tableau and the steps in the pivoting process to take advantage of the new information we have. We add an additional row and column to a tableau. The row is written above the column labels and contains the vector $\mathbf{c}^\mathsf{T} = [c_1 \quad c_2 \ldots c_s]$. The column is written to the left of the column denoting the basic variables and contains the vector

$$\mathbf{c}_B = \begin{bmatrix} c_{i_1} \\ c_{i_2} \\ \vdots \\ c_{i_m} \end{bmatrix}$$

The entries of $\mathbf{c}_B$ can be determined by copying the values of c_j from the new top row corresponding to the basic variables of the tableau.

The pivoting step can be changed to compute the entries of the objective row by using (12) rather than by using elementary row operations. Recall that the entries of the objective row are $z_j - c_j$.

We now rework an example using these new ideas.

EXAMPLE 1.   Consider the linear programming problem from Example 3 in Section 2.5:

$$\text{Maximize}\quad z = x_1 - 2x_2 - 3x_3 - x_4 - x_5 + 2x_6$$

subject to

$$
\begin{aligned}
x_1 + 2x_2 + 2x_3 + x_4 + x_5 + x_7 &= 12 \\
x_1 + 2x_2 + x_3 + x_4 + 2x_5 + x_6 &= 18 \\
3x_1 + 6x_2 + 2x_3 + x_4 + 3x_5 + x_8 &= 24
\end{aligned}
$$

$$x_j \geq 0, \qquad j = 1, 2, \ldots, 8$$

We have   _and p Max_ $z' = -(x_7 + x_8)$

$$
A = \begin{bmatrix}
1 & 2 & 2 & 1 & 1 & 0 & 1 & 0 \\
1 & 2 & 1 & 1 & 2 & 1 & 0 & 0 \\
3 & 6 & 2 & 1 & 3 & 0 & 0 & 1
\end{bmatrix},
$$

$$
b = \begin{bmatrix} 12 \\ 18 \\ 24 \end{bmatrix}, \quad \text{and} \quad
c = \begin{bmatrix} 1 \\ -2 \\ -3 \\ -1 \\ -1 \\ 2 \\ 0 \\ 0 \end{bmatrix}
$$

Note that this problem is in the form of equation (5). The matrix **A** contains the identity matrix by taking columns 7, 6, and 8 in this order. We have denoted the artificial variables by x_7 and x_8 rather than y_1 and y_2 to be consistent with (5).

The initial tableau for the auxiliary problem of the two-phase method is shown in Tableau 3.1a where the objective row has not been filled in. Recall from Section 2.5 that we had to eliminate the initial basic variables x_7 and x_8 from the objective function by substitution. This substitution procedure is replaced by the procedure of computing the objective row

Tableau 3.1a

c_j		0	0	0	0	0	0	−1	−1	
c_B		x_1	x_2	x_3	x_4	x_5	x_6	x_7	x_8	x_B
−1	x_7	1	2	2	1	1	0	1	0	12
0	x_6	1	2	1	1	2	1	0	0	18
−1	x_8	3	6	2	1	3	0	0	1	24

as $z_j - c_j$. We have

$$\mathbf{c}_B^T = \begin{bmatrix} -1 & 0 & -1 \end{bmatrix}$$

since $i_1 = 7$, $i_2 = 6$, and $i_3 = 8$. The entry in column 1 of the objective row is, by (11),

$$z_1 - c_1 = \begin{bmatrix} -1 & 0 & -1 \end{bmatrix} \begin{bmatrix} 1 \\ 1 \\ 3 \end{bmatrix} - 0 = -4$$

In the same way we compute the other entries obtaining the full initial tableau (Tableau 3.1b). Note that it is the same as Tableau 2.20 (Section 2.5).

Tableau 3.1b

$\mathbf{c}_B$		0 x_1	0 x_2	0 x_3	0 x_4	0 x_5	0 x_6	-1 x_7	-1 x_8	x_B
-1	x_7	1	2	2	1	1	0	1	0	12
0	x_6	1	2	1	1	2	1	0	0	18
-1	x_8	3	⑥	2	1	3	0	0	1	24
		-4	-8	-4	-2	-4	0	0	0	-36

The value of the last entry in the objective row is computed as

$$z = \mathbf{c}_B^T \mathbf{x}_B$$

where $\mathbf{x}_B$ is the last column of the tableau.

For this tableau,

$$\mathbf{B} = \begin{bmatrix} 1 & 0 & 0 \\ 0 & 1 & 0 \\ 0 & 0 & 1 \end{bmatrix} \quad \text{and} \quad \mathbf{B}^{-1} = \begin{bmatrix} 1 & 0 & 0 \\ 0 & 1 & 0 \\ 0 & 0 & 1 \end{bmatrix}$$

since

$$i_1 = 7 \quad \text{and} \quad \mathbf{A}_{i_1} = \begin{bmatrix} 1 \\ 0 \\ 0 \end{bmatrix}, \qquad i_2 = 6 \quad \text{and} \quad \mathbf{A}_{i_2} = \begin{bmatrix} 0 \\ 1 \\ 0 \end{bmatrix},$$

$$i_3 = 8 \quad \text{and} \quad \mathbf{A}_{i_3} = \begin{bmatrix} 0 \\ 0 \\ 1 \end{bmatrix}$$

The entering and departing variables are determined as previously. We find x_2 is the entering variable and x_8 is the departing variable. In the next tableau (Tableau 3.2), x_7, x_6, and x_2 will be the basic variables, in

Tableau 3.2

		0	0	0	0	0	0	-1	-1	
$\mathbf{c}_B$		x_1	x_2	x_3	x_4	x_5	x_6	x_7	x_8	$\mathbf{x}_B$
-1	x_7	0	0	$\boxed{\tfrac{4}{3}}$	$\tfrac{2}{3}$	0	0	1	$-\tfrac{1}{3}$	4
0	x_6	0	0	$\tfrac{1}{3}$	$\tfrac{2}{3}$	1	1	0	$-\tfrac{1}{3}$	10
0	x_2	$\tfrac{1}{2}$	1	$\tfrac{1}{3}$	$\tfrac{1}{6}$	$\tfrac{1}{2}$	0	0	$\tfrac{1}{6}$	4
		0	0	$-\tfrac{4}{3}$	$-\tfrac{2}{3}$	0	0	0	$\tfrac{4}{3}$	-4

this order, so that $i_1 = 7$, $i_2 = 6$, and $i_3 = 2$. Thus,

$$\mathbf{c}_B = \begin{bmatrix} -1 \\ 0 \\ 0 \end{bmatrix} \quad \text{and} \quad \mathbf{B} = \begin{bmatrix} 1 & 0 & 2 \\ 0 & 1 & 2 \\ 0 & 0 & 6 \end{bmatrix}$$

The objective row is computed after the usual procedure for pivoting is applied to find the other rows. The entries in the objective row are found by

$$z_j - c_j = \mathbf{c}_B^T \mathbf{t}_j - c_j$$

For example, the entry in the third column of Tableau 3.2 is

$$z_3 - c_3 = \mathbf{c}_B^T \mathbf{t}_3 - c_3 = \begin{bmatrix} -1 & 0 & 0 \end{bmatrix} \begin{bmatrix} \tfrac{4}{3} \\ \tfrac{1}{3} \\ \tfrac{1}{3} \end{bmatrix} - 0 = -\tfrac{4}{3}$$

We see that Tableau 3.2 is the same as Tableau 2.21 of Section 2.5. We continue the simplex algorithm by determining the entering and departing variables for Tableau 3.2 and pivoting to form Tableau 3.3.

Tableau 3.3

		0	0	0	0	0	0	-1	-1	
$\mathbf{c}_B$		x_1	x_2	x_3	x_4	x_5	x_6	x_7	x_8	$\mathbf{x}_B$
0	x_3	0	0	1	$\tfrac{1}{2}$	0	0	$\tfrac{3}{4}$	$-\tfrac{1}{4}$	3
0	x_6	0	0	0	$\tfrac{1}{2}$	1	1	$-\tfrac{1}{4}$	$-\tfrac{1}{4}$	9
0	x_2	$\boxed{\tfrac{1}{2}}$	1	0	0	$\tfrac{1}{2}$	0	$-\tfrac{1}{4}$	$\tfrac{1}{4}$	3
		0	0	0	0	0	0	1	1	0

Tableau 3.3 gives an optimal solution to the auxiliary problem. The next step is to form the initial tableau for Phase 2. The entries in the objective row of this tableau may be computed as $z_j - c_j$. Remember to put the coefficients of the original objective function at the top of the tableau. Carrying out these steps, we obtain Tableau 3.4.

Tableau 3.4

$\mathbf{c_B}$		1 x_1	−2 x_2	−3 x_3	−1 x_4	−1 x_5	2 x_6	$\mathbf{x_B}$
−3	x_3	0	0	1	½	0	0	3
2	x_6	0	0	0	½	1	1	9
−2	x_2	②	1	0	0	½	0	3
		−2	0	0	½	2	0	3

Pivoting, we obtain Tableau 3.5, which yields an optimal solution to the problem. The reader should check that he or she understands how the entries were determined.

Tableau 3.5

$\mathbf{c_B}$		1 x_1	−2 x_2	−3 x_3	−1 x_4	−1 x_5	2 x_6	$\mathbf{x_B}$
−3	x_3	0	0	1	½	0	0	3
2	x_6	0	0	0	½	1	1	9
1	x_1	1	2	0	0	1	0	6
		0	4	0	½	4	0	15

We now have defined enough tools to complete our discussion of artificial variables. Recall that we had taken a problem in canonical form

$$\text{Maximize} \quad z = \mathbf{c}^T\mathbf{x}$$

subject to

$$\mathbf{Ax} = \mathbf{b}$$

$$\mathbf{x} \geq \mathbf{0}$$

where $\mathbf{b} \geq \mathbf{0}$. We introduced artificial variables into each of the constraint equations. For Phase 1 we used a different objective function, namely,

$$\text{Minimize} \quad z = y_1 + y_2 + \cdots + y_m$$

where y_i, $i = 1, 2, \ldots, m$ are the artificial variables. Suppose that at the end of Phase 1, the minimum of this objective function is zero but that there are artificial variables which remain basic (at value zero) in the final optimal tableau. We now proceed as follows:

Phase 2

The initial tableau for Phase 2 is the final tableau of Phase 1 with the following modifications:

(a) Delete the columns from the final tableau of Phase 1 which are labeled with the nonbasic artificial variables.

(b) Replace the row above the column labels with the coefficients of the original objective function, assigning 0 as a cost for each of the basic artificial variables.

(c) Form the vector $\mathbf{c}_B$ from the new row of objective function coefficients.

(d) Calculate the entries in the new objective row as $z_j - c_j = \mathbf{c}_B^T \mathbf{t}_j - c_j$.

As we proceed with the simplex method for Phase 1 we must insure that the remaining artificial variables do not take on positive values. This would happen when one of these variables remained basic and the pivotal elimination gave a positive entry in $\mathbf{x}_B$ for the position labeled by the artificial variable. Suppose that x_k is to be the entering variable and that the rows labeled by artificial variables are $i_1, i_2, \ldots, i_p$. Denote the kth column of the current tableau by

$$\mathbf{t}_k = \begin{bmatrix} t_{1k} \\ t_{2k} \\ \vdots \\ t_{mk} \end{bmatrix}$$

It can be shown (Exercise 19) that if

$$t_{i_1 k} \geq 0, \quad t_{i_2 k} \geq 0, \quad \ldots, \quad t_{i_p k} \geq 0$$

none of the artificial variables which are basic will take on a positive value. If, however, we have

$$t_{i_r k} < 0$$

for some r, $r = 1, 2, \ldots, p$, then the usual simplex procedure could cause the artificial variable which labels row i_r to take on a positive value. Consequently, we must modify the usual simplex procedure. The new procedure for selecting a departing variable is this:

If at least one of the entries in the entering variable column corresponding to a row labeled by an artificial variable is negative, choose one of these artificial variables as the departing variable. Otherwise, use the usual simplex procedure.

EXAMPLE 2. In Section 2.5 we showed that the problem given in Example 5 ended with an artificial variable in the basis. We rework this example using the two-phase method. The original problem in canonical form is

$$\text{Maximize} \quad z = x_1 + 2x_2 + x_3$$

subject to

$$3x_1 + x_2 - x_3 = 15$$
$$8x_1 + 4x_2 - x_3 = 50$$
$$2x_1 + 2x_2 + x_3 = 20$$
$$x_1 \geq 0, \quad x_2 \geq 0, \quad x_3 \geq 0$$

The new problem for Phase 1 is

$$\text{Minimize} \quad z = y_1 + y_2 + y_3$$

subject to

$$3x_1 + x_2 - x_3 + y_1 \qquad\qquad = 15$$
$$8x_1 + 4x_2 - x_3 \qquad + y_2 \qquad = 50$$
$$2x_1 + 2x_2 + x_3 \qquad\qquad + y_3 = 20$$
$$x_j \geq 0, \quad j = 1, 2, 3; \qquad y_i \geq 0, \quad i = 1, 2, 3$$

We rewrite the objective function as

$$\text{Maximize} \quad z = -y_1 - y_2 - y_3$$

to have a maximization problem.

We now have the following sequence of tableaux (Tableaux 3.6–3.9) for Phase 1. The objective row of the initial tableau is computed by using $z_j - c_j$. Since the initial basic variables are y_1, y_2, and y_3, we have

$$\mathbf{c}_B = \begin{bmatrix} -1 \\ -1 \\ -1 \end{bmatrix}$$

and

$$z_1 - c_1 = \begin{bmatrix} -1 & -1 & -1 \end{bmatrix} \begin{bmatrix} 3 \\ 8 \\ 2 \end{bmatrix} - 0 = -13$$

Similarly,

$$z_2 - c_2 = [-1 \quad -1 \quad -1]\begin{bmatrix} 1 \\ 4 \\ 2 \end{bmatrix} - 0 = -7$$

and

$$z_3 - c_3 = 1$$

For the basic variables y_1, y_2, y_3 we have

$$z_4 - c_4 = [-1 \quad -1 \quad -1]\begin{bmatrix} 1 \\ 0 \\ 0 \end{bmatrix} - (-1) = 0$$

and similarly, $z_5 - c_5 = 0$ and $z_6 - c_6 = 0$.
The value of the objective function is

$$[-1 \quad -1 \quad -1]\begin{bmatrix} 15 \\ 50 \\ 20 \end{bmatrix} = -85$$

The results of these computations are shown in Tableaux 3.6–3.9.

Tableau 3.6

	c_B		0	0	0	-1	-1	-1	
			x_1	x_2	x_3	y_1	y_2	y_3	x_B
←	-1	y_1	③	1	-1	1	0	0	15
	-1	y_2	8	4	-1	0	1	0	50
	-1	y_3	2	2	1	0	0	1	20
			-13	-7	1	0	0	0	-85

Tableau 3.7

	c_B		0	0	0	-1	-1	-1	
			x_1	x_2	x_3	y_1	y_2	y_3	x_B
	0	x_1	1	$\frac{1}{3}$	$-\frac{1}{3}$	$\frac{1}{3}$	0	0	5
←	-1	y_2	0	$\frac{4}{3}$	⑤/③	$-\frac{8}{3}$	1	0	10
	-1	y_3	0	$\frac{4}{3}$	$\frac{5}{3}$	$-\frac{2}{3}$	0	1	10
			0	$-\frac{8}{3}$	$-\frac{10}{3}$	$\frac{13}{3}$	0	0	-20

Tableau 3.8

↓

c_B		0 x_1	0 x_2	0 x_3	-1 y_1	-1 y_2	-1 y_3	X_B
0	x_1	1	$\frac{3}{5}$	0	$-\frac{1}{5}$	$\frac{1}{5}$	0	7
0	x_3	0	$\frac{4}{5}$	1	$-\frac{8}{5}$	$\frac{3}{5}$	0	6
-1	y_3	0	0	0	②	-1	1	0
		0	0	0	-1	2	0	0

← (arrow pointing to y_3 row)

Tableau 3.9

c_B		0 x_1	0 x_2	0 x_3	-1 y_1	-1 y_2	-1 y_3	X_B
0	x_1	1	$\frac{3}{5}$	0	0	$\frac{1}{10}$	$\frac{1}{10}$	7
0	x_3	0	$\frac{4}{5}$	1	0	$-\frac{1}{5}$	$\frac{4}{5}$	6
-1	y_1	0	0	0	1	$-\frac{1}{2}$	$\frac{1}{2}$	0
		0	0	0	0	$\frac{3}{2}$	$\frac{1}{2}$	0

Thus, Phase 1 has an optimal solution with $x_1 = 7$, $x_3 = 6$, $y_1 = 0$, and value of 0 for the objective function. The artificial variable y_1 appears as a basic variable in the optimal solution.

The initial tableau for Phase 2 is shown in Tableau 3.10. The columns corresponding to y_2 and y_3 have been deleted and a cost of 0 has been assigned to y_1. The objective row has been filled in, using $z_j - c_j$.

Tableau 3.10

↓

c_B		1 x_1	2 x_2	1 x_3	0 y_1	X_B
1	x_1	1	$\frac{3}{5}$	0	0	7
1	x_3	0	④/5	1	0	6
0	y_1	0	0	0	1	0
		0	$-\frac{3}{5}$	0	0	13

← (arrow pointing to x_3 row)

$$z_j - c_j = C_B \cdot t_j - C_j \quad / \quad z = C_b X_B$$

We now apply the simplex method to Tableau 3.10. In selecting the departing variable we make sure the entry in the row labeled by y_1 and the pivotal column is nonnegative. If it is not, we will choose y_1 as the departing variable. We get Tableau 3.11.

Tableau 3.11

		1	2	1	0	
c_B		x_1	x_2	x_3	y_1	x_B
1	x_1	1	0	$-\frac{3}{4}$	0	$\frac{5}{2}$
2	x_2	0	1	$\frac{5}{4}$	0	$\frac{15}{2}$
0	y_1	0	0	0	1	0
		0	0	$\frac{3}{4}$	0	$\frac{35}{2}$

An optimal solution to the original problem is therefore

$$\beta^{-1} = \begin{bmatrix} 1 & -3/4 & 0 \\ 0 & 5/4 & 0 \\ 0 & 0 & 1 \end{bmatrix}$$

$$x_1 = \frac{5}{2}$$

$$x_2 = \frac{15}{2}$$

$$x_3 = 0$$

which gives $\frac{35}{2}$ as the optimal value of the objective function.

Solution to Dual Problem from Final Tableau of Primal

One of our objectives in this section is to describe how to find an optimal solution to the dual problem from the final tableau for the primal problem. We discuss the easiest case first, namely, when the primal problem is in canonical form

$$\text{Maximize} \quad z = c^T x$$

subject to

$$Ax = b$$

$$x \geq 0$$

where A contains the identity matrix and $b \geq 0$. In particular this situation arises when the primal problem is given in standard form and has been converted to canonical form by adding slack variables. There are two easily given descriptions for finding an optimal solution to the dual of the problem given above. The dual problem is

$$\text{Minimize} \quad z' = \mathbf{b}^T\mathbf{w}$$

subject to

$$\mathbf{A}^T\mathbf{w} \geq \mathbf{c}$$

$$\mathbf{w} \text{ unrestricted}$$

An optimal solution

$$\mathbf{w} = \begin{bmatrix} w_1 \\ w_2 \\ \vdots \\ w_m \end{bmatrix}$$

to it is given by

$$\mathbf{w}^T = \mathbf{c}_B^T \mathbf{B}^{-1} \tag{16}$$

where $\mathbf{B}$ is the matrix consisting of certain columns of the initial tableau. The columns which are used in $\mathbf{B}$ correspond to the basic variables of the final tableau of the primal problem. We can find $\mathbf{B}^{-1}$ from the *final* tableau as follows. From our assumptions about $\mathbf{A}$ and $\mathbf{b}$ we may infer that the columns labeled by the initial basic variables in the initial tableau form the $m \times m$ identity matrix when properly ordered. The columns in the final tableau with the same labels as the initial basic variables and arranged in the same order give $\mathbf{B}^{-1}$.

An optimal solution to the dual problem is also given by

$$w_j = c_{i_j} + (z_{i_j} - c_{i_j}) \tag{17}$$

where the subscript i_j ranges over the indices of the *initial basic variables*. Of course, c_{i_j} is the entry above the label of the i_jth column and $z_{i_j} - c_{i_j}$ is the corresponding entry in the objective row. This second description shows that if an initial basic variable is a slack or artificial variable, the value of the corresponding dual variable is the entry in the i_jth column of the objective row of the final tableau of the primal problem. This fact follows, since $c_{i_j} = 0$ for any slack or artificial variable.

EXAMPLE 3. Consider as our primal problem the linear programming problem

$$\text{Maximize} \quad z = 8x_1 + 9x_2 + 4x_3$$

subject to

$$x_1 + x_2 + 2x_3 \leq 2$$
$$2x_1 + 3x_2 + 4x_3 \leq 3$$
$$7x_1 + 6x_2 + 2x_3 \leq 8$$
$$x_1 \geq 0, \quad x_2 \geq 0, \quad x_3 \geq 0$$

Introducing the slack variables x_4, x_5, x_6, our primal problem becomes

$$\text{Maximize} \quad z = 8x_1 + 9x_2 + 4x_3$$

subject to

$$
\begin{aligned}
x_1 + x_2 + 2x_3 + x_4 &= 2 \\
2x_1 + 3x_2 + 4x_3 + x_5 &= 3 \\
7x_1 + 6x_2 + 2x_3 + x_6 &= 8 \\
x_1 \geq 0, \quad x_2 \geq 0, \quad x_3 \geq 0
\end{aligned}
$$

Solving this problem by the simplex method we are led to the sequence of Tableaux 3.12–3.14.

The dual problem in this example is

$$\text{Minimize} \quad z' = 2w_1 + 3w_2 + 8w_3$$

subject to

$$
\begin{aligned}
w_1 + 2w_2 + 7w_3 &\geq 8 \\
w_1 + 3w_2 + 6w_3 &\geq 9 \\
2w_1 + 4w_2 + 2w_3 &\geq 4 \\
w_1 \geq 0, \quad w_2 \geq 0, \quad w_3 \geq 0
\end{aligned}
$$

The solution to this problem is found from Tableau 3.14. In Tableau 3.14 the basic variables are x_4, x_2, x_1, in that order. Therefore, reading from Tableau 3.12, we find

$$
\mathbf{A}_4 = \begin{bmatrix} 1 \\ 0 \\ 0 \end{bmatrix}, \quad
\mathbf{A}_2 = \begin{bmatrix} 1 \\ 3 \\ 6 \end{bmatrix}, \quad
\mathbf{A}_1 = \begin{bmatrix} 1 \\ 2 \\ 7 \end{bmatrix}
$$

and

$$
\mathbf{B} = \begin{bmatrix} 1 & 1 & 1 \\ 0 & 3 & 2 \\ 0 & 6 & 7 \end{bmatrix}
$$

Since the initial basic variables are x_4, x_5, and x_6, in that order, we find the columns of $\mathbf{B}^{-1}$ under the labels x_4, x_5, and x_6 in Tableau 3.14. Thus,

$$
\mathbf{B}^{-1} = \begin{bmatrix} 1 & -\frac{1}{9} & -\frac{1}{9} \\ 0 & \frac{7}{9} & -\frac{2}{9} \\ 0 & -\frac{2}{3} & \frac{1}{3} \end{bmatrix}
$$

Tableau 3.12

↓ *Phase 2 for primary*

c_B		8 x_1	9 x_2	4 x_3	0 x_4	0 x_5	0 x_6	x_B
0	x_4	1	1	2	1	0	0	2
0	x_5	2	③	4	0	1	0	3
0	x_6	7	6	2	0	0	1	8
		-8	-9	-4	0	0	0	0

$z_j = c_B \cdot t_j$

Tableau 3.13

↓

c_B		8 x_1	9 x_2	4 x_3	0 x_4	0 x_5	0 x_6	x_B
0	x_4	$\frac{1}{3}$	0	$\frac{2}{3}$	1	$-\frac{1}{3}$	0	1
9	x_2	$\frac{2}{3}$	1	$\frac{4}{3}$	0	$\frac{1}{3}$	0	1
0	x_6	③	0	-6	0	-2	1	2
		-2	0	8	0	3	0	9

Tableau 3.14

c_B		8 x_1	9 x_2	4 x_3	0 x_4	0 x_5	0 x_6	x_B
0	x_4	0	0	$\frac{4}{3}$	1	$-\frac{1}{9}$	$-\frac{1}{9}$	$\frac{7}{9}$
9	x_2	0	1	$\frac{8}{3}$	0	$\frac{7}{9}$	$-\frac{2}{9}$	$\frac{5}{9}$
8	x_1	1	0	-2	0	$-\frac{2}{3}$	$\frac{1}{3}$	$\frac{2}{3}$
		0	0	4	0	$\frac{5}{3}$	$\frac{2}{3}$	$\frac{31}{3}$

Then an optimal solution to the dual problem is, by (16),

$w^T = C B^{-1}$

$$\mathbf{w}^T = [0 \quad 9 \quad 8] \begin{bmatrix} 1 & -\frac{1}{9} & -\frac{1}{9} \\ 0 & \frac{7}{9} & -\frac{2}{9} \\ 0 & -\frac{2}{3} & \frac{1}{3} \end{bmatrix} = [0 \quad \frac{5}{3} \quad \frac{2}{3}]$$

If (17) is used, an optimal value of w_1 is

$$c_4 - (z_4 - c_4) = 0 + 0 = 0$$

since x_4 was the first initial basic variable. Likewise, an optimal value of

w_2 comes from the second initial basic variable x_5 and is given as

$$w_2 = c_5 + (z_5 - c_5) = 0 + \tfrac{5}{3} = \tfrac{5}{3}$$

Finally,

$$w_3 = c_6 + (z_6 - c_6) = 0 + \tfrac{2}{3} = \tfrac{2}{3}$$

Thus, this solution to the dual problem yields the value $\tfrac{31}{3}$ for the dual objective function which checks with the value of the primal objective function. The duality theorem (Theorem 3.6) assures us that these solutions are indeed optimal, since they yield the same value for the objective functions.

Now let us consider finding a solution to the dual of an arbitrary general linear programming problem. As we discussed in earlier sections, such a problem can always be converted to one in the following form:

$$
\left.
\begin{array}{c}
\text{Maximize} \quad z = \mathbf{c}^T\mathbf{x} \\[4pt]
\text{subject to} \\[4pt]
\mathbf{A}^{(1)}\mathbf{x} \le \mathbf{b}^{(1)} \\[4pt]
\mathbf{A}^{(2)}\mathbf{x} \ge \mathbf{b}^{(2)} \\[4pt]
\mathbf{A}^{(3)}\mathbf{x} = \mathbf{b}^{(3)} \\[4pt]
\mathbf{x} \ge \mathbf{0} \\[4pt]
\mathbf{b}^{(1)} \ge \mathbf{0}, \quad \mathbf{b}^{(2)} > \mathbf{0}, \quad \mathbf{b}^{(3)} \ge \mathbf{0}
\end{array}
\right\} \qquad (18)
$$

This conversion does not change the set of feasible solutions but may change the objective function of the original problem by multiplying it by -1. A problem in the form above is ready to be solved by the simplex method after slack variables and artificial variables are introduced.

After multiplying the second set of constraints of (18) by -1, we may set up the dual problem. It is

$$
\left.
\begin{array}{c}
\text{Minimize} \quad z' = \begin{bmatrix} \mathbf{b}^{(1)T} & -\mathbf{b}^{(2)T} & \mathbf{b}^{(3)T} \end{bmatrix} \begin{bmatrix} \mathbf{w}^{(1)} \\ \mathbf{w}^{(2)} \\ \mathbf{w}^{(3)} \end{bmatrix} \\[18pt]
\text{subject to} \\[18pt]
\begin{bmatrix} \mathbf{A}^{(1)T} & -\mathbf{A}^{(2)T} & \mathbf{A}^{(3)T} \end{bmatrix} \begin{bmatrix} \mathbf{w}^{(1)} \\ \mathbf{w}^{(2)} \\ \mathbf{w}^{(3)} \end{bmatrix} \ge \mathbf{c} \\[18pt]
\mathbf{w}^{(1)} \ge \mathbf{0}, \quad \mathbf{w}^{(2)} \ge \mathbf{0}, \quad \mathbf{w}^{(3)} \text{ unrestricted}
\end{array}
\right\} \qquad (19)
$$

We seek the solution to (19) using the final tableau from the solution to (18). If the two-phase method is used, the columns labeled by artificial variables must not be discarded at the end of Phase 1. They must be kept throughout Phase 2 and must be modified accordingly by each pivoting step. However, the entries in the objective row labeled by artificial variables must *not* be used when checking the optimality criterion. This means that there may be negative numbers in the objective row when an optimal solution is reached, but these numbers will only be in columns labeled by artificial variables.

Assuming that all the columns labeled by artificial variables are available, an optimal solution to the dual problem (19) can be obtained from (16) or (17). If (16) is used, then $\mathbf{B}^{-1}$ is automatically available, as described above, in the final tableau of Phase 2 from the columns labeled by the initial basic variables. We then compute

$$\hat{\mathbf{w}}^T = \mathbf{c}_B^T \mathbf{B}^{-1}$$

but $\hat{\mathbf{w}}$ is *not* an optimal solution to the dual problem (19). Because the second set of constraints in (18) was multiplied by -1 to find the dual but was not changed to use the simplex algorithm, those entries in $\hat{\mathbf{w}}$ corresponding to the second set of constraints in (18) must be multiplied by -1. The vector thus formed is $\mathbf{w}$, an optimal solution to the dual problem (19).

If (17) is used, we must distinguish between the two-phase and the big M methods.

If the two-phase method is used, then the cost of an artificial variable is 0, so that

$$\hat{w}_i = 0 + (z_i - 0) = z_i$$

With the big M method the cost of an artificial variable c_i is $-M$, and $z_i - c_i$ will be of the form $k + M$, so that

$$\hat{w}_i = -M + (k + M) = k$$

In either case we must, as above, multiply each of the $\hat{w}_i$ by -1, where i runs through the indices of the second set of constraints in (18). The set of values $w_1, w_2, \ldots, w_n$ thus obtained is an optimal solution to the dual problem.

EXAMPLE 4. Consider as our primal problem the linear programming problem

$$\text{Maximize} \quad z = 3x_1 + 2x_2 + 5x_3$$

subject to

$$x_1 + 3x_2 + 2x_3 \le 15$$
$$2x_2 - x_3 \ge 5$$
$$2x_1 + x_2 - 5x_3 = 10$$
$$x_1 \ge 0, \quad x_2 \ge 0, \quad x_3 \ge 0$$

The dual problem is given as

$$\text{Minimize} \quad z' = 15w_1 - 5w_2 + 10w_3$$

subject to

$$w_1 \qquad\quad + 2w_3 \geq 3$$
$$3w_1 - 2w_2 + \; w_3 \geq 2$$
$$2w_1 + \; w_2 - 5w_3 \geq 5$$
$$w_1 \geq 0, \quad w_2 \geq 0, \quad v_3 \;\; \text{unrestricted}$$

Introducing the slack variables x_4 and x_5 and the artificial variables y_1 and y_2, we can formulate the auxiliary problem to the primal problem as

$$\text{Minimize} \quad z = y_1 + y_2$$

subject to

$$x_1 + 3x_2 + 2x_3 + x_4 \qquad\qquad\qquad = 15$$
$$2x_2 - \; x_3 \qquad - x_5 + y_1 \qquad = 5$$
$$2x_1 + \; x_2 - 5x_3 \qquad\qquad + y_2 = 10$$
$$x_i \geq 0, \quad i = 1, 2, 3, 4; \qquad y_1 \geq 0, y_2 \geq 0$$

From this statement of the primal problem we can construct our initial tableau (Tableau 3.15).

The initial basic variables are x_4, y_1, and y_2. At the end of Phase 1 we have Tableau 3.16.

Tableau 3.15 *ausc. prob.*

c_B		0 x_1	0 x_2	0 x_3	0 x_4	0 x_5	-1 y_1	-1 y_2	$\mathbf{x_B}$
0	x_4	1	3	2	1	0	0	0	15
-1	y_1	0	2	-1	0	-1	1	0	5
-1	y_2	2	1	-5	0	0	0	1	10
		-2	-3	6	0	1	0	0	-15

Tableau 3.16

c_B		0 x_1	0 x_2	0 x_3	0 x_4	0 x_5	-1 y_1	-1 y_2	$\mathbf{x_B}$
0	x_4	0	0	$\frac{23}{4}$	1	$\frac{5}{4}$	$-\frac{5}{4}$	$-\frac{1}{4}$	$\frac{15}{4}$
0	x_2	0	1	$-\frac{1}{2}$	0	$-\frac{1}{2}$	$\frac{1}{2}$	0	$\frac{5}{2}$
0	x_1	1	0	$-\frac{9}{4}$	0	$\frac{1}{4}$	$-\frac{1}{4}$	$\frac{1}{2}$	$\frac{15}{4}$
		0	0	0	0	0	1	1	0

value y = 0 means art. var. = 0

Tableau 3.17

	c_B	3 x_1	2 x_2	-5 x_3	0 x_4	0 x_5	0 y_1	0 y_2	x_B
← 0	x_4	0	0	$\frac{23}{4}$	1	$\frac{5}{4}$	$-\frac{5}{4}$	$-\frac{1}{2}$	$\frac{15}{4}$
2	x_2	0	1	$-\frac{1}{2}$	0	$-\frac{1}{2}$	$\frac{1}{2}$	0	$\frac{5}{2}$
3	x_1	1	0	$-\frac{9}{4}$	0	$\frac{1}{4}$	$-\frac{1}{4}$	$\frac{1}{2}$	$\frac{15}{4}$
		0	0	$-\frac{51}{4}$	0	$-\frac{1}{4}$	$\frac{1}{4}$	$\frac{3}{2}$	$\frac{65}{4}$

Converting to Phase 2 but keeping the columns labeled with artificial variables, we construct Tableau 3.17.

Now we perform more iterations of the simplex method, including the results of the pivoting step on the columns labeled by artificial variables but ignoring these columns when applying the optimality criterion. We obtain the final tableau (Tableau 3.18). Using (16) to obtain an optimal solution to the dual problem, we first compute

$$\hat{\mathbf{w}}^T = \mathbf{c}_B^T \mathbf{B}^{-1}$$

or

$$\hat{\mathbf{w}}^T = \begin{bmatrix} 5 & 2 & 3 \end{bmatrix} \begin{bmatrix} \frac{4}{23} & -\frac{5}{23} & -\frac{2}{23} \\ \frac{2}{23} & \frac{9}{23} & -\frac{1}{23} \\ \frac{9}{23} & -\frac{17}{23} & \frac{7}{23} \end{bmatrix}$$

$$= \begin{bmatrix} \frac{51}{23} & -\frac{58}{23} & \frac{9}{23} \end{bmatrix}$$

As before,

$$w_1 = \hat{w}_1, \qquad w_2 = -\hat{w}_2, \quad \text{and} \quad w_3 = \hat{w}_3$$

Tableau 3.18

	c_B	3 x_1	2 x_2	5 x_3	0 x_4	0 x_5	0 y_1	0 y_2	x_B
5	x_3	0	0	1	$\frac{4}{23}$	$\frac{5}{23}$	$-\frac{5}{23}$	$-\frac{2}{23}$	$\frac{15}{23}$
2	x_2	0	1	0	$\frac{2}{23}$	$-\frac{9}{23}$	$\frac{9}{23}$	$-\frac{1}{23}$	$\frac{65}{23}$
3	x_1	1	0	0	$\frac{9}{23}$	$\frac{17}{23}$	$-\frac{17}{23}$	$\frac{7}{23}$	$\frac{120}{23}$
		0	0	0	$\frac{51}{34}$	$\frac{58}{23}$	$-\frac{58}{23}$	$\frac{9}{23}$	$\frac{565}{23}$

Thus, an optimal solution to the dual problem is

$$\mathbf{w}^T = [\tfrac{51}{23} \quad \tfrac{58}{23} \quad \tfrac{9}{23}]$$

and the value of the dual objective function is

$$z' = 15(\tfrac{51}{23}) - 5(\tfrac{58}{23}) + 10(\tfrac{9}{23}) = \tfrac{565}{23}$$

We see from Tableau 3.18 that an optimal solution to the primal problem is

$$x_1 = \tfrac{120}{23}, \quad x_2 = \tfrac{65}{23}, \quad x_3 = \tfrac{15}{23}$$

and the value of the objective function is $\tfrac{565}{23}$, which is the same as the value for the dual problem.

We can also solve the primal problem using the big M method, obtaining the final Tableau 3.19. Using (17), we can obtain an optimal solution to the dual problem from Tableau 3.19 as follows. The first initial basic variable is x_4; we then find

$$w_1 = \hat{w}_1 = c_4 + (z_4 - c_4) = 0 + \tfrac{51}{23} = \tfrac{51}{23}$$

The second initial basic variable is y_1, so that

$$\hat{w}_2 = -M + (M - \tfrac{58}{23}) = -\tfrac{58}{23}$$

But since the second constraint was multiplied by -1 in forming the dual, we must multiply $\hat{w}_2$ by -1 to obtain the value of the dual variable at the optimum. We have

$$w_2 = -\hat{w}_2 = \tfrac{58}{23}$$

Proceeding as in the case of the first dual variable, we find

$$w_3 = \hat{w}_3 = -M + M + \tfrac{9}{23} = \tfrac{9}{23}$$

Substituting these values into the objective function, we get

$$z' = 15(\tfrac{51}{23}) - 5(\tfrac{58}{23}) + 90(\tfrac{9}{23}) = \tfrac{565}{23}$$

which is the same as the value for the primal problem.

Tableau 3.19

c_B		3	2	5	0	0	$-M$	$-M$	
		x_1	x_2	x_3	x_4	x_5	y_1	y_2	$\mathbf{x}_B$
5	x_3	0	0	1	$\tfrac{4}{23}$	$\tfrac{5}{23}$	$-\tfrac{5}{23}$	$-\tfrac{2}{23}$	$\tfrac{15}{23}$
2	x_2	0	1	0	$\tfrac{2}{23}$	$-\tfrac{9}{23}$	$\tfrac{9}{23}$	$-\tfrac{1}{23}$	$\tfrac{65}{23}$
3	x_1	1	0	0	$\tfrac{9}{23}$	$\tfrac{17}{23}$	$-\tfrac{17}{23}$	$\tfrac{7}{23}$	$\tfrac{120}{23}$
		0	0	0	$\tfrac{51}{23}$	$\tfrac{58}{23}$	$M - \tfrac{58}{23}$	$M + \tfrac{9}{23}$	$\tfrac{565}{23}$

3.2 Exercises

In Exercises 1 and 2 fill in *all* the missing entries in the given simplex tableaux.

1.

c_B		-1 x_1	2 x_2	-6 x_3	0 x_4	0 x_5	5 x_6	x_B
2	x_2	-3	1	-1	-2	0	0	2
5	x_6	2	0	0	-1	0	1	3
0	x_5	6	0	7	6	1	0	1
		5	0	4	-9	0	0	19

2.

c_B		4 x_1	$\frac{5}{3}$ x_2	$\frac{4}{3}$ x_3	3 x_4	-1 x_5	0 x_6	$-\frac{2}{3}$ x_7	x_B
-1	x_5	0	$\frac{4}{3}$	$\frac{2}{3}$	0	1	0	$-\frac{1}{3}$	4
0	x_6	0	$\frac{1}{3}$	$\frac{2}{3}$	1	0	1	$-\frac{1}{3}$	10
4	x_1	1	$\frac{1}{3}$	$\frac{1}{6}$	$\frac{1}{2}$	0	0	$\frac{1}{6}$	4
		0				0	0		

In Exercises 3 and 4 we give the original objective function of a linear programming problem and the final tableau at the end of Phase 1. Find the initial tableau for Phase 2 and solve the resulting linear programming problem.

3. Maximize $z = 2x_1 + 3x_2 + 5x_3 + x_4.$

c_B		x_1	x_2	x_3	x_4	x_5	x_6	x_7	y_1	y_2	x_B
0	x_1	1	$\frac{1}{2}$	1	0	3	0	-1	0	0	$\frac{1}{2}$
0	x_4	0	1	$-\frac{1}{4}$	1	-2	0	-1	0	0	4
0	x_6	0	2	$-\frac{2}{3}$	0	$-\frac{1}{2}$	1	0	0	0	$\frac{3}{2}$
-1	y_1	0	$-\frac{3}{2}$	-2	0	-3	0	0	1	0	0
-1	y_2	0	-4	1	0	2	0	-2	0	1	0
		0	$\frac{11}{2}$	1	0	1	0	2	0	0	0

4. Maximize $z = 3x_1 + x_2 + 3x_3.$

c_B		x_1	x_2	x_3	x_4	x_5	x_6	y_1	x_B
0	x_1	1	1	0	2	0	$\frac{2}{3}$	0	2
0	x_3	0	-1	1	-1	0	-1	0	$\frac{5}{2}$
0	x_5	0	2	0	3	1	1	0	$\frac{2}{3}$
-1	y_1	0	$-\frac{3}{2}$	0	$-\frac{1}{2}$	0	-2	1	0
		0	$\frac{3}{2}$	0	$\frac{1}{2}$	0	2	0	0

5. Verify the entries in Tableaux 3.3 and 3.4.

In Exercises 6–12 solve the given linear programming problem calculating $z_j - c_j$ as described in this section and using the new format for the tableaux.

6. Maximize $z = 2x_1 + x_2 + 3x_3$

subject to

$$2x_1 - x_2 + 3x_3 \le 6$$
$$x_1 + 3x_2 + 5x_3 \le 10$$
$$2x_1 \qquad + x_3 \le 7$$
$$x_1 \ge 0, \quad x_2 \ge 0, \quad x_3 \ge 0$$

7. Maximize $z = x_1 + x_2 + x_3 + x_4$

subject to

$$x_1 + 2x_2 - x_3 + 3x_4 \le 12$$
$$x_1 + 3x_2 + x_3 + 2x_4 \le 8$$
$$2x_1 - 3x_2 - x_3 + 2x_4 \le 7$$
$$x_1 \ge 0, \quad x_2 \ge 0, \quad x_3 \ge 0, \quad x_4 \ge 0$$

8. Minimize $z = 8x_1 + 6x_2 + 11x_3$

subject to

$$5x_1 + x_2 + 3x_3 \le 4$$
$$5x_1 + x_2 + 3x_3 \ge 2$$
$$2x_1 + 4x_2 + 7x_3 \le 5$$
$$2x_1 + 4x_2 + 7x_3 \ge 3$$
$$x_1 + x_2 + x_3 = 1$$
$$x_1 \ge 0, \quad x_2 \ge 0, \quad x_3 \ge 0$$

(Exercise 14 will require the use of the columns corresponding to artificial variables.)

9. Minimize $z = 4x_1 + x_2 + x_3 + 3x_4$

subject to

$$2x_1 + x_2 + 3x_3 + x_4 \ge 12$$
$$3x_1 + 2x_2 + 4x_3 \qquad = 5$$
$$2x_1 - x_2 + 2x_3 + 3x_4 = 8$$
$$3x_1 + 4x_2 + 3x_3 + x_4 \ge 16$$
$$x_1 \ge 0, \quad x_2 \ge 0, \quad x_3 \ge 0, \quad x_4 \ge 0$$

10. Maximize $z = 2x_1 + x_2 + x_3 + x_4$

   subject to

$$x_1 + 2x_2 + x_3 + 2x_4 \leq 7$$
$$x_1 + 2x_2 + x_3 + 2x_4 \geq 3$$
$$2x_1 + 3x_2 - x_4 - 4x_4 \leq 10$$
$$x_1 + x_2 + x_3 + x_4 = 1$$
$$x_1 \geq 0, \quad x_2 \geq 0, \quad x_3 \geq 0, \quad x_4 \geq 0$$

11. Maximize $z = 3x_1 + x_2 + 3x_3 + x_4$

   subject to

$$x_1 + x_2 + 4x_3 + x_4 \leq 6$$
$$2x_1 + 6x_3 + 2x_4 \geq 8$$
$$20x_1 + 2x_2 + 47x_3 + 11x_4 \leq 48$$
$$x_1 \geq 0, \quad x_2 \geq 0, \quad x_3 \geq 0, \quad x_4 \geq 0$$

12. Maximize $z = 2x_1 + x_2 + 3x_4$

   subject to

$$9x_1 + 14x_2 - 6x_3 - 6x_4 \leq 2$$
$$x_1 + x_2 - x_3 - x_4 \geq -1$$
$$-20x_1 - 5x_2 + 5x_3 + 13x_4 = 11$$
$$5x_1 + 10x_2 - 2x_3 + 14x_4 = 6$$
$$x_1 \geq 0, \quad x_2 \geq 0, \quad x_3 \geq 0, \quad x_4 \geq 0$$

13. For each of the linear programming problems in Exercises 6 and 9 find the matrices **B** and $\mathbf{B}^{-1}$ from each of the tableaux which arose in the solution of the problem.

14. For each of the primal linear programming problems in Exercises 6 and 8 find an optimal solution to the dual problem using the final tableau determined in solving the primal problem.

15. Solve Example 5 in Section 2.1 (page 64).

16. Solve Project 2 in Section 2.1 (page 77).

17. Solve Project 3 in Section 2.1 (page 77).

18. Consider the primal linear programming problem

$$\text{Maximize} \quad z = \mathbf{c}^T\mathbf{x}$$

   subject to

$$\mathbf{Ax} \leq \mathbf{b}$$
$$\mathbf{x} \geq \mathbf{0}$$

Suppose that $\mathbf{x}_0$ is an optimal solution to this problem. Show that the dual objective function takes the value $\mathbf{c}_B^T \mathbf{x}_0$ for $\mathbf{w} = (\mathbf{B}^{-1})^T \mathbf{c}_B$. (Thus, by the duality theorem, if $\mathbf{w}$ is a feasible solution to the dual problem, it is an optimal solution.)

19. Verify the remarks preceding Example 2 regarding Phase 2.

✳3.3 THE DUAL SIMPLEX METHOD

The jth dual constraint of the dual problem is

$$z_j = a_{1j}w_1 + a_{2j}w_2 + \cdots + a_{mj}w_m \geq c_j \tag{1}$$

We have already seen that the entry in the jth column of the objective row of a tableau for the primal problem is $z_j - c_j$, the difference between the left and right sides of equation (1). If the entry $z_j - c_j$ is negative, then the feasible solution to the primal problem represented by the tableau is not optimal. Moreover, (1) does not hold. Thus, whenever the primal problem is not optimal, the dual problem is not feasible, and conversely.

When we use the simplex algorithm on a primal problem we begin with a feasible but nonoptimal solution. Each iteration of the simplex algorithm finds a feasible solution which is closer to optimality, and this procedure continues until an optimal solution is reached. In the meantime, what is happening to the dual problem? If the primal problem has a solution which is feasible and nonoptimal, the solution determined for the dual problem is infeasible. That is, at least one of the nonnegativity constraints fails to be satisfied. As the simplex method progresses, the solutions determined for the dual problem are all infeasible until the optimal solution is attained for the primal problem. The dual solution corresponding to the optimal primal solution is both optimal and feasible. The goal for the primal problem when using the simplex method is to achieve optimality. The goal for a corresponding method for the dual problem is to achieve feasibility, that is, to have all the nonnegativity constraints satisfied.

The dual simplex method handles problems for which it is easy to obtain an initial basic solution that is infeasible but satisfies the optimality criterion. That is, the initial tableau has nonnegative entries in the objective row but some negative entries in the right-hand column. The following example will be used as we present our description of the dual simplex algorithm.

EXAMPLE 1. Consider the following linear programming problem:

$$\text{Maximize} \quad z = -x_1 - 2x_2$$

subject to

$$x_1 - 2x_2 + x_3 \geq 4$$

$$2x_1 + x_2 - x_3 \geq 6$$

$$x_1 \geq 0, \quad x_2 \geq 0, \quad x_3 \geq 0$$

We change each constraint to an $\leq$ inequality and then introduce slack variables x_4 and x_5. The result is a problem in canonical form:

$$\text{Maximize} \quad z = -x_1 - 2x_2$$

subject to

$$-x_1 + 2x_2 - x_3 + x_4 \qquad = -4$$

$$-2x_1 - x_2 + x_3 \qquad + x_5 = -6$$

$$x_j \geq 0, \qquad j = 1, 2, \ldots, 5$$

The initial tableau for the simplex algorithm is given in Tableau 3.20. It has x_4 and x_5 as the initial basic variables. The solution which this tableau represents is

$$x_1 = 0, \quad x_2 = 0, \quad x_3 = 0, \quad x_4 = -4, \quad x_5 = -6$$

Tableau 3.20

		-1	-2	0	0	0	
c_B		x_1	x_2	x_3	x_4	x_5	x_B
0	x_4	-1	2	-1	1	0	-4
0	x_5	-2	-1	1	0	1	-6
		1	2	0	0	0	0

This solution is not feasible, and here $z = 0$. The entries in the objective row show that the optimality criterion is satisfied.

The dual simplex method consists of two parts: a feasibility criterion which tells us whether the current solution (which satisfies the optimality criterion) is feasible, and a procedure for getting a new solution which removes some of the infeasibilities of the current solution and consequently drives the current solution toward a feasible solution. The dual simplex method consists of the following steps:

1. Find an initial basic solution such that all entries in the objective row are nonnegative and at least one basic variable has a negative value. (Tableau 3.20 represents this step for our example.)
2. Select a departing variable by examining the basic variables and choosing the most negative one. This is the departing variable and the row it labels is the pivotal row.

3. Select an entering variable. This selection depends on the ratios of the objective row entries to the corresponding pivotal row entries. The ratios are only formed for those entries of the pivotal row which are negative. If all entries in the pivotal row are nonnnegative, the problem has no feasible solution. Among all the ratios (which must all be nonpositive), select the maximum ratio. The column for which this ratio occurred is the pivotal column and the corresponding variable is the entering variable. In case of ties among the ratios, choose one column arbitrarily.

4. Perform pivoting to obtain a new tableau. The objective row can be computed as $z_j - c_j = \mathbf{c}_B^T \mathbf{t}_j - c_j$, where $\mathbf{t}_j$ is the jth column of the new tableau.

5. The process stops when a basic solution which is feasible (all variables ≥ 0) is obtained.

A flowchart for the dual simplex method is given in Figure 3.5 and a structure diagram in Figure 3.6.

Continuing with our example, we do Step 2 of the dual simplex algorithm. We see that $x_5 = -6$ is the most negative basic variable, so that x_5 is the departing variable. The ratios of the entries in the objective row

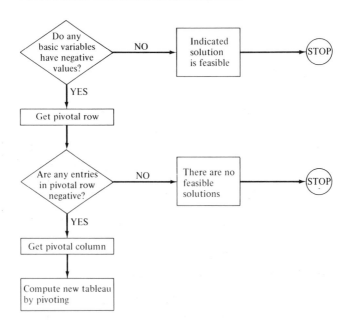

Figure 3.5 Flowchart for dual simplex algorithm.

Input is a tableau which satisfies the optimality criterion.

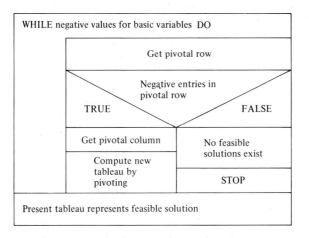

Figure 3.6 Structure diagram for dual simplex algorithm.

to corresponding negative entries in the pivotal row are

$$\text{Column 1:} \quad -1/2$$

$$\text{Column 2:} \quad -2$$

The maximum ratio is $-1/2$, so that x_1 is the entering variable. We repeat Tableau 3.20, with the entering and departing variables and the pivot labeled (Tableau 3.20a).

Tableau 3.20a

c_B		-1 x_1	-2 x_2	0 x_3	0 x_4	0 x_5	$\mathbf{x_B}$
0	x_4	-1	2	-1	1	0	-4
0	x_5	-2	-1	1	0	1	-6
		1	2	0	0	0	0

We now perform a pivotal elimination to get Tableau 3.21. The basic solution given by Tableau 3.21 is

$$x_1 = 3, \quad x_2 = 0, \quad x_3 = 0, \quad x_4 = -1, \quad x_5 = 0$$

This solution is still optimal (the objective row has nonnegative entries)

Tableau 3.21

$\downarrow$

c_B		-1 x_1	-2 x_2	0 x_3	0 x_4	0 x_5	x_B
0	x_4	0	$\frac{5}{3}$	$\boxed{-\frac{3}{2}}$	1	$-\frac{1}{2}$	-1
-1	x_1	1	$\frac{1}{3}$	$-\frac{1}{2}$	0	$-\frac{1}{2}$	3
		0	$\frac{3}{2}$	$\frac{1}{2}$	0	$\frac{1}{2}$	-3

(handwritten: $\theta-n$ 0 $5/3$ ... -1)

but is infeasible. However, it is less infeasible in that only one variable is negative.

For the next iteration of the dual simplex algorithm, x_4 is the departing variable, since it is the only negative basic variable. Forming the ratios of the entries in the objective row to the corresponding negative entries of the pivotal row, we have

$$\text{Column 3:} \qquad \frac{\frac{1}{2}}{-\frac{3}{2}} = -1/3$$

$$\text{Column 5:} \qquad \frac{\frac{1}{2}}{-\frac{1}{2}} = -1$$

The maximum ratio is $-1/3$, so that x_3 is the entering variable. Pivoting, we now obtain Tableau 3.22.

Tableau 3.22

c_B		-1 x_1	-2 x_2	0 x_3	0 x_4	0 x_5	x_B
0	x_3	0	$-\frac{5}{3}$	1	$-\frac{2}{3}$	$\frac{1}{3}$	$\frac{2}{3}$
-1	x_1	1	$-\frac{1}{3}$	0	$-\frac{1}{3}$	$-\frac{1}{3}$	$\frac{10}{3}$
		0	$\frac{7}{3}$	0	$\frac{1}{3}$	$\frac{1}{3}$	$-\frac{10}{3}$

The basic solution represented by Tableau 3.22 is

$$x_1 = \tfrac{10}{3}, \quad x_2 = 0, \quad x_3 = \tfrac{2}{3}, \quad x_4 = 0, \quad x_5 = 0$$

This solution satisfies the optimality criterion, *and* it is feasible since all the variables have nonnegative values.

Much of our discussion of the simplex algorithm centered on finding an initial basic feasible solution to a linear programming problem in arbitrary form. We developed the concept of artificial variables to provide a method for constructing a starting point for such a problem.

In the above example we were fortunate in finding an initial basic solution to the given problem which satisfied the optimality criterion but was not feasible. In general, it is difficult to find such a starting point for the dual simplex algorithm. However, a common use of the dual simplex algorithm is to restore feasibility when an additional constraint is added to a linear programming problem for which an optimal solution has already been obtained (see Chapter 4). In this case the algorithm can start with the optimal solution; a special method for finding an initial basic solution is not needed. Since we will only have use for the dual simplex algorithm in this situation, we will not discuss any procedures for finding an initial basic solution.

3.3 Exercises

In Exercises 1–5 the given tableau represents a solution to a linear programming problem which satisfies the optimality criterion, but is infeasible. Use the dual simplex method to restore feasibility.

1.

c_B		5 x_1	6 x_2	0 x_3	0 x_4	0 x_5	X_B
5	x_1	1	0	$\frac{1}{8}$	$-\frac{1}{8}$	0	$\frac{17}{4}$
6	x_2	0	1	$-\frac{1}{12}$	$\frac{5}{12}$	0	$\frac{19}{6}$
0	x_5	0	0	$-\frac{1}{8}$	$-\frac{7}{8}$	1	$-\frac{1}{4}$
		0	0	$\frac{1}{8}$	$\frac{15}{8}$	0	$\frac{161}{4}$

2.

c_B		5 x_1	6 x_2	0 x_3	0 x_4	0 x_5	0 x_6	X_B
5	x_1	1	0	0	-1	1	0	4
6	x_2	0	1	0	1	$-\frac{2}{3}$	0	$\frac{10}{3}$
0	x_3	0	0	1	7	-8	0	2
0	x_6	0	0	0	0	$-\frac{1}{3}$	1	$-\frac{1}{3}$
		0	0	0	1	1	0	40

3.

c_B		4 x_1	5 x_2	3 x_3	0 x_4	0 x_5	0 x_6	0 x_7	X_B
3	x_3	$\frac{3}{4}$	0	1	$\frac{1}{4}$	0	0	0	$\frac{5}{2}$
0	x_5	$\frac{11}{16}$	0	0	$-\frac{3}{16}$	1	$-\frac{1}{4}$	0	$\frac{17}{8}$
5	x_2	$\frac{9}{16}$	1	0	$-\frac{1}{16}$	0	$\frac{1}{4}$	0	$\frac{19}{8}$
0	x_7	$-\frac{1}{4}$	0	0	$-\frac{1}{4}$	0	0	1	$-\frac{1}{2}$
		$\frac{17}{16}$	0	0	$\frac{7}{16}$	0	$\frac{5}{4}$	0	$\frac{155}{8}$

4.

c_B		5 x_1	6 x_2	0 x_3	0 x_4	0 x_5	0 x_6	x_B
5	x_1	1	0	0	0	1	0	4
6	x_2	0	1	0	$\frac{1}{3}$	$-\frac{2}{3}$	0	$\frac{10}{3}$
0	x_3	0	0	1	-1	-8	0	2
0	x_6	0	0	0	$\frac{1}{3}$	$\frac{2}{3}$	1	$-\frac{1}{3}$
		0	0	0	2	1	0	40

5.

c_B		7 x_1	3 x_2	0 x_3	0 x_4	0 x_5	0 x_6	x_B
0	x_3	0	0	1	$-\frac{1}{4}$	$-\frac{17}{4}$	0	$\frac{19}{2}$
7	x_1	1	0	0	$\frac{1}{8}$	$-\frac{3}{8}$	0	$\frac{1}{4}$
3	x_2	0	1	0	0	1	0	2
0	x_6	0	0	0	$\frac{1}{8}$	$-\frac{3}{8}$	1	$-\frac{23}{4}$
		0	0	0	$\frac{7}{8}$	$\frac{3}{8}$	0	$\frac{31}{4}$

6. (Computer project) Compare the structure diagrams for the simplex algorithm and the dual simplex algorithm. How is the duality exemplified by these diagrams?

3.4 THE REVISED SIMPLEX METHOD

The revised simplex method makes use of some of the notation and ideas we developed in Section 3.2. There we considered a linear programming problem in canonical form

$$\left. \begin{array}{l} \text{Maximize} \quad \mathbf{z} = \mathbf{c}^T\mathbf{x} \\ \text{subject to} \\ \mathbf{A}\mathbf{x} = \mathbf{b} \\ \mathbf{x} \geq \mathbf{0} \end{array} \right\} \tag{1}$$

In this section we confine our attention to the case where the canonical form above was obtained using only slack variables, not artificial variables. A more general procedure for dealing with artificial variables is available, but will not be discussed in this book.

For each tableau in the simplex algorithm we defined the matrix

$$\mathbf{B} = [\mathbf{A}_{i_1} \quad \mathbf{A}_{i_2} \cdots \mathbf{A}_{i_m}]$$

where A_{i_r} is the i_r column of the constraint matrix A and i_r is the index of the rth basic variable in the list at the left side of the tableau. The values of the basic variables $x_{i_1}, x_{i_2}, \ldots, x_{i_m}$ were represented by the vector

$$\mathbf{x}_B = \begin{bmatrix} x_{i_1} \\ x_{i_2} \\ \vdots \\ x_{i_m} \end{bmatrix}$$

and we showed that

$$\mathbf{x}_B = \mathbf{B}^{-1}\mathbf{b} \quad \text{or} \quad \mathbf{b} = \mathbf{B}\mathbf{x}_B \tag{2}$$

We also defined

$$\mathbf{c}_B = \begin{bmatrix} c_{i_1} \\ c_{i_2} \\ \vdots \\ c_{i_m} \end{bmatrix}$$

so that

$$z = \mathbf{c}_B^T\mathbf{x}_B \quad \text{or} \quad z - \mathbf{c}_B^T\mathbf{x}_B = 0 \tag{3}$$

We can combine equations (2) and (3) into one matrix equation by writing

$$\begin{bmatrix} 1 & -\mathbf{c}_B^T \\ 0 & \mathbf{B} \end{bmatrix} \begin{bmatrix} z \\ \mathbf{x}_B \end{bmatrix} = \begin{bmatrix} 0 \\ \mathbf{b} \end{bmatrix} \tag{4}$$

The coefficient matrix in (4) is $(m + 1) \times (m + 1)$, and the vectors are $(m + 1) \times 1$. We denote the coefficient matrix in (4) by $\mathbf{M}$; that is,

$$\mathbf{M} = \begin{bmatrix} 1 & -\mathbf{c}_B^T \\ 0 & \mathbf{B} \end{bmatrix}$$

By multiplying the matrices together, the reader may verify that

$$\mathbf{M}^{-1} = \begin{bmatrix} 1 & \mathbf{c}_B^T\mathbf{B}^{-1} \\ 0 & \mathbf{B}^{-1} \end{bmatrix}$$

Hence, the solution represented by any tableau can be succinctly stated as

$$\begin{bmatrix} z \\ \mathbf{x}_B \end{bmatrix} = \mathbf{M}^{-1} \begin{bmatrix} 0 \\ \mathbf{b} \end{bmatrix}$$

$$= \begin{bmatrix} 1 & \mathbf{c}_B^T\mathbf{B}^{-1} \\ 0 & \mathbf{B}^{-1} \end{bmatrix} \begin{bmatrix} 0 \\ \mathbf{b} \end{bmatrix} \tag{5}$$

$$= \begin{bmatrix} \mathbf{c}_B^T\mathbf{B}^{-1}\mathbf{b} \\ \mathbf{B}^{-1}\mathbf{b} \end{bmatrix}$$

The revised simplex method exploits the form of the solution in (5) by only working with the matrix $\mathbf{M}^{-1}$ instead of the entire tableau. In fact, because $\mathbf{M}^{-1}$ has a particularly simple form, we need only consider $\mathbf{B}^{-1}$. Now the initial tableau is constructed so that $\mathbf{B} = \mathbf{I}_m$, where $\mathbf{I}_m$ denotes the $m \times m$ identity matrix. Thus, initially, $\mathbf{B}^{-1} = \mathbf{I}_m$. The revised simplex method uses a procedure to find the $\mathbf{B}^{-1}$ for the next iteration from information about the entering and departing variables along with the current $\mathbf{B}^{-1}$. We start by writing

$$(\mathbf{B}^{-1})_{\text{new}} = \mathbf{E}\mathbf{B}^{-1} \tag{6}$$

where $\mathbf{E}$ is an $m \times m$ matrix which can be obtained as follows:

(a) Start with $\mathbf{I}_m$. Suppose x_p is the entering variable and x_{i_q} is the departing variable. We have previously shown that $\mathbf{t}_p$, the pth column of the current simplex tableau, which is the pivotal column, is given by

$$\mathbf{t}_p = \mathbf{B}^{-1}\mathbf{A}_p,$$

where $\mathbf{A}_p$ is the pth column of the original matrix $\mathbf{A}$. Denote the entries of the pivotal column of the current tableau by

$$\mathbf{t}_p = \begin{bmatrix} t_{1p} \\ t_{2p} \\ \vdots \\ t_{mp} \end{bmatrix}$$

(b) Replace the qth column of $\mathbf{I}_m$ by the vector

$$\begin{bmatrix} -t_{1p}/t_{qp} \\ -t_{2p}/t_{qp} \\ \vdots \\ 1/t_{qp} \\ \vdots \\ -t_{mp}/t_{qp} \end{bmatrix} \quad \leftarrow q\text{th entry}$$

This modification of the identity matrix which we have constructed is our $\mathbf{E}$. Notice that we never have to numerically invert a matrix (a procedure which may require some care); we simply obtain a sequence of matrices which are $\mathbf{B}^{-1}$ for each of our tableaux.

The revised simplex method consists of the following steps:

1. Determine the entering variable x_p by choosing the most negative $z_j - c_j$, $j = 1, 2, \ldots, s$. Pick randomly if there are ties. Recall that $z_j - c_j$ may be computed as

$$z_j - c_j = \mathbf{c}_B^{\mathsf{T}}\mathbf{t}_j - c_j = \mathbf{c}_B^{\mathsf{T}}\mathbf{B}^{-1}\mathbf{A}_j - c_j$$

In terms of a matrix product, we may write

$$z_j - c_j = \begin{bmatrix} 1 & \mathbf{c}_B^T \mathbf{B}^{-1} \end{bmatrix} \begin{bmatrix} -c_j \\ \mathbf{A}_j \end{bmatrix}$$

2. Determine the departing variable x_{i_q}. This is the variable with the minimum θ-ratio. The i_rth basic variable has a θ-ratio

$$x_{i_r} / t_{rp}$$

where x_{i_r} is the entry in $\mathbf{x}_B$ on the right-hand side of the tableau and where $t_{rp} > 0$. To find the θ-ratios, we may compute

$$\mathbf{t}_p = \begin{bmatrix} t_{1p} \\ t_{2p} \\ \vdots \\ t_{mp} \end{bmatrix} = \mathbf{B}^{-1} \mathbf{A}_p$$

and

$$\mathbf{x}_B = \mathbf{B}^{-1} \mathbf{b}$$

We use only those entries in $\mathbf{x}_B$ which correspond to positive entries in $\mathbf{t}_p$ to form the set of θ-ratios.

3. Determine the new $\mathbf{B}^{-1}$ as described above.
4. Determine the new basic feasible solution and objective function value. From equations (5) and (6), $(\mathbf{x}_B)_{\text{new}} = (\mathbf{B}^{-1})_{\text{new}} \mathbf{b} = \mathbf{E}\mathbf{B}^{-1}\mathbf{b} = \mathbf{E}\mathbf{x}_B$. Thus, if $\mathbf{x}_B$ is available in storage, then

$$(\mathbf{x}_B)_{\text{new}} = \mathbf{E}\mathbf{x}_B$$

This formula is computationally faster than equation (5), since $\mathbf{E}$ is sparse (has many zeros).

As in the simplex method, if none of the $z_j - c_j$ is negative, an optimal solution has been achieved. If none of the entries in the pivotal column $\mathbf{t}_p$ is positive, the optimal solution is unbounded.

EXAMPLE 1. Consider the linear programming problem in canonical form which came from the model for the sawmill:

$$\text{Maximize} \quad z = 120x_1 + 100x_2$$

subject to

$$2x_1 + 2x_2 + x_3 \qquad = 8$$

$$5x_1 + 3x_2 \qquad + x_4 = 15$$

$$x_j \geq 0, \qquad j = 1, 2, 3, 4$$

For this problem

$$A = \begin{bmatrix} 2 & 2 & 1 & 0 \\ 5 & 3 & 0 & 1 \end{bmatrix}, \quad b = \begin{bmatrix} 8 \\ 15 \end{bmatrix}, \quad \text{and} \quad c = \begin{bmatrix} 120 \\ 100 \\ 0 \\ 0 \end{bmatrix} \quad c_B = \begin{bmatrix} 0 \\ 0 \end{bmatrix}$$

The slack variables x_3 and x_4 are our initial basic variables, so that $i_1 = 3$ and $i_2 = 4$. Consequently,

$$B = \begin{bmatrix} 1 & 0 \\ 0 & 1 \end{bmatrix}$$

$$B^{-1} = \begin{bmatrix} 1 & 0 \\ 0 & 1 \end{bmatrix} \quad \text{and} \quad c_B = \begin{bmatrix} 0 \\ 0 \end{bmatrix}$$

Therefore,

$$M^{-1} = \begin{bmatrix} 1 & c_B^T B^{-1} \\ 0 & B^{-1} \end{bmatrix} = \begin{bmatrix} 1 & 0 & 0 \\ 0 & 1 & 0 \\ 0 & 0 & 1 \end{bmatrix}$$

so that

$$\begin{bmatrix} z \\ x_B \end{bmatrix} = M^{-1} \begin{bmatrix} 0 \\ b \end{bmatrix} = \begin{bmatrix} 1 & 0 & 0 \\ 0 & 1 & 0 \\ 0 & 0 & 1 \end{bmatrix} \begin{bmatrix} 0 \\ 8 \\ 15 \end{bmatrix} = \begin{bmatrix} 0 \\ 8 \\ 15 \end{bmatrix} \tag{7}$$

We first determine the entering variable by computing

$$z_j - c_j = c_B A_j - c_j = 0 - c_j$$

$$z_j - c_j = \begin{bmatrix} 1 & c_B^T B^{-1} \end{bmatrix} \begin{bmatrix} -c_j \\ A_j \end{bmatrix} = \begin{bmatrix} 1 & 0 & 0 \end{bmatrix} \begin{bmatrix} -c_j \\ A_j \end{bmatrix} = -c_j$$

Therefore,

$$z_1 - c_1 = -c_1 = -120 \quad = \text{entering variable}$$

$$z_2 - c_2 = -c_2 = -100$$

$$z_3 - c_3 = -c_3 = 0 \qquad x_3$$

$$z_4 - c_4 = -c_4 = 0 \qquad x_4$$

and x_1 ($p = 1$) is the entering variable.

To find the departing variable, we form

$$t_1 = B^{-1}A_1 = \begin{bmatrix} 1 & 0 \\ 0 & 1 \end{bmatrix} \begin{bmatrix} 2 \\ 5 \end{bmatrix} = \begin{bmatrix} 2 \\ 5 \end{bmatrix}$$

and copy x_B from (7). All the entries of t_1 are positive, so we compute

$$\min \{ \tfrac{8}{2}, \tfrac{15}{5} \} = 3$$

The minimum θ-ratio occurs for $x_{i_2} = x_4$ ($q = 2$) and thus, x_4 becomes the departing variable. To compute E, we replace the second column of

I_2 by

$$\begin{bmatrix} -\frac{2}{5} \\ \frac{1}{5} \end{bmatrix}$$

(handwritten, top right) $EA = \begin{bmatrix} 1 & -2/5 \\ 0 & 1/5 \end{bmatrix}\begin{bmatrix} 2 & 2 & 1 & 0 \\ 5 & 3 & 0 & 1 \end{bmatrix} = \begin{bmatrix} 0 & 4/5 & 1 & -2/5 \\ 1 & 3/5 & 0 & 1/5 \end{bmatrix} = T$

Therefore,

$$E = \begin{bmatrix} 1 & -\frac{2}{5} \\ 0 & \frac{1}{5} \end{bmatrix}$$

and

$$(\mathbf{B}^{-1})_{\text{new}} = \mathbf{E}\mathbf{B}^{-1} = \begin{bmatrix} 1 & -\frac{2}{5} \\ 0 & \frac{1}{5} \end{bmatrix}\begin{bmatrix} 1 & 0 \\ 0 & 1 \end{bmatrix} = \begin{bmatrix} 1 & -\frac{2}{5} \\ 0 & \frac{1}{5} \end{bmatrix}$$

Now we have $i_1 = 3$, $i_2 = 1$, so that

$$\mathbf{c}_B = \begin{bmatrix} 0 \\ 120 \end{bmatrix}$$

Therefore,

$$\mathbf{M}^{-1} = \begin{bmatrix} 1 & 0 & 24 \\ 0 & 1 & -\frac{2}{5} \\ 0 & 0 & \frac{1}{5} \end{bmatrix}$$

and the current solution is

$$\begin{bmatrix} z \\ \mathbf{x}_B \end{bmatrix} = \mathbf{M}^{-1}\begin{bmatrix} 0 \\ \mathbf{b} \end{bmatrix} = \begin{bmatrix} 1 & 0 & 24 \\ 0 & 1 & -\frac{2}{5} \\ 0 & 0 & \frac{1}{5} \end{bmatrix}\begin{bmatrix} 0 \\ 8 \\ 15 \end{bmatrix} = \begin{bmatrix} 360 \\ 2 \\ 3 \end{bmatrix} \qquad (8)$$

Next we compute

$$z_j - c_j = \begin{bmatrix} 1 & \mathbf{c}_B^T\mathbf{B}^{-1} \end{bmatrix}\begin{bmatrix} -c_j \\ \mathbf{A}_j \end{bmatrix} = \begin{bmatrix} 1 & 0 & 24 \end{bmatrix}\begin{bmatrix} -c_j \\ \mathbf{A}_j \end{bmatrix}$$

so that

$$z_1 - c_1 = \begin{bmatrix} 1 & 0 & 24 \end{bmatrix}\begin{bmatrix} -120 \\ 2 \\ 5 \end{bmatrix} = 0$$

$$z_2 - c_2 = \begin{bmatrix} 1 & 0 & 24 \end{bmatrix}\begin{bmatrix} -100 \\ 2 \\ 3 \end{bmatrix} = -28 \quad \text{entering variable}$$

$$z_3 - c_3 = \begin{bmatrix} 1 & 0 & 24 \end{bmatrix}\begin{bmatrix} 0 \\ 1 \\ 0 \end{bmatrix} = 0$$

$$z_4 - c_4 = \begin{bmatrix} 1 & 0 & 24 \end{bmatrix}\begin{bmatrix} 0 \\ 0 \\ 1 \end{bmatrix} = 24$$

$$E^{-1}6 = \begin{bmatrix} 1 & -2/5 \\ 0 & 4/5 \end{bmatrix}\begin{bmatrix} 4 \\ 15 \end{bmatrix} = \begin{bmatrix} 6 \\ 3 \end{bmatrix}$$

$$\theta\text{-}Min\left(\frac{6}{5/4}, \frac{3}{3/5}\right) = min\left(\frac{30}{4}, \frac{5}{3}\right) = \frac{5}{3}min$$

202   Further Topics in Linear Programming

The entering variable is x_2 ($p = 2$). To find the departing variable, we compute

$$t_2 = B^{-1}A_2 = \begin{bmatrix} 1 & -\frac{2}{5} \\ 0 & \frac{1}{5} \end{bmatrix}\begin{bmatrix} 2 \\ 3 \end{bmatrix} = \begin{bmatrix} \frac{4}{5} \\ \frac{3}{5} \end{bmatrix}$$

and copy x_B from (8). All the entries of t_2 are positive, so that we compute

$$\min\left\{\frac{2}{\frac{4}{5}}, \frac{3}{\frac{3}{5}}\right\} = \frac{2}{\frac{4}{5}} = \frac{5}{2}$$

The minimum θ-ratio occurs for $x_{i_1} = x_3$ ($q = 1$) and thus, x_3 becomes the departing variable. We compute E by replacing the first column of I_2 by

$$\begin{bmatrix} 1/\frac{4}{5} \\ -\frac{3/5}{4/5} \end{bmatrix} = \begin{bmatrix} \frac{5}{4} \\ -\frac{3}{4} \end{bmatrix}$$

We obtain

$$E = \begin{bmatrix} \frac{5}{4} & 0 \\ -\frac{3}{4} & 1 \end{bmatrix}$$

and

$$(B^{-1})_{new} = \begin{bmatrix} \frac{5}{4} & 0 \\ -\frac{3}{4} & 1 \end{bmatrix}\begin{bmatrix} 1 & -\frac{2}{5} \\ 0 & \frac{1}{5} \end{bmatrix} = \begin{bmatrix} \frac{5}{4} & -\frac{1}{2} \\ -\frac{3}{4} & \frac{1}{2} \end{bmatrix}$$

Now we have $i_1 = 2$ and $i_2 = 1$, so that

$$c_B = \begin{bmatrix} 100 \\ 120 \end{bmatrix}$$

Therefore,

$$M^{-1} = \begin{bmatrix} 1 & 35 & 10 \\ 0 & \frac{5}{4} & -\frac{1}{2} \\ 0 & -\frac{3}{4} & \frac{1}{2} \end{bmatrix}$$

and the current solution is

$$\begin{bmatrix} z \\ x_B \end{bmatrix} = \begin{bmatrix} 1 & 35 & 10 \\ 0 & \frac{5}{4} & -\frac{1}{2} \\ 0 & -\frac{3}{4} & \frac{1}{2} \end{bmatrix}\begin{bmatrix} 0 \\ 8 \\ 15 \end{bmatrix} = \begin{bmatrix} 430 \\ \frac{5}{2} \\ \frac{3}{2} \end{bmatrix} \tag{9}$$

Next we compute

$$z_j - c_j = [1 \quad 35 \quad 10]\begin{bmatrix} -c_j \\ A_j \end{bmatrix}$$

We have

$$z_1 - c_1 = 0$$
$$z_2 - c_2 = 0$$
$$z_3 - c_3 = 35$$
$$z_4 - c_4 = 10$$

Since $z_j - c_j \geq 0$ for all j, the solution given by (9) is optimal.

3.4 Exercises

In Exercises 1 and 2 calculate the matrix $\mathbf{E}$ from the given information.

1. The pivotal column $\mathbf{t}_p$ is

$$\mathbf{t}_p = \begin{bmatrix} 1 \\ -2 \\ 0 \\ 3 \end{bmatrix}$$

and the departing variable labels the fourth row ($q = 4$).

2. The pivotal column $\mathbf{t}_p$ is

$$\mathbf{t}_p = \begin{bmatrix} \frac{1}{2} \\ \frac{3}{4} \\ -\frac{1}{3} \\ 0 \\ \frac{5}{12} \end{bmatrix}$$

and the departing variable labels the second row ($q = 2$).

In Exercises 3 and 4, find the new $\mathbf{B}^{-1}$ from the given information.

3. The current $\mathbf{B}^{-1}$ is

$$\mathbf{B}^{-1} = \begin{bmatrix} 1 & 0 & 2 \\ -1 & 1 & 3 \\ 0 & 2 & 1 \end{bmatrix}$$

and the pivotal column is given by

$$\mathbf{t}_p = \begin{bmatrix} 1 \\ \frac{3}{2} \\ 2 \end{bmatrix} \quad \text{and} \quad \mathbf{b} = \begin{bmatrix} 1 \\ 3 \\ 4 \end{bmatrix}$$

4. The current $\mathbf{B}^{-1}$ is

$$\mathbf{B}^{-1} = \begin{bmatrix} 1 & -1 & 2 & 1 \\ 0 & 1 & 0 & 1 \\ -1 & 3 & 1 & -3 \\ 2 & -1 & 2 & 4 \end{bmatrix}$$

and the pivotal column is given by

$$\mathbf{t}_p = \begin{bmatrix} 0 \\ 4 \\ -2 \\ \frac{2}{3} \end{bmatrix} \quad \text{and} \quad \mathbf{b} = \begin{bmatrix} 1 \\ 1 \\ 4 \\ 2 \end{bmatrix}$$

In Exercises 5–9 solve the given linear programming problem using the revised simplex method.

5. Exercise 6, Section 3.2 (page 188).

6. Maximize $z = x_1 + 2x_2 + 3x_3 + x_4$

subject to

$$2x_1 + x_2 + x_3 + 2x_4 \leq 18$$
$$3x_1 + 5x_2 + 2x_3 + 3x_4 \leq 24$$
$$3x_1 + 2x_2 + x_3 + x_4 \leq 12$$
$$x_j \geq 0, \quad j = 1, 2, 3, 4$$

7. Maximize $z = 2x_1 + 3x_2 + x_3 + x_4 + 2x_5$

subject to

$$2x_1 + x_2 - 3x_3 + x_4 + x_5 \leq 10$$
$$x_1 + 4x_2 + x_4 + 2x_5 \leq 20$$
$$3x_1 + 4x_4 + 2x_5 \leq 15$$
$$x_j \geq 0, \quad j = 1, 2, \ldots, 5$$

8. Maximize $z = 3x_1 + 2x_2 + 4x_5 + x_6 + 2x_8$

subject to

$$3x_1 + x_2 + x_3 + x_4 + 2x_5 + 3x_6 + x_8 \leq 12$$
$$2x_1 + x_2 + 2x_4 + 5x_6 + x_7 + 2x_8 \leq 15$$
$$3x_1 + 2x_2 + x_3 + 3x_5 + x_7 + 3x_8 \leq 18$$
$$x_j \geq 0, \quad j = 1, 2, \ldots, 8$$

9. Maximize $z = 2x_1 + x_2 + 3x_3 + x_6 + 2x_7 + 3x_8$

subject to

$$2x_1 + x_2 + x_4 + 3x_5 + x_7 \leq 24$$
$$x_1 + 3x_3 + x_4 + x_5 + 2x_6 + 3x_8 \leq 30$$
$$5x_1 + 3x_2 + 3x_4 + 2x_5 + x_7 + 5x_8 \leq 18$$
$$3x_1 + 2x_2 + x_3 + x_6 + 3x_8 \leq 20$$

10. If

$$M = \begin{bmatrix} 1 & -c_B^T \\ 0 & B \end{bmatrix}$$

verify that

$$M^{-1} = \begin{bmatrix} 1 & c_B^T B^{-1} \\ 0 & B^{-1} \end{bmatrix}$$

11. Consider the standard linear programming problem

$$\text{Maximize} \quad z = c^T x$$

subject to

$$Ax \leq b$$

$$x \geq 0$$

(a) Show that this problem can be written in canonical form as

$$\text{Maximize} \quad z = c^T x + (c')^T x'$$

subject to

$$[A \mid I] \begin{bmatrix} x \\ x' \end{bmatrix} = b$$

$$\begin{bmatrix} x \\ x' \end{bmatrix} \geq 0$$

(*Hint:* x' will be a vector of slack variables.)

(b) Show that the initial simplex tableau represents the matrix equation

$$\begin{bmatrix} 1 & -c^T & -(c')^T \\ 0 & A & I \end{bmatrix} \begin{bmatrix} z \\ x \\ x' \end{bmatrix} = \begin{bmatrix} 0 \\ b \end{bmatrix}$$

(*Hint:* Proceed as in the derivation of equation (4).)

(c) Equation (5) shows that the solution represented by any tableau is obtained by multiplying the vector

$$\begin{bmatrix} 0 \\ b \end{bmatrix}$$

by M^{-1}. Show that the system of equations represented by any tableau is

$$\begin{bmatrix} 1 & c_B^T B^{-1} A - c^T & c_B^T B^{-1} - (c')^T \\ 0 & B^{-1} A & B^{-1} \end{bmatrix} \begin{bmatrix} z \\ x \\ x' \end{bmatrix} = \begin{bmatrix} c_B^T B^{-1} b \\ B^{-1} b \end{bmatrix}$$

(d) Show from part c that at an optimal solution to the problem in part a, $c_B^T B^{-1} A \geq c^T$ and $c_B^T B^{-1} \geq (c')^T$.

12. (a) Find the dual of the linear programming problem in Exercise 11a.

(b) Show that $\mathbf{w} = (\mathbf{B}^{-1})^T\mathbf{c}_B$ is a feasible solution to the dual problem. (*Hint:* Use Exercise 11d.)

(c) Using Exercise 9 in Section 3.2, explain why $\mathbf{w} = (\mathbf{B}^{-1})^T\mathbf{c}_B$ is an optimal solution to the dual problem.

13. (**Computer project**) Construct a flowchart or structure diagram for the revised simplex method.

3.5 SENSITIVITY ANALYSIS

In the Introduction it was pointed out that solving a linear programming problem is just one part of mathematically modeling a situation. After the problem is solved, one must ask whether the solution makes sense in the actual situation. It is also very likely that the numbers which are used for the linear programming problem are not known exactly. In most cases they will be estimates of the true numbers, and many times they will not be very good estimates. Consequently, it is desirable to have ways of measuring the sensitivity of the solution to changes in the values which specify the problem. Of course, one way to proceed would be to recompute the solution using different values. However, the tableau representing an optimal solution contains the information we need to measure the sensitivity. Using the final tableau saves completely repeating the calculations for a different set of values.

There are five things that can be singled out as subject to variation in defining a linear programming problem:

1. One or more of the objective function coefficients can change.
2. One or more of the resource values (components of **b**) can change.
3. One or more of the entries in the constraint matrix **A** can change.
4. A variable might need to be added. This may happen if management wants information about the effects of making an additional product.
5. Addition of a constraint might be necessary, especially if the solution to the original problem is somewhat unreasonable. This situation also occurs when some of the variables are constrained to take integer values.

In this section we examine the first two possibilities. In Chapter 4 we will discuss the fifth case. Cases 3 and 4 are discussed in more advanced texts. (See the Further Readings at the end of the chapter.)

Another approach to this study is to assume a change in each entry of the objective function coefficient vector, for example, and to assume that the amount of the change depends on a parameter. This leads to the study of *parametric programming*. We do not study this topic here but refer the interested reader to the Further Readings at the end of this chapter.

We assume that the original problem has been converted to the form

$$\text{Maximize} \quad z = \mathbf{c}^T\mathbf{x}$$

subject to

$$\mathbf{Ax} = \mathbf{b}$$

$$\mathbf{x} \geq 0$$

and that an optimal solution to the problem has been obtained. We further assume that we have available the final tableau for the simplex method.

As changes are made in the problem statement, there are several things that may happen to the old optimal solution. It may remain both feasible and optimal, so that no further calculations are necessary. In fact, some computer codes for linear programming problems will automatically compute a range of values for **b** and **c** in which the solution found will remain optimal. The solution to the problem may remain feasible but become nonoptimal. In this case a few iterations of the simplex algorithm will restore the optimality. The optimal solution to the original problem, being a basic feasible solution, provides an initial basic feasible solution for these iterations. On the other hand, the solution to the given problem may remain optimal, as judged by the optimality criterion, but become infeasible. In this case a few iterations of the dual simplex algorithm will usually restore the feasibility.

We now examine Cases 1 and 2.

Change in the Objective Function

Suppose c_k changes to $\hat{c}_k = c_k + \Delta c_k$. The old optimal solution must remain feasible, since neither **A** nor **b** was changed. The optimality criterion is stated in terms of

$$z_j - c_j = \mathbf{c}_B^T \mathbf{t}_j - c_j$$

where $\mathbf{t}_j$ is the jth column of the final tableau (see Section 3.2). If k is the index of one of the basic variables, then $\mathbf{c}_B$ changes and every $z_j - c_j$ must be recomputed. If x_k is a nonbasic variable, $\mathbf{c}_B$ is unchanged and only $z_k - c_k$ changes. In this latter case we have

$$z_k - \hat{c}_k = (z_k - c_k) - \Delta c_k$$

Therefore,

$$z_k - \hat{c}_k \geq 0$$

if and only if

$$z_k - c_k \geq \Delta c_k \tag{1}$$

That is, the profit coefficient of the kth variable, c_k, can be increased by as much as $z_k - c_k$ and the solution will remain optimal. However, making this increase in c_k will not change the value of the objective function, since $x_k = 0$ in the optimal solution.

Now suppose that x_k is a basic variable in the optimal solution. Suppose $k = i_r$, so that the new value of $\mathbf{c}_B$ is

$$
\hat{\mathbf{c}}_B = \begin{bmatrix} c_{i_1} \\ c_{i_2} \\ \vdots \\ c_{i_r} + \Delta c_{i_r} \\ \vdots \\ c_{i_m} \end{bmatrix} = \mathbf{c}_B + \Delta c_{i_r} \begin{bmatrix} 0 \\ 0 \\ \vdots \\ 1 \\ \vdots \\ 0 \end{bmatrix} \leftarrow r\text{th entry}
$$

Let $\mathbf{e}_r$ denote the $m \times 1$ matrix which is all zeros except for a 1 in the rth row. Then we may write

$$
\hat{\mathbf{c}}_B = \mathbf{c}_B + \Delta c_{i_r} \mathbf{e}_r
$$

Now for all values of j except $j = i_r$, we have _[handwritten: $j \neq k$]_

$$
\hat{z}_j - c_j = \hat{\mathbf{c}}_B^T \mathbf{t}_j - c_j = \mathbf{c}_B^T \mathbf{t}_j + \Delta c_{i_r} \mathbf{e}_r^T \mathbf{t}_j - c_j = z_j - c_j + t_{rj} \Delta c_{i_r} \geq \quad (2)
$$

The reader may show, using a similar argument, that

$$
\hat{z}_{i_r} - \hat{c}_{i_r} = 0
$$

Recall that basic variables are determined only by the constraints, not by the objective function. Furthermore, for each basic variable x_{i_k} we must have $\hat{z}_{i_k} - c_{i_k} = 0$. This follows from (2), since $t_{ri_k} = 0$ when $k \neq r$. Consequently, the old optimal solution remains optimal when the objective coefficient of a basic variable is changed if and only if for all nonbasic variables x_j, _[handwritten: $z_k - c_k = 0$]_

$$
z_j - c_j + t_{rj} \Delta c_{i_r} \geq 0
$$

or

[handwritten: sign restrictions]

$$
z_j - c_j \geq -t_{rj} \Delta c_{i_r} \tag{3}
$$

If $t_{rj} = 0$, the inequality in (3) holds for all changes Δc_{i_r}. If $t_{rj} > 0$, we can divide each side of (3) by $-t_{rj}$, reversing the inequality, to obtain

[handwritten: b)]
$$
\Delta c_{i_r} \geq -\frac{z_j - c_j}{t_{rj}} \quad \text{[handwritten: for all } j \neq i_n = k] \tag{4}
$$

for those j for which x_j is nonbasic and $t_{rj} > 0$. If $t_{rj} < 0$, we again divide both sides of (3) by $-t_{rj}$, but do not reverse the inequality, and obtain

$$c) \quad \Delta c_{i_r} \leq - \frac{z_j - c_j}{t_{rj}} \quad \text{for all } j \neq i_r = k \quad (5)$$

for those j for which x_j is nonbasic and $t_{rj} < 0$. Combining (4) and (5), we find that the old optimal solution remains optimal if the change in c_{i_r} satisfies

$$\text{all } X_j \text{ is now basic} \quad \max_{j} \left\{ - \frac{z_j - c_j}{t_{rj}} \bigg| t_{rj} > 0 \right\} \leq \Delta c_{i_r} \leq \min_{\text{all } X_j} \left\{ - \frac{z_j - c_j}{t_{rj}} \bigg| t_{rj} < 0 \right\} \quad (6)$$

$\text{is a nonbasic variable}$

where the index j runs over all nonbasic variables. If there are no j such that $t_{rj} > 0$, then the left side of (6) gives no restriction on Δc_{i_r}. It may take on arbitrarily large negative values. Likewise, if there are no j such that $t_{rj} < 0$, the right side of (6) gives no restriction on Δc_{i_r}.

If Δc_k does not satisfy the inequality in (1) when x_k is a nonbasic variable, or if Δc_k does not satisfy the inequalities in (6) when x_k is a basic variable, the solution represented by the final tableau is no longer optimal. Some iterations of the simplex method will restore optimality.

EXAMPLE 1. Consider the linear programming problem from Example 2 in Section 2.5:

$$\text{Maximize} \quad z = x_1 - 2x_2 - 3x_3 - x_4 - x_5 + 2x_6$$

subject to

$$x_1 + 2x_2 + 2x_3 + x_4 + x_5 \qquad = 12$$
$$x_1 + 2x_2 + x_3 + x_4 + 2x_5 + x_6 = 18$$
$$3x_1 + 6x_2 + 2x_3 + x_4 + 3x_5 \qquad = 24$$

An optimal solution is given by Tableau 3.23.

We see that c_2 can be increased by $4 = z_2 - c_2$, c_4 can be increased by $\frac{1}{2} = z_4 - c_4$, or c_5 can be increased by $4 = z_5 - c_5$ without changing the optimality of the solution.

$X_1 = X_{i_3}$ is 3rd basic variable

$X_3 = X_{i_1}$ is 1st

Tableau 3.23

		1	-2	-3	-1	-1	2	
c_B		x_1	x_2	x_3	x_4	x_5	x_6	x_B
-3	x_3	0	0	1	$\frac{1}{2}$	0	0	3
2	x_6	0	0	0	$\frac{1}{2}$	1	1	9
1	x_1	1	2	0	0	1	0	6
		0	4	0	$\frac{1}{2}$	4	0	15

— no upper bound

$-t_{35}$

To determine the range for a change in c_1, we compute

$$\max\left\{-\frac{z_j - c_j}{t_{3j}}\;\middle|\; t_{3j} > 0\right\} = \max\{-\tfrac{4}{2}, -\tfrac{4}{1}\} = -2$$

and

$$\min\left\{-\frac{z_j - c_j}{t_{3j}}\;\middle|\; t_{3j} < 0\right\}$$

Since there are no j for which $t_{3j} < 0$, there is no upper bound for the change in c_1. That is, if

$$-2 \le \Delta c_1 < \infty$$

the current solution remains optimal.

To determine the range for a change in c_3 for which the solution remains optimal, we compute

$$\max\left\{-\frac{z_j - c_j}{t_{1j}}\;\middle|\; t_{1j} > 0\right\} = \max\left\{\frac{-\tfrac{1}{2}}{\tfrac{1}{2}}\right\} = -1$$

Again there are no values of j for which $t_{1j} < 0$, so that the change in c_3 is not bounded above. We have that the solution remains optimal if

$$-1 \le \Delta c_3 < \infty$$

Finally, checking the range for a change in c_6, we see that the solution remains optimal if

$$-1 \le \Delta c_6 < \infty$$

Summarizing our results, the optimal solution

$$x_1 = 6, \quad x_2 = 0, \quad x_3 = 3, \quad x_4 = 0, \quad x_5 = 0, \quad x_6 = 9$$

remains optimal for a change Δc_k in the kth coefficient of the objective function Δc_k if Δc_k satisfies the appropriate inequality listed in Table 3.2. This assumes only one coefficient is changed.

Table 3.2

k	Δc_k	k	Δc_k
1	$-2 \le \Delta c_1 < \infty$	4	$-\infty < \Delta c_4 \le \tfrac{1}{2}$
2	$-\infty < \Delta c_2 \le 4$	5	$-\infty < \Delta c_5 \le 4$
3	$-1 \le \Delta c_3 < \infty$	6	$-1 \le \Delta c_6 < \infty$

EXAMPLE 2.   The canonical form of the sawmill problem

$$\text{Maximize} \quad z = 120x_1 + 100x_2$$

subject to

$$2x_1 + 2x_2 \le 8$$

$$5x_1 + 3x_2 \le 15$$

$$x_1 \ge 0, \quad x_2 \ge 0$$

has a final tableau as shown in Tableau 3.24.

Tableau 3.24

		120	100	0	0	
c_B		x_1	x_2	x_3	x_4	x_B
100	x_2	0	1	$\frac{5}{4}$	$-\frac{1}{2}$	$\frac{5}{2}$
120	x_1	1	0	$-\frac{3}{4}$	$\frac{1}{2}$	$\frac{3}{2}$
		0	0	35	10	430

The owner of the sawmill would like to know how much it is possible to increase the price (and hence the profit) on each of the outputs individually and still make the same amount of each type of board.

The change in c_1 which does not affect optimality is

$$\frac{-10}{\frac{1}{2}} = -20 \le \Delta c_1 \le \frac{140}{3} = \frac{-35}{-\frac{3}{4}}$$

Likewise, the bounds for the change in c_2 are

$$\frac{-35}{\frac{5}{4}} = -28 \le \Delta c_2 \le 20 = \frac{-10}{-\frac{1}{2}}$$

These computations show that if the owner increases prices so that the profit on 1000 board feet of finish-grade lumber is $166\frac{2}{3}$ (instead of $120) or so that on the same amount of construction-grade lumber it is $120 (instead of $100), then the owner can still make 1500 board feet of finish-grade and 2500 board feet of construction-grade lumber to maximize the profit. If the owner chooses to increase c_2 to $120, the final tableau will be Tableau 3.25 and the profit will be $480.

Tableau 3.25

		120	120	0	0	
c_B		x_1	x_2	x_3	x_4	x_B
120	x_2	0	1	$\frac{5}{4}$	$-\frac{1}{2}$	$\frac{5}{2}$
120	x_1	1	0	$-\frac{3}{4}$	$\frac{1}{2}$	$\frac{3}{2}$
		0	0	60	0	480

Changes in the Resource Vector

Suppose b_k changes to $\hat{b}_k = b_k + \Delta b_k$. We assume that the final basic variables remain as the basic variables for the new problem. However, their values may change. To compute their new values, we proceed as follows. Let

$$\hat{\mathbf{b}} = \begin{bmatrix} b_1 \\ b_2 \\ \vdots \\ b_k + \Delta b_k \\ \vdots \\ b_m \end{bmatrix} = \begin{bmatrix} b_1 \\ b_2 \\ \vdots \\ b_k \\ \vdots \\ b_m \end{bmatrix} + \Delta b_k \begin{bmatrix} 0 \\ 0 \\ \vdots \\ 1 \\ \vdots \\ 0 \end{bmatrix} \leftarrow k\text{th entry}$$

We may write $\hat{\mathbf{b}} = \mathbf{b} + \Delta b_k \mathbf{e}_k$. Then

$$\hat{\mathbf{x}}_B = \mathbf{B}^{-1}\hat{\mathbf{b}} = \mathbf{B}^{-1}\mathbf{b} + \Delta b_k \mathbf{B}^{-1}\mathbf{e}_k$$

and

$$\hat{\mathbf{x}}_B = \mathbf{x}_B + \Delta b_k \mathbf{B}^{-1}\mathbf{e}_k$$

Now $\mathbf{B}^{-1}\mathbf{e}_k$ is the kth column of the matrix $\mathbf{B}^{-1}$. This column appears in the final simplex tableau in the column which held the kth column of the identity matrix in the initial tableau. If the new solution $\hat{\mathbf{x}}_B$ is to be feasible, Δb_k must be chosen so that

$$\mathbf{x}_B + \Delta b_k \mathbf{B}^{-1}\mathbf{e}_k \geq 0$$

EXAMPLE 3. Looking again at the sawmill problem, we see that a change Δb_1 in the availability of the saw yields

$$\hat{\mathbf{x}}_B = \begin{bmatrix} \frac{5}{2} \\ \frac{3}{2} \end{bmatrix} + \Delta b_1 \begin{bmatrix} \frac{5}{4} \\ -\frac{3}{4} \end{bmatrix}$$

If $\hat{\mathbf{x}}_B$ is to remain feasible, Δb_1 must satisfy

$$\frac{5}{2} + \frac{5}{4}\Delta b_1 \geq 0$$
$$\frac{3}{2} - \frac{3}{4}\Delta b_1 \geq 0$$

or

$$-2 \leq \Delta b_1 \leq 2$$

Likewise, we find that if the change Δb_2 in the available hours of the plane satisfies

$$-3 \leq \Delta b_2 \leq 5$$

the solution remains feasible (and optimal).

If the availability of the saw is increased from 8 to 12 hours a week, the final tableau of the problem becomes Tableau 3.26. By applying the dual simplex method we may restore feasibility. We then obtain Tableau 3.27.

Tableau 3.26

		120	100	0	0	
c_B		x_1	x_2	x_3	x_4	x_B
100	x_2	0	1	$\frac{5}{4}$	$-\frac{1}{2}$	$\frac{15}{2}$
120	x_1	1	0	$\left(-\frac{3}{4}\right)$	$\frac{1}{2}$	$-\frac{3}{2}$
		0	0	35	10	570

Tableau 3.27

		120	100	0	0	
c_B		x_1	x_2	x_3	x_4	x_B
100	x_2	$\frac{5}{3}$	1	0	$\frac{1}{3}$	5
0	x_3	$-\frac{4}{3}$	0	1	$-\frac{2}{3}$	2
		$\frac{140}{3}$	0	0	$\frac{100}{3}$	500

We see that the mill operator should only make construction-grade lumber if the saw is available 12 hours per day. Five thousand board feet can be made with this much sawing time.

3.5 Exercises

1. Consider the linear programming problem

$$\text{Maximize} \quad z = x_1 + 2x_2 + x_3 + x_4$$

subject to

$$2x_1 + x_2 + 3x_3 + x_4 \leq 8$$
$$2x_1 + 3x_2 \qquad + 4x_4 \leq 12$$
$$3x_1 + x_2 + 2x_3 \qquad \leq 18$$
$$x_j \geq 0, \quad 1 \leq j \leq 4$$

After adding slack variables x_5, x_6, and x_7 and solving by the simplex method, we obtain the final tableau shown below.

		1	2	1	1	0	0	0	
c_B		x_1	x_2	x_3	x_4	x_5	x_6	x_7	x_B
1	x_3	$\frac{4}{9}$	0	1	$-\frac{1}{9}$	$\frac{1}{3}$	$-\frac{1}{9}$	0	$\frac{4}{3}$
2	x_2	$\frac{2}{3}$	1	0	$\frac{4}{3}$	0	$\frac{1}{3}$	0	4
0	x_7	$\frac{13}{9}$	0	0	$-\frac{10}{9}$	$-\frac{2}{3}$	$-\frac{1}{9}$	1	$\frac{34}{3}$
		$\frac{7}{9}$	0	0	$\frac{14}{9}$	$\frac{1}{3}$	$\frac{5}{9}$	0	$\frac{28}{3}$

(a) For each of the cost coefficients c_j, $1 \le j \le 4$, find the range of values for Δc_j for which the above solution remains optimal.

(b) For each of the resources b_i, $1 \le i \le 3$, find the range of values for Δb_i for which the above solution remains optimal.

2. What will be the optimal solution to the problem in Exercise 1
   (a) if c_1 is changed to 3?
   (b) if b_2 is changed to 26?
   (c) if c_3 is changed to $\frac{1}{2}$?
   (d) if b_3 is changed to 127?

3. Resolve the linear programming problem in Example 2, Section 2.5 (page 127), keeping the columns for the artificial variables in the Tableau 2.18.
   (a) For each of the cost coefficients c_j, $1 \le j \le 6$, find the range of values for Δc_j for which the solution remains optimal.
   (b) For each of the resources b_i, $1 \le i \le 3$, find the range of values for Δb_i for which the solution remains optimal.

4. What will be the optimal solution to the problem in Exercise 3
   (a) if c_2 is changed to 5?
   (b) if c_3 is changed to $-\frac{7}{2}$?
   (c) if b_1 is changed to 30?
   (d) if b_2 is changed to 25?

5. Consider the agricultural problem in Exercise 4, Section 2.1 (page 72).
   (a) Suppose the farmer is able to hire additional workers who can devote 60 hours to working the crops. How much of each crop should be planted?
   (b) Suppose the farmer decides to only plant 10 acres. How much of each crop should be planted?
   (c) Suppose the price received for oats increases by $1/acre, so that the profit per acre of oats is now $21. How much of each crop should the farmer plant?
   (d) What increase in profit for soybeans would induce the farmer to plant soybeans?

6. Consider the investment problem in Exercise 8, Section 2.1 (page 73).

(a) Suppose the rate of return on the electronics stock increases to 5.2%. What is the optimal investment policy?

(b) Suppose the rate of return on the utility stock decreases to 7%. What is the optimal investment policy?

(c) Suppose the amount invested in the bond is at least $90,000. What is the optimal investment policy?

7. The text discusses changing only one component of the resource vector at a time. Consider now the situation where

$$\hat{\mathbf{b}} = \mathbf{b} + \Delta\mathbf{b} \quad \text{and} \quad \Delta\mathbf{b} = \begin{bmatrix} \Delta b_1 \\ \Delta b_2 \\ \vdots \\ \Delta b_m \end{bmatrix}$$

That is, several components of the resource vector are changed. Following the text discussion, show that

$$\hat{\mathbf{x}}_B = \mathbf{x}_B + \mathbf{B}^{-1}\Delta\mathbf{b}$$

and that the solution $\hat{\mathbf{x}}_B$ is a feasible solution (and hence optimal) if and only if $\mathbf{x}_B + \mathbf{B}^{-1}\Delta\mathbf{b} \geq \mathbf{0}$.

3.5 Project

A tractor manufacturer in a developing nation subcontracts the task of making air filters for the tractors to a small company. The filter consists of a cylindrical main chamber with a cylindrical exit duct mounted on top of it. The specifications for the filters are as follow:

1. In order to keep the dust efficiency within permissible limits, the diameter of the main chamber and the exit duct should not exceed 16 cm and 6.5 cm, respectively.

2. To keep the pressure drop across the air cleaner small enough to prevent excessive power loss, the main chamber diameter and exit duct diameter should not be less than 9.5 cm and 3.5 cm, respectively.

3. The main chamber is to be 24 cm tall, and the exit duct is to be 6.5 cm long.

4. To maintain acceptable weight and durability, each filter must contain at least 1600 cm² of metal.

5. At least 50 air filters must be supplied each month.

As is typical in such countries, industrial materials such as sheet metal are not available in unlimited supply. The government has allocated 9.65 m² of metal each month to the filter manufacturer.

A cross-sectional view of the filter is shown in Figure 3.7. Assume that the intake port and other interior structures need 40% of the total metal used for the main chamber and exit duct. Also assume that unusable scrap accounts for 15% of the total metal used for the main chamber and exit duct.

(a) Set up a linear programming model for this situation to meet the objective of minimizing the amount of sheet metal used per air cleaner.

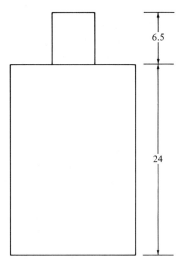

Figure 3.7

(b) Solve the model.
(c) Perform a sensitivity analysis on the model.
(d) Comment on the solution, noting the meanings of the slack variables, the dual variables, and the ranges for ''resource'' and ''cost'' components.

3.6 COMPUTER ASPECTS (Optional)

The development of linear programming and its applications has paralleled the development of the digital computer. The first linear programming problem to be solved on a computer dealt with a U.S. Air Force problem regarding the development and support of aircraft subject to strategic and physical requirements. It was solved in January 1952, on a machine at the National Bureau of Standards. The large programs (called **LP codes** or **LP systems**) which are available today to solve linear programming problems owe their existence to the current generation of computers with their extremely large auxiliary storage units (drums and disks) and their fast arithmetic operations. These larger LP codes can theoretically handle problems involving as many as 285,000 variables and 16,000 constraints in reasonable (but not short) amounts of time.

An LP system is typically provided by the computer manufacturer as an option when leasing a large system, or it can be leased from one of the many software development companies. Such a system may represent an investment of 10–20 person-years of development time. Typical of the commercially available systems are UNIVAC's Mathematical Programming System, FMPS; IBM's Mathematical Programming System Extended, MPSX/370; Control Data Corporation's OPHELIE; Haverly Systems' LP/360; and Bonner and Moore Software Systems' Functional Mathematical Programming System, FMPS.

In this section we will describe typical features of an LP code to give the reader some idea of what to look for in the course of solving a linear programming problem. The following example will be used as an illustration.

EXAMPLE 1. A health food store packages two types of snack foods, Chewy and Nutty, by mixing sunflower seeds, raisins, and peanuts. For the coming week the store has available in bulk quantities 90 kg of sunflower seeds, 100 kg of raisins, and 60 kg of peanuts. Invoices show that they paid $135 for the sunflower seeds, $180 for the raisins, and $60 for the peanuts.

Chewy consists of 2 parts sunflower seeds, 2 parts peanuts, and 6 parts raisins. Nutty consists of 3 parts sunflower seeds and 7 parts peanuts. Chewy sells for $2.00/kg, and Nutty sells for $1.50/kg. Determine a mixing scheme which will maximize the store's profits, assuming that its entire production of Chewy and Nutty will be sold.

Input

To solve a general linear programming problem, clearly the LP code will need to know the type of problem (maximize or minimize), the coefficients of the objective function c, the coefficients of the constraints A, the right-hand sides of the constraints b, and the relation ($\leq$, $=$, $\geq$) for each constraint. There are several tradeoffs that can be made between ease of use for the problem solver and ease of programming and standardization for the programmer. Some codes assume that all problems are minimization problems, that all entries in b are nonnegative, and that all constraints are equalities. Thus, the user of these codes must put the model into a particular form by including slack variables and multiplying the objective function and each constraint by -1 if necessary. In this case the code provides the artificial variables where necessary.

In larger problems the majority of the entries of A will be zero. In this case the user would not want to have to input all these zeros, so most codes provide for entering only the nonzero elements of A and assume all other entries are zero. Consequently, the input must identify the constraint and variable to which the coefficient belongs. This specification is accomplished by asking the user to assign a name (perhaps limited to six or eight characters) to each constraint and each variable. Then each nonzero coefficient can be entered by giving the names of the constraint and variable to which it belongs. A drawback to this method is that the computer will interpret a misspelled name as a new variable or constraint. Some codes try to protect against such errors by keeping the name in two pieces and flagging all input where one piece of the name agrees but the other does not. Naming the variables and constraints also provides more descriptive output, especially when the names are chosen as mnemonics for the quantities that the variables and constraints represent.

The objective function is either entered separately by giving the name of the variable and its objective function coefficient or it is entered as part of the constraints. In the latter case the coefficients are given the system-reserved constraint name OBJECT.

The right-hand sides are entered by giving the constraint name and the coefficient value. If the constraints have not been changed into a particular form, the type of relation for each constraint must also be specified.

Most of the existing codes are written in a combination of assembly language and FORTRAN, so that all input is bound by the formatting rules of these languages. Among other things this means that the names or coefficients must be punched in specified columns of the input cards, with no deviation allowed. It also means that all numbers must be given with decimal points.

Besides specifying the problem, many codes allow for the user to specify various error-correcting features and output formats, which we will discuss later. They also provide the capability of identifying the problem to the code and user by allowing the user to give a title to the problem which will be printed as a heading on each page of output.

A typical job setup for an LP code is shown in Figure 3.8. Note that the various types of data are separated by the data control cards ROWS, COLUMNS, and RHS. Control cards for computer center accounting and program library access must precede the cards shown in the figure.

EXAMPLE 1 (continued).   Using mnemonic labels for the variables, we can write the mathematical model of the situation as

Maximize

$$z = 2 \times \text{CHEWY} - (1.5)(.2) \times \text{CHEWY} - (1.8)(.6) \times \text{CHEWY}$$
$$- (1)(.2) \times \text{CHEWY} + 1.5 \times \text{NUTTY}$$
$$- (1.5)(.3) \times \text{NUTTY} - (1)(.7) \times \text{NUTTY}$$

or

Maximize $z = .42 \times \text{CHEWY} + .35 \times \text{NUTTY}$

subject to

RAISIN: $6 \times \text{CHEWY} \leq 100$

PEANUT: $2 \times \text{CHEWY} + 7 \times \text{NUTTY} \leq 60$

SUN: $2 \times \text{CHEWY} + 3 \times \text{NUTTY} \leq 90$

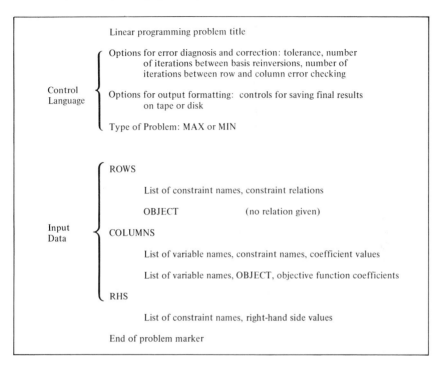

Figure 3.8

The input format for this example for the widely available LP code MPS/90, which comes from the Sperry-Univac program library, is shown in Figure 3.9.

Algorithm

Virtually all LP codes designed for production, rather than teaching, use the revised simplex method. This method has several desirable features, including the ability to handle a large number of variables. The real limit on the size of a problem is the number of constraints (see Section 2.9). Other features will be described when we discuss error detection and correction.

Most of the large LP codes provide an option for computing B^{-1} which is based upon a procedure from numerical linear algebra called **LU factorization**. That is, B is written as LU, the product of a lower triangular matrix L and an upper triangular matrix U. The inverses of upper and lower triangular matrices are easily calculated. Then $B^{-1} = U^{-1}L^{-1}$. Many large linear programming models have **sparse** matrices (ones with few nonzero entries). The matrix representations can then be highly compressed and L^{-1} and U^{-1} can be calculated in core, with special

```
0001          PROGRAM MIXTURE
0002          AMINMAX='MAX'
0003          CALL INPUT
0004          CALL SETUP
0005          CALL BCDOUT
0006          CALL PICTURE
0007          CALL TRANCOL
0008          CALL TRANROW
0009          CALL OPTIMIZE
0010          CALL SOLUTION
0011          STOP
0012          END
```

COMPILATION COMPLETE   0 ERRORS

```
NAME
ROWS
 N PROFIT
 L RAISIN
 L PEANUT
 L SUN
COLUMNS
     CHEWY      PROFIT       .42000    RAISIN        6.00000
     CHEWY      PEANUT      2.00000    SUN           2.00000
     NUTTY      PROFIT       .35000    PEANUT        7.00000
     NUTTY      SUN         3.00000
RHS
     RHS1       RAISIN    100.00000    PEANUT       60.00000
     RHS1       SUN        90.00000
ENDATA
```

	CHEWY	NUTTY	RHS1
AT	*LO*	*LO*	
PROFIT	.42000	.35000	.00000
RAISIN	6.00000	.00000	100.00000
PEANUT	2.00000	7.00000	60.00000
SUN	2.00000	3.00000	90.00000

Figure 3.9

routines for sparse matrices, resulting in significant time savings. For this reason, more and more codes will provide an **LU**-factorization option.

The revised simplex algorithm with iterative B^{-1} calculation is usually programmed to check itself at specified intervals. Between checks it follows the description we gave in Section 3.4. The check involves computing the next $\mathbf{B}^{-1}$ in a manner different from the one we described. The matrix $\mathbf{B}$ can be constructed from the list of basic variables and the original problem as it was read in and stored. Then a very good method of numerically inverting $\mathbf{B}$, such as the **LU**-factorization method described above, is used. This procedure of occasionally recomputing $\mathbf{B}^{-1}$ from the given problem serves to produce a more accurate basic feasible solution. However, in general the procedure is expensive in terms of computation time and must be used sparingly.

As was indicated in Section 2.4, most LP codes provide several options for handling degeneracy when it occurs. Usually, the default option is to ignore it.

Generalized Upper Bounding

Large commercially available mathematical programming systems typically include a procedure called generalized upper bounding (GUB, for short). It is usual for a large linear programming problem to have a substantial number of constraints which deal with bounds on certain variables or sums of variables, or with the balance of materials between two stages of a process. The special structure of these constraints allows them to be treated in a way different from that of a general constraint. The GUB procedure may be used on a problem whose constraints have the form shown in Figure 3.10. The GUB constraints are shaded.

The GUB procedure allows the reduction of the number of basic variables from one corresponding to each constraint to one corresponding to each general constraint (r variables in Figure 3.10). Thus, a problem with 1000 constraints, 800 of which are GUB constraints, can be solved with about the same effort as a problem with 200 general constraints.

Models to which the GUB procedure may be applied arise in a number of areas, including production resource allocation (e.g., forest manage-

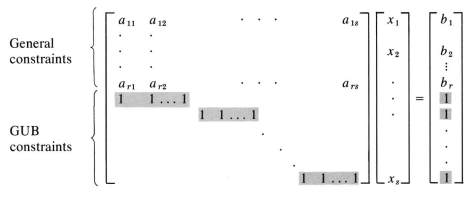

Figure 3.10

ment, commercial fishing strategies), multiproduct blend, and large-scale models of economic sectors.

Output

We have seen in Sections 3.3 and 3.5 that there is a large amount of information about the solution to a linear programming which is contained in the final tableau. A typical output from an LP code will summarize this information in a useful form. Some codes also allow the option of having the output saved on disk or tape for future input to a report generator or to rerun the problem with slight variations.

The output will give the problem heading, the number of iterations used, and the optimal value of the objective function. Then usually a list of all the variables originally specified for the problem is produced. Those variables which are basic are labeled as such. The value of each variable is given along with the value of $z_j - c_j$ from the objective row.

The next part of the output lists the constraints and notes whether each is slack or binding. The value of the slack variable (if the computer code had to provide it) is also given. Since each constraint corresponds to a dual variable, the value of each dual variable, which may be interpreted as the marginal cost of each right-hand side (or resource) value, is given by most codes.

Finally, ranges of values for each of the objective function coefficients and each of the right-hand side values are given. These ranges come from the sensitivity analysis of the optimal solution. One must interpret each range as giving the values which that particular coefficient may take, assuming that

1. no other coefficients are changed, and
2. the computed optimal value remains optimal.

EXAMPLE 1 (continued).   The output from MPS/90 for our example is shown in Figure 3.11.

Error Detection and Correction

Throughout this chapter we have used rational numbers such as $\frac{2}{3}$ and $-\frac{8}{11}$ while solving linear programming problems using the simplex or revised simplex algorithms. A computer, however, will convert these numbers to decimal representation and round off in the process. If the computer typically carries seven digits, $\frac{2}{3}$ becomes .6666667 and $-\frac{8}{11}$ becomes $-.7272727$. After many calculations the errors made through roundoff and truncation tend to accumulate. It is possible for the computer to produce a "solution" to a linear programming problem which is not feasible because of the roundoff error.

```
OPTIMIZE    OBJ = PROFIT    RHS = RHS1
TIME =  0.13          *MAXIMIZE*

ITER   OBJ VAL     NON OPT   VEC IN   VEC OUT    PI/DJ
  1    7.00000        2         5        2      -.42000
  2    8.33333                  6        3      -.35000

OPTIMALITY REACHED


SECTION 1 = ROWS

NUMBER   ROW      AT   ACTIVITY    SLACK ACTIVITY   LOWER LIMIT   UPPER LIMIT   PI VALUE
  1     PROFIT    BS     8.33333      -8.33333        NONE          NONE        1.00000
  2     RAISIN    UL   100.00000        .00000        NONE        100.00000    -.05333
  3     PEANUT    UL    60.00000        .00000        NONE         60.00000    -.05000
  4     SUN       BS    44.76190       45.23810       NONE         90.00000     .00000


SECTION 2 - COLUMNS

NUMBER   COLUMN   AT   ACTIVITY     INPUT COST      LOWER LIMIT   UPPER LIMIT   DJ VALUE
  5     CHEWY     BS    16.66667       .42000          .00000        NONE        .00000
  6     NUTTY     BS     3.80952       .35000          .00000        NONE        .00000
```

Figure 3.11

Fortunately, the revised simplex method allows us to easily detect such errors as they accumulate. The key to this detection is that the revised simplex algorithm keeps the problem as it was entered and only changes the $\mathbf{B}^{-1}$ matrix. Suppose after some iterations the algorithm claims that $\hat{\mathbf{x}}_B = \hat{\mathbf{B}}^{-1}\mathbf{b}$ is a feasible solution. Since at each iteration $\mathbf{B}^{-1}$ is computed from the previous value, roundoff errors could have occurred, making what we have recorded as $\mathbf{B}^{-1}$ different from the theoretical $\mathbf{B}^{-1}$.

However, we may take the original constraints in canonical form as $\mathbf{Ax} = \mathbf{b}$ and set all the nonbasic variables to zero. This yields the system

$$\mathbf{Bx}_B = \mathbf{B}\begin{bmatrix} x_{i_1} \\ x_{i_2} \\ \vdots \\ x_{i_m} \end{bmatrix} = \mathbf{b}$$

We then compute $\mathbf{B}\hat{\mathbf{x}}_B$ and compare it with $\mathbf{b}$. If the difference between the two values exceeds a preset tolerance, we can invoke an error-correcting routine. This process is generally referred to as **checking row sums**. Normally, the tolerance is set at 10^{-6}, although most LP systems allow the user to change its value.

Column sums may also be checked by recalling that the entries in the objective row should be

$$\mathbf{c}_B^T \mathbf{t}_j - c_j = \mathbf{c}_B^T \mathbf{B}^{-1}\mathbf{A}_j - c_j$$

The error-correcting routine which is used when row or column sum errors are detected involves what is usually referred to as **reinverting the basis.** The matrix $\mathbf{B}$ can be formed exactly for the present list of basic variables from the original problem. Then a very accurate method from numerical analysis is used to calculate $\mathbf{B}^{-1}$. From this a better value of $\mathbf{x}_B$ can be found, as $\mathbf{x}_B = \mathbf{B}^{-1}\mathbf{b}$.

In some codes this method of reinverting the basis is done at specified intervals irrespective of whether the row and column sums check. The length of this interval is commonly either fixed at every 75 iterations or is determined automatically as a function of the size of the problem. If $\mathbf{A}$ is $m \times n$, then a typical choice for the interval length is either $[m/5]$ or $3[\sqrt{n}]$. In any case, the user has the option of overriding the specification of the interval length by the code by choosing his or her own interval length.

Scaling

One thing the user can do to minimize the errors generated by the algorithm is to make sure the problem is appropriately scaled. This means that all the numbers in the constraint coefficient matrix $\mathbf{A}$ should be about the same size. If the coefficients of one constraint are more than 100 times the size of those of another constraint, severe error propagation may result. However, the large coefficients can be divided by an appropriate number to make them about the right size.

EXAMPLE 2. If two constraints in a linear problem are

$$2x_1 + .12x_2 + x_3 - 3x_4 = 7 \qquad (1)$$

and

$$75x_1 + 250x_2 + 121x_3 + 314x_4 = 500 \qquad (2a)$$

the second can be divided by 100 to yield

$$.75x_1 + 2.5x_2 + 1.21x_3 + 3.14x_4 = 5 \qquad (2b)$$

Constraints (1) and (2b) should be used as input to the revised simplex algorithm.

In the same way, if the coefficients of one variable are significantly different in magnitude from those of all other variables, the units of the variable should be changed to bring the size of the coefficients in line. This means that the corresponding objective row coefficient will also have to be changed.

EXAMPLE 3. Consider the linear programming problem

$$\text{Maximize} \quad z = 23x_1 + 2x_2 + 3x_3$$

subject to

$$140x_1 + 5x_2 - x_3 \leq 10$$
$$210x_1 + x_2 + 3x_3 \leq 14$$
$$x_1 \geq 0, \quad x_2 \geq 0, \quad x_3 \geq 0$$

Replacing x_1 by $x_1' = 100x_1$, we obtain the following problem:

$$\text{Maximize} \quad z = .23x_1' + 2x_2 + 3x_3$$

subject to

$$1.4x_1' + 5x_2 - x_3 \leq 10$$
$$2.1x_1' + x_2 + 3x_3 \leq 14$$
$$x_1' \geq 0, \quad x_2 \geq 0, \quad x_3 \geq 0$$

Thus, if the units of x_1 are grams/kilogram, the units of x_1' will be 100 grams/kilogram.

Restarting

Most large codes allow the user the option of saving the computations which have been done at a certain point. One scheme allows the user to specify the maximum number of iterations permitted. When that maximum number is reached, the current $\mathbf{B}^{-1}$, list of basic variables, and original problem statements are dumped onto tape or disk for restarting at a later time. Another scheme allows the final solution, $\mathbf{B}^{-1}$, and list of basic variables to be saved along with the original problem. This latter scheme is particularly useful when the restarting procedure allows additional constraints or variables to be put into the problem. In this case the dual simplex method is used to restore feasibility if necessary.

The restart procedures are especially useful when analysis shows that the original model produced results which are inappropriate to the situation which was modeled.

Further Readings

Beale, E. M. L. *Mathematical Programming in Practice*. Pitman, London, 1968.

Dantzig, George B., and R. M. Van Slyke. "Generalized Upper Bounding Techniques for Linear Programming." *Journal of Computer and System Sciences,* 1 (1967), 213–26.

Gass, Saul I. *Linear Programming*. 4th ed. McGraw-Hill, New York, 1975.

Geoffrion, A. M., "Elements of Large-scale Mathematical Programming." *Management Science,* 16 (1970), 652–91.

Murty, Katta. *Linear and Combinatorial Programming*. Wiley, New York, 1976.

Orchard-Hays, William. *Advanced Linear Programming Computing Techniques*. McGraw-Hill, New York, 1968.

Salkin, Harvey S., and Jahar Saha, eds. *Studies in Linear Programming*. North-Holland, Amsterdam, 1975.

Thesen, Arne. *Computer Methods in Operations Research*. Academic Press, New York, 1978.

Wagner, Harvey. *Principles of Operations Research*. Prentice-Hall, Englewood Cliffs, N.J., 1969.

White, William W. "A Status Report on Computing Algorithms for Mathematical Programming." *ACM Computing Surveys,* 5 (1973), 135–66.

CHAPTER
012345

INTEGER PROGRAMMING

In the linear programming problems considered so far, the variables have been permitted to assume all nonnegative real values. However, there are many problems where the variables must assume only integer values. For example, it would be meaningless to have an answer calling for the manufacture of half a table or for the chartering of 1.2 airplanes. In some problems, such as the transportation problem with integer values for supply and demand, the simplex method will yield integer answers; however, in many other problems it will not. In this chapter we formulate a number of problems which require integer variables and present three algorithms for solving these integer programming problems.

4.1 EXAMPLES

EXAMPLE 1. (**The Transportation Problem**) Suppose a manufacturer making one product has m factories and n warehouses. The demand at the jth warehouse is d_j, $j = 1, 2, \ldots, n$, and the supply available from the ith factory is s_i, $i = 1, 2, \ldots, m$. The cost of shipping one unit from the ith factory to the jth warehouse is c_{ij}. Our problem is to determine the amount x_{ij} of the product to be sent from the ith factory to the jth warehouse.

If we assume that the total supply at least equals the total demand,

$$\sum_{i=1}^{m} s_i \geq \sum_{j=1}^{n} d_j$$

so that our problem is feasible, then the mathematical model is

$$\text{Minimize} \quad z = \sum_{i=1}^{m} \sum_{j=1}^{n} c_{ij} x_{ij}$$

subject to

$$\sum_{j=1}^{n} x_{ij} \leq s_i, \quad i = 1, 2, \ldots, m$$

$$\sum_{i=1}^{m} x_{ij} \geq d_j, \quad j = 1, 2, \ldots, n$$

$$x_{ij} \geq 0 \quad \text{and} \quad x_{ij} \text{ integral,}$$

$$i = 1, 2, \ldots, m; \quad j = 1, 2, \ldots, n$$

If this problem is converted to standard form, then the only entries in the constraint matrix are 1's, -1's, and 0's. In this case the simplex method will automatically yield integer solutions if the s_i and d_j are integers. However, the simplex method is a rather poor way of solving the transportation problem. In Chapter 5 we present a special algorithm for this problem which is rather efficient. This algorithm was developed because the transportation model arises repeatedly in practice.

EXAMPLE 2. **(The Knapsack Problem)** Consider the problem faced by a hiker who cannot carry more than k pounds of equipment. She has n items which she is considering bringing. To each item she assigns a relative value c_j with the most important items having the highest values. Let a_j be the weight of the jth item. The hiker's problem is to decide which of the n items to carry; she will choose those which maximize the total relative value subject to the weight limitation.

To construct the mathematical model, let $x_j = 1$ if the jth item is chosen and let $x_j = 0$ if the jth item is not chosen. Then the model is

$$\text{Maximize} \quad z = \sum_{j=1}^{n} c_j x_j$$

subject to

$$\sum_{j=1}^{n} a_j x_j \leq k$$

$$x_j = 0 \quad \text{or} \quad 1, \quad j = 1, 2, \ldots, n$$

Note that by limiting the value of x_j to 0 or 1, the left-hand side of the constraint represents just the weight of the items which are chosen. This type of an integer programming problem is called a **zero–one programming problem.**

EXAMPLE 3. (**The Assignment Problem**) Suppose n people P_1, P_2, ..., P_n are being considered for n jobs J_1, J_2, ..., J_n. Using various criteria, including past performance, aptitude, and job ability, we specify a value c_{ij} which would accrue if the ith person is given the jth job. We assume that each person is assigned to exactly one job and that each job is assigned to exactly one person. Our problem is to assign the people to the jobs so that the total value of the assignment is maximized.

To construct the mathematical model, define the variables x_{ij} so that

$$x_{ij} = \begin{cases} 1 & \text{if } P_i \text{ is assigned to } J_j \\ 0 & \text{otherwise} \end{cases}$$

Then the mathematical model is

$$\text{Maximize} \quad z = \sum_{i=1}^{n} \sum_{j=1}^{n} c_{ij} x_{ij}$$

subject to

$$\sum_{i=1}^{n} x_{ij} = 1, \qquad j = 1, 2, \ldots, n \tag{1}$$

$$\sum_{j=1}^{n} x_{ij} = 1, \qquad i = 1, 2, \ldots, n \tag{2}$$

$$x_{ij} = 0 \quad \text{or} \quad 1, \quad i, j = 1, 2, \ldots, n$$

Under the condition that x_{ij} has a value of either 0 or 1, exactly one of the summands in equation (1) can be nonzero, and likewise, exactly one of the summands in (2) can be nonzero. The constraint (1) says that job j is assigned to exactly one person; (2) says that person i is assigned to exactly one job. Just as in the transportation problem, the simplex algorithm will yield a zero-one solution to the assignment problem. However, there is a special algorithm which efficiently handles this problem; it will be discussed in Chapter 5.

EXAMPLE 4. (**The Traveling Salesman Problem**) A traveling salesman has to visit each of n cities C_1, C_2, ..., C_n. He must start from his home office in city C_1 and return to C_1 after visiting each city exactly once. The order in which he visits cities C_2, C_3, ..., C_n does not matter. He knows the distance between each pair of cities and wants to choose a route which minimizes the total distance traveled.

To formulate the mathematical model, let c_{ij} be the distance between C_i and C_j. Let the variable x_{ij} be defined by

$$x_{ij} = 1 \quad \text{if the route includes traveling from } C_i \text{ to } C_j$$

$$= 0 \quad \text{otherwise}$$

The condition that the route must go to exactly one city after leaving C_i may be written

$$\sum_{j=1}^{n} x_{ij} = 1, \qquad i = 1, 2, \ldots, n$$

The condition that the route goes through every city exactly once can be phrased by saying that each city must be reached from exactly one city, or

$$\sum_{i=1}^{n} x_{ij} = 1, \qquad j = 1, 2, \ldots, n$$

Our mathematical model is then

$$\text{Minimize} \quad z = \sum_{i=1}^{n} \sum_{j=1}^{n} c_{ij} x_{ij}$$

subject to

$$\sum_{i=1}^{n} x_{ij} = 1, \qquad j = 1, 2, \ldots, n \tag{3}$$

$$\sum_{j=1}^{n} x_{ij} = 1, \qquad i = 1, 2, \ldots, n \tag{4}$$

$$x_{ij} = 0 \quad \text{or} \quad 1, \qquad i, j = 1, 2, \ldots, n$$

Consider the feasible solution for this problem when $n = 12$:

$$x_{12} = x_{23} = x_{34} = x_{45} = x_{56} = x_{61} = 1$$

and

$$x_{78} = x_{89} = x_{9,10} = x_{10,11} = x_{11,12} = x_{12,7} = 1$$

and

$$x_{ij} = 0 \quad \text{for all other values of } i \text{ and } j$$

This solution is feasible, since each index from 1 to 12 occurs exactly once in the first position and exactly once in the second position. However, it is not an acceptable solution, since there are two disconnected subtours. We must design a way to eliminate disconnected routes from our set of feasible solutions.

We introduce $n - 1$ new variables $u_2, u_3, \ldots, u_n$ and $(n - 1)^2 - (n - 1)$ new constraints. The constraints are

$$u_i - u_j + nx_{ij} \leq n - 1, \qquad i, j = 2, 3, \ldots, n, \qquad \text{and} \quad i \neq j \tag{5}$$

$$u_i \geq 0 \text{ and integral}, \qquad i = 2, 3, \ldots, n$$

$u_i \leq m, \ u_j \geq 2$

Before we had $2n$ constraints and n^2 variables; these variables had values either 0 or 1 and n of them x_{ii} were always 0. We now have

$$2n + (n - 1)^2 - (n - 1) = n^2 - n + 2$$

linear constraints and

$$n^2 + n - 1$$

integer-valued variables.

We now show that the constraints (3), (4), and (5) do not permit disconnected routes and still include all routes satisfying the original problem statement. First we assume there is a subtour; that is, the route leads back to C_1 before visiting all the cities. Then there must be another subtour, since each city is visited exactly once. This subtour will start and end at some city in the list C_2, C_3, ..., C_n; it will not include C_1; and it will include $r \le n - 1$ cities. The r variables x_{ij} which describe this subtour will be equal to 1. We add up the r constraints (5) which correspond to these nonzero x_{ij}. This new constraint is satisfied by any solution which satisfies (5). As we take the sum to form this new constraint, we have $-u_j$ when the route enters city C_j and $+u_j$ when it leaves. Since the route enters and leaves each of the r cities exactly once, the u_j's cancel out in the sum. Thus, the new constraint is

$$nr \le (n - 1)r$$

which is a contradiction to our assumption that there was a subtour of length $r \le n - 1$.

For example, if we had the subtour starting at C_4,

$$C_4 \rightarrow C_5 \rightarrow C_3 \rightarrow C_2 \rightarrow C_4$$

so that

$$x_{45} = x_{53} = x_{32} = x_{24} = 1$$

then we would form our new constraint by adding the constraints

$$u_4 - u_5 + nx_{45} \le n - 1$$
$$u_5 - u_3 + nx_{53} \le n - 1$$
$$u_3 - u_2 + nx_{32} \le n - 1$$
$$u_2 - u_4 + nx_{24} \le n - 1$$

and obtain

$$4n \le 4(n - 1)$$

We have now shown that constraints (3), (4), and (5) allow no subtours. Now we show that these constraints do not exclude any potential routes.

To do this we show that each u_i can be assigned a nonnegative integer value for any route and that these values satisfy the constraints given in (5).

Let t_i be the position in the route at which C_i is visited. Thus, $t_1 = 1$ for C_1. Now consider the route which starts $C_1 \to C_4 \to C_6 \to C_2 \to \cdots$, then $t_1 = 1$, $t_4 = 2$, $t_6 = 3$, $t_2 = 4$. Let $u_i = t_i$ for $i = 2, 3, \ldots, n$. We show that for each i and j, (5) holds. Either $x_{ij} = 1$ or $x_{ij} = 0$. If $x_{ij} = 1$, then C_j is visited immediately after C_i, so that

$$t_j = t_i + 1$$

Substituting into (5), we have

$$u_i - u_j + nx_{ij} = t_i - (t_i + 1) + n = n - 1$$

as we needed. If $x_{ij} = 0$, then since $u_i \leq n$ and $u_j \geq 2$, we have

$$u_i - u_j \leq n - 2 \leq n - 1$$

so that (5) holds.

We have shown that a model for the traveling salesman problem is

$$\text{Minimize} \quad z = \sum_{i=1}^{n} \sum_{j=1}^{n} c_{ij} x_{ij}$$

subject to

$$\sum_{i=1}^{n} x_{ij} = 1, \qquad j = 1, 2, \ldots, n$$

$$\sum_{j=1}^{n} x_{ij} = 1, \qquad i = 1, 2, \ldots, n$$

$$u_i - u_j + nx_{ij} \leq n - 1, \qquad i, j = 2, 3, \ldots, n, \qquad \text{and} \quad i \neq j$$

$$x_{ij} = 0 \text{ or } 1, \qquad i, j = 1, 2, \ldots, n$$

$$u_i \geq 0 \text{ and integral}, \qquad i = 2, 3, \ldots, n$$

EXAMPLE 5. (Stock Cutting Problem) A plumber can buy plastic pipe in 6-foot and 12-foot lengths. The current job requires eight 4-foot lengths, five 5-foot lengths, and three 7-foot lengths. The plumber wants to figure out how many of each of the two stock lengths should be bought to minimize waste.

We determine all the possible ways the stock lengths can be cut to yield the necessary lengths. A 6-foot piece can be cut to give

One 4-foot length and 2 feet scrap   (Cutting pattern 1)
One 5-foot length and 1 foot scrap   (Cutting pattern 2)

A 12-foot piece can be cut to give

One 4-foot piece and 8 feet scrap	(Cutting pattern 3)
Two 4-foot pieces and 4 feet scrap	(Cutting pattern 4)
Three 4-foot pieces	(Cutting pattern 5)
One 4-foot piece, one 5-foot piece, and 3 feet scrap	(Cutting pattern 6)
One 4-foot piece, one 7-foot piece, and 1 foot scrap	(Cutting pattern 7)
One 5-foot piece and 7 feet scrap	(Cutting pattern 8)
Two 5-foot pieces and 2 feet scrap	(Cutting pattern 9)
One 7-foot piece and 5 feet scrap	(Cutting pattern 10)
One 7-foot piece and one 5-foot piece	(Cutting pattern 11)

Let piece 1 be of length $l_1 = 4$ feet, let piece 2 be of length $l_2 = 5$ feet, and let piece 3 be of length $l_3 = 7$ feet.

Let

a_{ij} = Number of pieces of length l_i in cutting pattern j

b_i = Number of pieces of length l_i which are needed

c_j = Waste in cutting pattern j

x_j = Number of times cutting pattern j is used

Our mathematical model is

$$\text{Minimize} \quad z = \mathbf{c}^T \mathbf{x}$$

subject to

$$\mathbf{Ax} = \mathbf{b}$$

$$\mathbf{x} \geq \mathbf{0} \text{ and integral}$$

where

$$\mathbf{A} = \begin{bmatrix} 1 & 0 & 1 & 2 & 3 & 1 & 1 & 0 & 0 & 0 & 0 \\ 0 & 1 & 0 & 0 & 0 & 1 & 0 & 1 & 2 & 0 & 1 \\ 0 & 0 & 0 & 0 & 0 & 0 & 1 & 0 & 0 & 1 & 1 \end{bmatrix}$$

$$\mathbf{c}^T = \begin{bmatrix} 2 & 1 & 8 & 4 & 0 & 3 & 1 & 7 & 2 & 5 & 0 \end{bmatrix}$$

and

$$\mathbf{b} = \begin{bmatrix} 8 \\ 5 \\ 3 \end{bmatrix}$$

EXAMPLE 6. (**Fixed Charge Problem**) A manufacturing corporation makes n products, and naturally the board of directors wants to minimize manufacturing costs. Each unit of product j which is made costs c_j

dollars to produce (raw materials, labor, direct machine costs, etc.). Moreover, if any units of product j are made, there is a fixed cost of k_j dollars, which represents the initial cost of setting up the production and distribution process.

Let x_j be the number of units of product j that are made. Suppose that the production process is constrained by a system of inequalities such as that in Exercises 1 or 2 in Section 2.1.

Our objective function is

$$\text{Minimize} \quad z = \sum_{j=1}^{n} (c_j x_j + k_j y_j)$$

where

$$y_j = \begin{cases} 1 & \text{if} \quad x_j > 0 \\ 0 & \text{if} \quad x_j = 0 \end{cases} \tag{6}$$

The constraints in (6) are nonlinear functions of x_j and y_j. These constraints are not defined by hyperplanes as they must be for a linear programming problem.

However, this problem can be cast as a problem where some of the variables are restricted to be integers and others may take any value. Such problems are called **mixed integer programming problems**. In this form the new objective function for our problem will be linear.

Suppose we know an upper bound on the number of units of x_j which can be produced. That is, suppose we know numbers M_j such that

$$x_j \leq M_j, \qquad j = 1, 2, \ldots, n$$

We may reformulate the definition of y_j in (6) as

$$\left. \begin{array}{c} y_j \geq \dfrac{x_j}{M_j} \\[2mm] y_j = 0 \text{ or } 1 \end{array} \right\} \tag{7}$$

If $x_j > 0$, then $y_j \geq x_j/M_j > 0$ implies that $y_j = 1$. If $x_j = 0$, then $y_j \geq x_j/M_j \geq 0$, so that $y_j = 0$ or 1. But since we are minimizing, the objective function will be smaller if $y_j = 0$. Therefore, at the minimum value of the objective function, if $x_j = 0$, then $y_j = 0$. The constraints given by (7) are now linear. We combine (7) with those constraints describing the production process to obtain our mixed integer programming model.

EXAMPLE 7.   (Either-Or Problem)   Suppose that we have a situation in which either the constraint

$$\sum_{j=1}^{n} a_{1j} x_j \leq b_1 \tag{8}$$

or the constraint

$$\sum_{j=1}^{n} a_{2j}x_j \le b_2 \tag{9}$$

holds. We can convert this condition to one in which the constraints are linear if we have available numbers M_1 and M_2 such that

$$\sum_{j=1}^{n} a_{1j}x_j - b_1 \le M_1$$

and

$$\sum_{j=1}^{n} a_{2j}x_j - b_2 \le M_2$$

for all feasible solutions.

Let y be a zero-one variable. Consider the problem

$$\sum_{j=1}^{n} a_{1j}x_j - b_1 \le M_1 y \tag{10}$$

$$\sum_{j=1}^{n} a_{2j}x_j - b_2 \le M_2(1 - y) \tag{11}$$

$$y = 0 \quad \text{or} \quad 1$$

If $y = 0$ in our new problem constraint, then (10) is the same as constraint (8) and constraint (11) is redundant, since it holds for all feasible solutions. If $y = 1$, then constraint (11) is the same as constraint (9), and constraint (10) is redundant.

We now examine a general integer programming problem and describe some methods for solving it.

We consider the following problem:

$$\text{Maximize} \quad z = c^T x$$

$$\text{subject to}$$

$$Ax \le b$$

$$x \ge 0$$

$$x_j \quad \text{integer if} \quad j \in I$$

where I is a subset of $\{1, 2, \ldots, n\}$. If $I = \{1, 2, \ldots, n\}$, then the problem is called a **pure integer programming problem**. If I is a proper subset of $\{1, 2, \ldots, n\}$, then the problem is called a **mixed integer programming problem**. In a pure integer programming problem every variable is required to be an integer. In a mixed integer programming problem only some of the variables are required to have integer values. Examples 6 and 7 are mixed integer programming problems. Examples 1–5 are pure integer programming problems. In Examples 2 and 3 the variables are restricted to the values 0 or 1.

One might attempt to solve an integer programming problem by treating it as a linear programming problem (that is, by not restricting the variables to integer values) and then rounding the answer to the nearest integer. Under extremely fortunate circumstances one might not have to round at all. But there are other situations where rounding will produce an incorrect answer.

EXAMPLE 8.   Consider the integer programming problem

$$\text{Maximize} \quad z = 7x + 8y$$

subject to

$$10x + 3y \le 52$$

$$2x + 3y \le 18$$

$$x \ge 0, \quad y \ge 0, \quad \text{integers}$$

If we ignore the restriction that x and y are integers, the simplex method gives the solution (verify)

$$x = 4\tfrac{1}{4}, \qquad y = 3\tfrac{1}{6}$$

with optimal value

$$z = 55\tfrac{1}{12}$$

If we round the values of x and y to the nearest integer values which are feasible, we get

$$x = 4, \qquad y = 3$$

and

$$z = 52$$

However, the solution

$$x = 3, \qquad y = 4$$

is also feasible, and the value of the objective function for this solution is

$$z = 53$$

4.1 Exercises

In Exercises 1–6 formulate the given problem as an integer programming problem.

1. **(Equipment purchasing problem)** A ribbon manufacturer is considering the purchase of two different types of printing machines which will be used to emboss designs on the ribbon. Machine A can print 100 meters per minute and requires 50 square meters of floor space, while machine B can print 200 meters per minute and requires 140 square meters of floor space. Suppose the manufacturer must print at least 600 meters per minute and has no more than 350 square meters of floor space. If a model A machine costs $22,000 and a model B machine costs $48,000, how many machines of each type should be bought to minimize the cost?

2. **(A production problem)** A chair manufacturer makes three different types of chairs, each of which must go through sanding, staining, and varnishing. In addition, the model with the vinyl-covered back and seat must go through an upholstering process. The following table gives the time required for each operation on each type of chair, the available time for each operation in hours per month, and the profit per chair for each model. How many chairs of each type should be made to maximize the total profit?

Model	Sanding (hours)	Staining (hours)	Varnishing (hours)	Upholstering (hours)	Profit ($)
A—solid back and seat	1.0	0.5	0.7	0	10
B—ladder back, solid seat	1.2	0.5	0.7	0	13
C—vinyl-covered back and seat	0.7	0.3	0.3	0.7	8
Total time available per month	600	300	300	140	

3. Pam Hardy currently has six favorite country and western songs. There are ten records available which contain different groups of these songs. Suppose the jth record costs c_j dollars. Set up a model which Pam could use to determine the cheapest selection of records to buy to get at least one version of each of her favorite songs.

4. Tommy Jones's mother is planning his tenth birthday party and will serve a variety of soft drinks which will be chosen from the list below.

Drink	Cola	Root beer	Cherry	Lemon	Orange	Grape	Ginger ale
Price per bottle (in cents)	69	59	62	62	65	55	65

From past experience it has been determined that at least 12 bottles of soft drinks are needed. Also, at least two bottles of ginger ale and at least two bottles of cola and no more than three fruit-flavored soft drinks are needed. How many bottles of each type should be bought to minimize the total cost?

5. A manager for a large corporation must prepare a list of projects which her group will complete over the next year. She has under consideration ten such projects but will not be able to do all of them because of limits on personnel and budget. She has assigned a weight to each project which represents to her the value of completing the project. The personnel, capital requirements, and weights for each project are given in the following table.

Project:	1	2	3	4	5	6	7	8	9	10
Person-weeks	250	195	200	70	30	40	100	170	40	120
Cost (in thousands of dollars)	400	300	350	100	70	70	250	250	100	200
Value of completion	70	50	60	20	10	20	30	45	10	40

The manager has available 1000 person-weeks and \$1,500,000 to allocate among the projects. Which projects should she choose to complete to maximize the value?

6. Each day at the Graphic Arts Co. the press operator is given a list of jobs to be done during the day. He must determine the order in which he does the jobs based on the amount of time it takes to change from one job setup to the next. Clearly he will arrange the jobs in an order which minimizes the total setup time. Assume that each day he starts the press from a rest state and returns it to that state at the end of the day. Suppose on a particular day he must do six jobs for which he estimates the changeover times given in the following table. What schedule of jobs should the operator use?

		To job j (minutes)						
	i \ j	1	2	3	4	5	6	rest
	1	0	10	5	15	10	20	5
	2	10	0	10	10	20	15	10
	3	5	5	0	5	10	10	15
From job i	4	8	10	3	0	9	14	10
	5	4	7	8	6	0	10	10
	6	10	5	10	15	10	0	8
	rest	7	7	9	12	10	8	0

4.1 Projects

1. Meg Watkins is trying to decide which college to attend. From among the applications she submitted she has been admitted to four schools. One is a large state university which is about 250 miles from her home. At this school she may live in a dormitory for two years, but then must find accommodations in the community. Another is a small private school about 1000 miles from her home which has an excellent reputation. At this school there are dormitory accommodations for all students. Meg, under pressure from her father, also applied to and was accepted by the private church-related school in her home town. Since she is a local student, the school would expect her to live at home.

Another possibility open to Meg is to go to the local community college for two years and then transfer to another school. The reputation of the community college has been improving over the last few years. The state university would carry forward her acceptance for two years and transfer all credits. The distant private school will also carry forward her acceptance but most likely will transfer 9 credits less than two full years of credit. The local private school has no formal statement on transfer from two-year schools.

The accompanying table gives the costs for attending each school and Meg's assessment of the worth (or utility) of having certain blocks of credit from each school.

	State	Distant private	Local private	Community college
Tuition	$700/year	$4000/year	$3000/year	—
Living				
On-campus	$2000/year	$2000/year	$50/month	$50/month
Off-campus	$1800/year			
Humanities	7	9	6	6
Social science	6	8	5	6
Science	8	5	4	3
Major	8	10	6	—

Set up a model which Meg could use to maximize the future worth of her education assuming that she can earn $2000/summer and that her father will provide $2500/year for her education and living expenses. You may wish to consider allowing Meg the option of working while going to school or going to summer school for certain courses.

2. Consider the problem of making change in a supermarket. Suppose that the cash register shows that your change is to be C cents. Initially, we assume $C < 100$. The coins that are available to make change have values

$$w_1 = 1, \quad w_2 = 5, \quad w_3 = 10, \quad w_4 = 25, \quad w_5 = 50$$

(a) Set up an integer programming problem for finding which combination of coins yields the correct change using the smallest number of coins.

(b) Construct a table giving the solution to part a for $C = 1, 2, \ldots, 99$.

(c) One algorithm for change making calls for giving as many of the largest-denomination coins as possible, then using the next-largest-denomination for the remaining amount, and so on. Does this algorithm give the correct answer for each $C = 1, 2, \ldots, 99$?

(d) Suppose our monetary system had coins with values

$$w_1 = 1, \quad w_2 = 5, \quad w_3 = 20, \quad w_4 = 25$$

Use the algorithm in part c to make up $C = 40$. Is this a minimal solution?

(e) Consider the problem of giving C dollars in change where $C = 1, 2, \ldots, 9$. Would it be advantageous to use a \$2 bill in addition to the \$1 and \$5 bills?

(f) What other situations have models similar to the change-making problem?

Further Reading

Chang, S. K., and A. Gill. "Algorithmic Solution of the Change-Making Problem." *Journal of the ACM,* 17 (1970), 113–22.

4.2 CUTTING PLANE METHODS

In this section we discuss one approach which has been used to solve integer programming problems. The algorithms for such problems are not as nice as those for linear programming problems in the sense that there is not one algorithm that works well for all integer programming problems. Among the difficulties with these algorithms is their inefficiency for even medium-sized problems. For example, computations for the traveling salesman problem (Example 4 in Section 4.1) become prohibitively long for over 200 cities.

Consider the pure integer programming problem

$$\text{Maximize} \quad z = c^T x \tag{1}$$

subject to

$$Ax = b \tag{2}$$

$$x \geq 0, \quad \text{integral} \tag{3}$$

where A is an $m \times n$ matrix, c is an $n \times 1$ column vector, and b is an $m \times 1$ column vector.

The **cutting plane algorithms** were developed by Ralph Gomory in 1958. The idea behind the algorithms is to start with the problem given by (1), (2), and (3), ignore the restriction that x must be a vector with integer components, and solve the resulting problem by the simplex method. If the resulting solution is integral, we are finished. Otherwise, we add a new constraint that "cuts off" (eliminates) some nonintegral solutions, including the one just obtained by the simplex method. However, the new constraint is carefully constructed so as not to eliminate any feasible integer solutions. We solve the new problem and repeat the simplex algorithm. By adding enough constraints, we eventually reach an optimal integer solution.

Suppose the problem has a feasible solution and that a finite optimal value exists. We may assume that $\mathbf{A}$, $\mathbf{B}$, and $\mathbf{c}$ have integer entries. The ith constraint as it appears in the final simplex tableau for the related linear programming problem is

$$\sum_{j=1}^{n} t_{ij} x_j = x_{Bi} \qquad (4)$$

where x_{Bi} is the value of the ith basic variable for the optimal solution of the related linear programming problem. We denote by $[a]$ the integer part of a. That is, $[a]$ is the largest integer K such that $K \leq a$. For example,

$$[\tfrac{3}{2}] = 1, \qquad [\tfrac{3}{5}] = 0$$
$$[-\tfrac{2}{3}] = -1, \qquad [-3\tfrac{1}{3}] = -4$$
$$[3] = 3, \qquad [-2] = -2$$

Since $[t_{ij}] \leq t_{ij}$ and $x_j \geq 0$, we can write from (4)

$$\sum_{j=1}^{n} [t_{ij}] x_j \leq x_{Bi} \qquad (5)$$

Any integer vector $\mathbf{x}$ which satisfies (4) must also satisfy (5). For such $\mathbf{x}$ the left-hand side of (5) is an integer. Thus, we may write the constraint (6) which $\mathbf{x}$ must also satisfy:

$$\sum_{j=1}^{n} [t_{ij}] x_j \leq [x_{Bi}] \qquad (6)$$

We can transform (6) into an equation by introducing the slack variable u_i:

$$\sum_{j=1}^{n} [t_{ij}] x_j + u_i = [x_{Bi}] \qquad (7)$$

Since u_i is the difference between two integers, we may require that u_i be an integer. We have shown that any feasible solution to the given integer programming problem will also satisfy (7) for some value of u_i.

Now assume that x_{Bi} is not an integer. We may write

$$[t_{ij}] + g_{ij} = t_{ij}$$

and

$$[x_{Bi}] + f_i = x_{Bi},$$

where $0 \le g_{ij} < 1$ and $0 < f_i < 1$. Thus, if

$$t_{ij} = \tfrac{9}{7}, \quad \text{then} \quad g_{ij} = \tfrac{2}{7}$$

and if

$$t_{ij} = -\tfrac{2}{3}, \quad \text{then} \quad g_{ij} = \tfrac{1}{3}$$

If we subtract (4) from (7), we have

$$\sum_{j=1}^{n} (-g_{ij})x_j + u_i = -f_i \tag{8}$$

Equation (8) is the **cutting plane constraint** to be added to the constraints in (2). We now proceed as follows:

Step 1. Solve the related linear programming problem obtained from the given integer programming problem by dropping the integrality requirements. If the solution is a vector with integer components, stop. Otherwise, go to Step 2.

Step 2. Generate a cutting plane constraint as in (8). A heuristic rule to use for choosing the constraint in the cutting plane construction is: choose the constraint in the final simplex tableau which gives the largest f_i.

Step 3. Consider the new integer programming problem which consists of the same objective function, the constraints in (2) and the cutting plane (8) which have been added in Step 2. Return to Step 1.

Since the coefficients of all the basic variables except the ith one in the list will be zero in (4), we may rewrite (4) as

$$x_{r_i} + \sum_{j \in N} t_{ij}x_j = x_{Bi} \tag{9}$$

where N is the set of indices of the nonbasic variables and where the variable labeling this ith constraint is x_{r_i}. Suppose that x_{Bi} is not an integer. The Gomory cutting plane obtained from (9) is

$$\sum_{j \in N} (-g_{ij})x_j + u_i = -f_i \tag{10}$$

where

$$[t_{ij}] + g_{ij} = t_{ij} \quad \text{and} \quad [x_{Bi}] + f_i = x_{Bi}$$

If we set the nonbasic variables x_j equal to zero in (10), we obtain $u_i = -f_i < 0$. Since the slack variable u_i is negative, we see that the optimal solution to the related linear programming problem whose ith constraint is given by (9) is no longer a feasible solution to the new linear programming problem with the added constraint (10). We have "cut off" the current optimal solution. The dual simplex method can now be used to solve the new problem; it will remove the infeasibility caused by adding the constraint (10).

We have shown that the cutting plane method fulfills the two criteria. It does not remove any integer vectors from the set of feasible solutions, but it does remove noninteger optimal solutions.

EXAMPLE 1. Consider the pure integer programming problem

$$\text{Maximize} \quad z = 5x_1 + 6x_2$$

subject to

$$10x_1 + 3x_2 \le 52$$

$$2x_1 + 3x_2 \le 18$$

$$x_1 \ge 0, \quad x_2 \ge 0, \quad \text{integers}$$

The final simplex tableau for the related linear programming problem is given in Tableau 4.1. Since the solution is nonintegral, we must add a

Tableau 4.1

c_B		5	6	0	0	
		x_1	x_2	x_3	x_4	x_B
5	x_1	1	0	$\frac{1}{8}$	$-\frac{1}{8}$	$\frac{17}{4}$
6	x_2	0	1	$-\frac{1}{12}$	$\frac{5}{12}$	$\frac{19}{6}$
		0	0	$\frac{1}{8}$	$\frac{15}{8}$	$\frac{161}{4}$

cutting plane constraint. We find $f_1 = \frac{1}{4}$, $f_2 = \frac{1}{6}$, so that we choose the first row for constructing the cutting plane. We have

$$[\tfrac{1}{8}] + \tfrac{1}{8} = \tfrac{1}{8}$$

and

$$[-\tfrac{1}{8}] + \tfrac{7}{8} = -\tfrac{1}{8}$$

The Gomory cutting plane is then

$$-\tfrac{1}{8}x_3 - \tfrac{7}{8}x_4 + u_1 = -\tfrac{1}{4}$$

Our new tableau is Tableau 4.2. We must now use the dual simplex method on Tableau 4.2 to restore feasibility. We obtain Tableau 4.3.

Since the solution is still nonintegral, we formulate another Gomory

Tableau 4.2

↓

c_B		5 x_1	6 x_2	0 x_3	0 x_4	0 u_1	$\mathbf{x_B}$
5	x_1	1	0	$\frac{1}{8}$	$-\frac{1}{8}$	0	$\frac{17}{4}$
6	x_2	0	1	$-\frac{1}{12}$	$\frac{5}{12}$	0	$\frac{19}{6}$
← 0	u_1	0	0	$\left(-\frac{1}{8}\right)$	$-\frac{7}{8}$	1	$-\frac{1}{4}$
		0	0	$\frac{1}{8}$	$\frac{15}{8}$	0	$\frac{161}{4}$

Tableau 4.3

c_B		5 x_1	6 x_2	0 x_3	0 x_4	0 u_1	$\mathbf{x_B}$
5	x_1	1	0	0	-1	1	4
6	x_2	0	1	0	1	$-\frac{2}{3}$	$\frac{10}{3}$
0	x_3	0	0	1	7	-8	2
		0	0	0	1	1	40

cutting plane using the x_2 row. We have

$$[1] + 0 = 1$$

$$[-\tfrac{2}{3}] + \tfrac{1}{3} = -\tfrac{2}{3}$$

$$[\tfrac{10}{3}] + \tfrac{1}{3} = \tfrac{10}{3}$$

The cutting plane equation is

$$0x_4 - \tfrac{1}{3}u_1 + u_2 = -\tfrac{1}{3}$$

We show the new tableau in Tableau 4.4.

Tableau 4.4

↓

c_B		5 x_1	6 x_2	0 x_3	0 x_4	0 u_1	0 u_2	$\mathbf{x_B}$
5	x_1	1	0	0	-1	1	0	4
6	x_2	0	1	0	1	$-\frac{2}{3}$	0	$\frac{10}{3}$
0	x_3	0	0	1	7	-8	0	2
← 0	u_2	0	0	0	0	$\left(-\frac{1}{3}\right)$	1	$-\frac{1}{3}$
		0	0	0	1	1	0	40

Tableau 4.5

c_B		5	6	0	0	0	0	
		x_1	x_2	x_3	x_4	u_1	u_2	x_B
5	x_1	1	0	0	-1	0	3	3
6	x_2	0	1	0	1	0	-2	4
0	x_3	0	0	1	7	0	-24	10
0	u_1	0	0	0	0	1	-3	1
		0	0	0	1	0	3	39

Using the dual simplex method on Tableau 4.4, we obtain Tableau 4.5. We have obtained the optimal integer solution

$$x_1 = 3, \quad x_2 = 4, \quad z = 39$$

Note that there is no slack in the fourth constraint.
    The first cutting plane

$$-\tfrac{1}{8}x_3 - \tfrac{7}{8}x_4 + u_1 = -\tfrac{1}{4}$$

can be expressed in terms of x_1 and x_2 by substituting

$$x_3 = 52 - 10x_1 - 3x_2$$

$$x_4 = 18 - 2x_1 - 3x_2$$

which come from the constraints of the original problem. We obtain

$$3x_1 + 3x_2 + u_1 = 22 \tag{11}$$

or

$$x_1 + x_2 \le \tfrac{22}{3}, \tag{12}$$

The second cutting plane

$$-\tfrac{1}{3}u_1 + u_2 = -\tfrac{1}{3}$$

gives

$$x_1 + x_2 \le 7 \tag{13}$$

In this calculation we used (11) to write u_1 in terms of x_1 and x_2. We sketch the original feasible region and the cutting planes in Figure 4.1.

    Several remarks are in order. For a linear programming problem the number of positive components in the solution vector is $\le m$, where A is $m \times s$. For an integer programming problem this is no longer true. In fact, the optimal value may not occur at an extreme point of the convex region defined by the constraints.
    The cutting plane algorithm which we just gave has several major

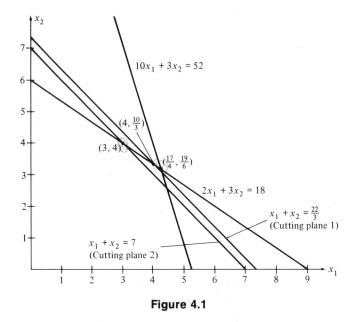

Figure 4.1

drawbacks. The integral answer does not appear until the very last step. This is unlike the simplex method where we can stop at any stage and obtain a feasible solution which is better than the previous feasible solutions. In the Gomory algorithm, if we stop before the end, we have no integral solution. A more recently developed method keeps all computations as integers and alleviates this difficulty. Computational difficulties with the cutting plane algorithm have also been encountered. There are some small problems which have taken an exceedingly large number of cutting planes to produce an integer answer.

Gomory's Method for Mixed Integer Programming

We can attempt to solve a mixed integer programming problem in the same way as a pure integer programming problem. We use the simplex method to obtain the optimal solution to the related linear programming problem. This will be a solution to the mixed integer programming problem if those variables which appear in the basis for the optimal solution that are required to have integer values actually do have such values. We suppose now that x_{r_i} is the ith basic variable in the list of basic variables for the optimal solution of the related linear programming problem, that x_{r_i} is required to be integral, and that x_{Bi} is not an integer. We saw that we may write the ith constraint (9) as

$$x_{r_i} + \sum_{j \text{ nonbasic}} t_{ij}x_j = x_{Bi} \tag{14}$$

Let $x_{Bi} = [x_{Bi}] + f_i$. We may write (14) as

$$x_{r_i} + \sum_{\substack{j \text{ nonbasic}}} t_{ij}x_j = [x_{Bi}] + f_i$$

or

$$\sum_{j \in N} t_{ij}x_j = f_i + ([x_{Bi}] - x_{r_i}) \tag{15}$$

We now divide the set of indices for the nonbasic variables into two subsets N^+ and N^-, where

$$N^+ = \{j \mid j \text{ is the index of a nonbasic variable and } t_{ij} \geq 0\}$$

$$N^- = \{j \mid j \text{ is the index of a nonbasic variable and } t_{ij} < 0\}$$

Then (15) may be written as

$$\sum_{j \in N^+} t_{ij}x_j + \sum_{j \in N^-} t_{ij}x_j = f_i + ([x_{Bi}] - x_{r_i}) \tag{16}$$

Using (16), we want to derive a cutting plane constraint that will cut off the current optimal solution to the related linear programming problem because in that solution x_{r_i} does not have an integer value. The cutting plane constraint must also have the property, as before, that no feasible solutions for the mixed integer problem should be cut off. We can interpret this condition as saying that if x satisfies (16) and x_{r_i} is an integer, then x must satisfy the cutting plane constraint.

We consider two cases: the right-hand side of (16) is positive, or it is negative. Assume, first, that $f_i + ([x_{Bi}] - x_{r_i}) < 0$. The quantity in parentheses is an integer by assumption and $0 < f_i < 1$. We see that $[x_{Bi}] - x_{r_i}$ must be a negative integer for all these assumptions to hold.

Thus, the largest value that the right-hand side of (16) can have while still remaining negative is $f_i - 1$. Our original constraint (14) implies, in this case,

$$\sum_{j \in N^+} t_{ij}x_j + \sum_{j \in N^-} t_{ij}x_j \leq f_i - 1 \tag{17}$$

We can make the left-hand side of (17) smaller by removing the first sum, since it represents a nonnegative quantity. We obtain

$$\sum_{j \in N^-} t_{ij}x_j \leq f_i - 1$$

Dividing by $f_i - 1$ and reversing the inequality, since $f_i - 1$ is negative, and then multiplying by f_i yields

$$\sum_{j \in N^-} \frac{f_i}{f_i - 1} t_{ij}x_j \geq f_i \tag{18}$$

Since the first sum in (17) is nonnegative, we may add it to (18) to obtain

$$\sum_{j \in N^+} t_{ij}x_j + \frac{f_i}{f_i - 1} \sum_{j \in N^-} t_{ij}x_j \geq f_i \tag{19}$$

For the other case, we assume $f_i + ([x_{Bi}] - x_{r_i}) \geq 0$. Using similar reasoning as in the first case, we see that $[x_{Bi}] - x_{r_i}$ must be nonnegative. Thus, our original constraint (14) implies

$$\sum_{j \in N^+} t_{ij}x_j + \sum_{j \in N^-} t_{ij}x_j \geq f_i \tag{20}$$

We may replace the second sum in (20) by any larger quantity and still maintain the inequality. We have

$$\sum_{j \in N^-} t_{ij}x_j \leq 0 \leq \frac{f_i}{f_i - 1} \sum_{j \in N^-} t_{ij}x_j$$

Consequently, (20) implies

$$\sum_{j \in N^+} t_{ij}x_j + \frac{f_i}{f_i - 1} \sum_{j \in N^-} t_{ij}x_j \geq f_i$$

which is the same as (19).

We have now shown that if (14) is a constraint in a mixed integer problem whose corresponding basic variable is supposed to have an integer value but does not, then (19) is satisfied by every vector x which satisfies (14), assuming x_{r_i} is an integer. From (19) we can construct the equation of the cutting plane. We reverse the inequality in (19) and add in a slack variable u_i, obtaining

$$-\sum_{j \in N^+} t_{ij}x_j - \frac{f_i}{f_i - 1} \sum_{j \in N^-} t_{ij}x_j + u_i = -f_i \tag{21}$$

The equation (21) is the Gomory cutting plane. For the current optimal solution, $x_j = 0$ if $j \in N^+$ or $j \in N^-$. Therefore, introducing the constraint (21) gives

$$u_i = -f_i < 0$$

for the current optimal solution, and hence it has been cut off, since u_i must be nonnegative.

Note that (21) may be written as

$$u_i = -f_i + \sum_{j \in N} d_j x_j$$

where

$$
d_j = \begin{cases} t_{ij} & \text{if} \quad t_{ij} \geq 0 \\[2mm] \dfrac{f_i}{f_i - 1} t_{ij} & \text{if} \quad t_{ij} < 0 \end{cases} \tag{22}
$$

It is possible that some of the nonbasic variables are also required to be integer-valued. We can refine the cutting plane defined by (21) if we use this information. The refinement will simply cut off more solutions that do not satisfy the integrability conditions. We write

$$
t_{ij} = [t_{ij}] + g_{ij}
$$

Then instead of the definitions of d_j given in (22), we have for the refined cutting plane

$$
d_j = \begin{cases} t_{ij} & \text{if} \quad t_{ij} \geq 0 \quad \text{and} \quad x_j \text{ may be nonintegral} \\[2mm] \dfrac{f_i}{f_i - 1} t_{ij} & \text{if} \quad t_{ij} < 0 \quad \text{and} \quad x_j \text{ may be nonintegral} \\[2mm] g_{ij} & \text{if} \quad g_{ij} \leq f_i \quad \text{and} \quad x_j \text{ must be integral} \\[2mm] \dfrac{f_i}{f_i - 1}(g_{ij} - 1) & \text{if} \quad g_{ij} > f_i \quad \text{and} \quad x_j \text{ must be integral} \end{cases} \tag{23}
$$

EXAMPLE 2.   Consider the mixed integer programming problem

$$
\text{Maximize} \quad z = 5x_1 + 6x_2
$$

subject to

$$
10x_1 + 3x_2 \leq 52
$$
$$
2x_1 + 3x_2 \leq 18
$$
$$
x_1 \geq 0, \quad \text{integral}
$$
$$
x_2 \geq 0
$$

The final simplex tableau for the related linear programming problem is the same as in Example 1 and is given in Tableau 4.1. Since the value for x_1 is not an integer, we must add a cutting plane constraint. The nonbasic variables are x_3 and x_4, the slack variables. Neither is constrained to be an integer. We therefore use (21) to define our cutting plane. In this case, $i = 1$ and

$$
x_{Bi} = 4 + \tfrac{1}{4}
$$

so that $f_1 = \frac{1}{4}$. From Tableau 4.1 we see that $N^+ = \{3\}$ and $N^- = \{4\}$. The cutting plane is

$$-t_{13}x_3 - \frac{f_1}{f_1 - 1} t_{14}x_4 + u_1 = -f_1$$

or

$$-\tfrac{1}{8}x_3 - \frac{\frac{1}{4}}{-\frac{3}{4}}(-\tfrac{1}{8})x_4 + u_1 = -\tfrac{1}{4}$$

or

$$-\tfrac{1}{8}x_3 - \tfrac{1}{24}x_4 + u_1 = -\tfrac{1}{4}$$

Putting this constraint into Tableau 4.1, we get Tableau 4.6.

Tableau 4.6

		5	6	0	0	0	
c_B		x_1	x_2	x_3	x_4	u_1	x_B
5	x_1	1	0	$\frac{1}{8}$	$-\frac{1}{8}$	0	$\frac{17}{4}$
6	x_2	0	1	$-\frac{1}{12}$	$\frac{5}{12}$	0	$\frac{19}{6}$
0	u_1	0	0	$-\frac{1}{8}$	$-\frac{1}{24}$	1	$-\frac{1}{4}$
		0	0	$\frac{1}{8}$	$\frac{15}{8}$	0	$\frac{161}{4}$

We use the dual simplex method to restore feasibility in Tableau 4.7. The solution which Tableau 4.7 represents,

$$x_1 = 4, \quad x_2 = \tfrac{10}{3}, \quad z = 40$$

satisfies the integrality conditions so that it is an optimal solution to the given mixed integer programming problem.

Tableau 4.7

		5	6	0	0	0	
c_B		x_1	x_2	x_3	x_4	u_1	x_B
5	x_1	1	0	0	$-\frac{1}{6}$	1	4
6	x_2	0	1	0	$\frac{4}{9}$	$-\frac{2}{3}$	$\frac{10}{3}$
0	x_3	0	0	1	$\frac{1}{3}$	-8	2
		0	0	0	$\frac{11}{6}$	1	40

EXAMPLE 3. Consider the mixed integer programming problem

$$\text{Maximize} \quad z = 4x_1 + 5x_2 + 3x_3$$

subject to

$$3x_1 \qquad + 4x_3 \le 10$$

$$2x_1 + x_2 + x_3 \le 7$$

$$3x_1 + 4x_2 + x_3 \le 12$$

$$x_1 \ge 0, \quad x_3 \ge 0, \quad \text{integral} \quad ,x_4 \ integer$$

$$x_2 \ge 0$$

The final tableau of the related linear programming problem is Tableau 4.8 (verify). Since x_1 and x_3 are constrained to be integer-valued, the solution represented by Tableau 4.8 is not feasible for the given mixed integer problem. We introduce a cutting plane from the first constraint in Tableau 4.8 ($i = 1$). Note that x_4 must also be integer-valued, since it is the difference of two integers. The set of indices of nonbasic variables is $N = \{1, 4, 6\}$. We calculate the value of d_j using (23) for each of these indices. The fractional part of $x_3 = \frac{5}{2}$ is $f_1 = \frac{1}{2}$.

$$X_{B_1} = \frac{5}{2} = [X_{B_1}] + f_1 = 2 + \frac{1}{2}$$

Tableau 4.8

		4	5	3	0	0	0	
c_B		x_1	x_2	x_3	x_4	x_5	x_6	x_B
3	x_3	$\frac{3}{4}$	0	1	$\frac{1}{4}$	0	0	$\frac{5}{2}$
0	x_5	$\frac{11}{16}$	0	0	$-\frac{3}{16}$	1	$-\frac{1}{4}$	$\frac{17}{8}$
5	x_2	$\frac{9}{16}$	1	0	$-\frac{1}{16}$	0	$\frac{1}{4}$	$\frac{19}{8}$
		$\frac{17}{16}$	0	0	$\frac{7}{16}$	0	$\frac{5}{4}$	$\frac{155}{8}$

For $j = 1$, x_1 must be integer-valued, and we have [*non-basic*]

$$t_{11} = \frac{3}{4} = 0 + \frac{3}{4} \quad \text{or} \quad g_{11} = \frac{3}{4} > f_1$$

Therefore, $= \frac{f_1}{f_1 - 1}(g_{11} - 1)$

$$d_1 = \frac{\frac{1}{2}}{\frac{1}{2} - 1}\left(\frac{3}{4} - 1\right) = \frac{\frac{1}{2}}{-\frac{1}{2}}\left(-\frac{1}{4}\right) = \frac{1}{4}$$

For $j = 4$, x_4 must be integer-valued, and we have

$$[\tfrac{1}{4}] = 0 + \tfrac{1}{4} \quad \text{or} \quad g_{14} = \tfrac{1}{4} < f_1$$

Therefore,

$$d_4 = \tfrac{1}{4}$$

For $j = 6$, x_6 may have any value, $t_{16} = 0$, and therefore

$$d_6 = 0$$

The cutting plane constraint is

$$u_1 = -\tfrac{1}{2} + \tfrac{1}{4}x_1 + \tfrac{1}{4}x_4$$

or

$$-\tfrac{1}{4}x_1 - \tfrac{1}{4}x_4 + u_1 = -\tfrac{1}{2}$$

We add this constraint to Tableau 4.8 to get Tableau 4.9. This tableau represents an optimal but infeasible solution to the related linear programming problem. We apply the dual simplex method to restore feasibility. The result is shown in Tableau 4.10, which yields the optimal solution

$$x_1 = 0, \quad x_2 = \tfrac{5}{2}, \quad x_3 = 2, \quad z = \tfrac{37}{2}$$

Tableau 4.9

		4	5	3	0	0	0	0	
c_B		x_1	x_2	x_3	x_4	x_5	x_6	u_1	x_B
3	x_3	$\frac{3}{4}$	0	1	$\frac{1}{4}$	0	0	0	$\frac{5}{2}$
0	x_5	$\frac{11}{16}$	0	0	$-\frac{3}{16}$	1	$-\frac{1}{4}$	0	$\frac{17}{8}$
5	x_2	$\frac{9}{16}$	1	0	$-\frac{1}{16}$	0	$\frac{1}{4}$	0	$\frac{19}{8}$
0	u_1	$-\frac{1}{4}$	0	0	$\left(-\frac{1}{4}\right)$	0	0	1	$-\frac{1}{2}$
		$\frac{17}{16}$	0	0	$\frac{7}{16}$	0	$\frac{5}{4}$	0	$\frac{155}{8}$

Tableau 4.10

		4	5	3	0	0	0	0	
c_B		x_1	x_2	x_3	x_4	x_5	x_6	u_1	x_B
3	x_3	$\frac{1}{2}$	0	1	0	0	0	1	2
0	x_5	$\frac{7}{8}$	0	0	0	1	$-\frac{1}{4}$	$-\frac{3}{4}$	$\frac{5}{2}$
5	x_2	$\frac{5}{8}$	1	0	0	0	$\frac{1}{4}$	$-\frac{1}{4}$	$\frac{5}{2}$
0	x_4	1	0	0	1	0	0	-4	2
		$\frac{5}{8}$	0	0	0	0	$\frac{5}{4}$	$\frac{7}{4}$	$\frac{37}{2}$

4.2 Exercises

In Exercises 1 and 2 assume that every variable is constrained to be an integer. Using the given final simplex tableau, find the equation of the cutting plane.

1.

c_B		$\begin{matrix}2\\x_1\end{matrix}$	$\begin{matrix}3\\x_2\end{matrix}$	$\begin{matrix}1\\x_3\end{matrix}$	$\begin{matrix}0\\x_4\end{matrix}$	$\begin{matrix}0\\x_5\end{matrix}$	x_B
0	x_4	0	0	$-\frac{1}{8}$	1	$-\frac{1}{2}$	$\frac{5}{2}$
3	x_2	0	1	$\frac{1}{2}$	0	1	$\frac{7}{8}$
2	x_1	1	0	$\frac{3}{8}$	0	$\frac{1}{4}$	$\frac{13}{8}$
		0	0	$\frac{5}{4}$	0	$\frac{7}{2}$	$\frac{47}{8}$

(handwritten: $\frac{7}{8}$, $-\frac{1}{2} \times 8^{+1} = -\frac{7}{8}$, $\frac{5}{8}$)

2.

c_B		$\begin{matrix}2\\x_1\end{matrix}$	$\begin{matrix}3\\x_2\end{matrix}$	$\begin{matrix}4\\x_3\end{matrix}$	$\begin{matrix}1\\x_4\end{matrix}$	$\begin{matrix}0\\x_5\end{matrix}$	$\begin{matrix}0\\x_6\end{matrix}$	$\begin{matrix}0\\x_7\end{matrix}$	x_B
3	x_2	$-\frac{5}{24}$	1	0	0	$\frac{1}{12}$	$\frac{1}{2}$	$\frac{9}{25}$	2
1	x_4	$\frac{11}{24}$	0	0	1	$-\frac{1}{3}$	1	$-\frac{3}{4}$	$\frac{7}{24}$
4	x_3	1	0	1	0	$\frac{3}{4}$	$\frac{11}{12}$	$\frac{13}{12}$	$\frac{1}{3}$
		$\frac{11}{6}$	0	0	0	$\frac{35}{12}$	$\frac{37}{6}$	$\frac{29}{6}$	$\frac{183}{8}$

(handwritten: $\frac{7}{24}$, $\frac{1}{3}$, $-\frac{3}{4}x_5 - \frac{1}{2}x_6 - \frac{1}{2}x_7 + U_1 = -\frac{1}{3}$)

In Exercises 3 and 4 solve the given integer programming problem by the cutting plane method and sketch the graph of the set of feasible solutions and the cutting planes.

3. Maximize $z = x + y$

subject to

$$2x + 3y \le 12$$

$$2x + y \le 6$$

$$x \ge 0, \quad y \ge 0, \quad \text{integers}$$

4. Maximize $z = x + 4y$

subject to

$$x + 6y \le 36$$

$$3x + 8y \le 60$$

$$x \ge 0, \quad y \ge 0, \quad \text{integers}$$

In Exercises 5–12 solve the given integer programming problem using the cutting plane algorithm.

5. Maximize $z = 4x + y$

subject to

$$3x + 2y \leq 5$$
$$2x + 6y \leq 7$$
$$3x + 7y \leq 6$$
$$x \geq 0, \quad y \geq 0, \quad \text{integers}$$

6. Maximize $z = x_2 + 4x_3$

subject to

$$3x_1 - 6x_2 + 9x_3 \leq 9$$
$$3x_1 + 2x_2 + x_3 \leq 7$$
$$x_1 \geq 0, \quad x_2 \geq 0, \quad x_3 \geq 0, \quad \text{integers}$$

7. Maximize $z = 5x + 2y$

subject to

$$6x - 15y \leq 24$$
$$6x + 10y \leq 69$$
$$3x + 10y \leq 60$$
$$x \geq 0, \quad y \geq 0, \quad \text{integers}$$

8. Maximize $z = x_1 + 2x_2 + x_3 + x_4$

subject to

$$2x_1 + x_2 + 3x_3 + x_4 \leq 8$$
$$2x_1 + 3x_2 + 4x_4 \leq 12$$
$$3x_1 + x_2 + 2x_3 \leq 18$$
$$x_j \geq 0, \quad j = 1, 2, 3, 4, \quad \text{integers}$$

9. Repeat Exercise 3 under the assumption that only x must be an integer.

10. Repeat Exercise 6 under the assumption that only x_1 and x_3 must be integers.

11. Repeat Exercise 8 under the assumption that only x_1 and x_2 must be integers.

12. Repeat Exercise 6 under the assumption that only x_2 must be an integer.

4.3 BRANCH AND BOUND METHODS

If the set of feasible solutions to the related linear programming problem of a mixed integer programming problem is bounded, then the integer-valued variables can take on only finitely many values in this region. We had previously discussed solving a linear programming problem by enumerating the extreme points of the set of feasible solutions and then choosing an optimal solution from among these. We dismissed this brute force method in favor of the simplex algorithm which listed only some of the extreme points and chose an order for the list in which the value of the objective function improved each time. However, using cutting planes and the simplex algorithm can involve very lengthy computations. Thus, it may be advantageous to again consider some enumerating technique.

We examine the possibility of cleverly enumerating the integer values which should be considered for a mixed integer programming problem. The cleverness is needed so that the task does not become overwhelming. One technique that is used is that of implicit enumeration. This involves generating a list of some of the feasible integral solutions and saving the best solution in the list for comparison with lists which will be generated subsequently.

A set S of feasible solutions to a mixed integer programming problem is said to be **implicitly enumerated** if S does not contain any solution which is better than the best currently known solution. Our strategy will be to partition the set of feasible solutions into several subsets and then to dismiss many of these subsets because they are implicitly enumerated. We can derive certain relations from the constraints of the problem and the value of the best current solution. These relations will be violated by any set of solutions which is implicitly enumerated; that is, these relations give conditions which are necessary for a solution to satisfy if it is to improve the value of the objective function. If many subsets of the set of feasible solutions can be implicitly enumerated, we can greatly limit the number of solutions that have to be explicitly examined.

It will be easiest to describe the enumeration technique if we initially limit ourselves to a zero-one programming problem. Consider such a problem and assume it has n variables. The set of all feasible solutions S can be partitioned into two subsets, S_0 and S_1, where

$$S_0 = \{\mathbf{x} \in S \mid x_1 = 0\}$$

and

$$S_1 = \{\mathbf{x} \in S \mid x_1 = 1\}$$

Likewise, both S_0 and S_1 can be partitioned into two subsets—say, S_0 is divided into S_{00} and S_{01}, and S_1 is divided into S_{10} and S_{11}. We define S_{00} by

$$S_{00} = \{\mathbf{x} \in S \mid x_1 = 0 \quad \text{and} \quad x_2 = 0\}$$

Similarly,

$$S_{01} = \{\mathbf{x} \in S \mid x_1 = 0 \quad \text{and} \quad x_2 = 1\}$$

$$S_{10} = \{\mathbf{x} \in S \mid x_1 = 1 \quad \text{and} \quad x_2 = 0\}$$

$$S_{11} = \{\mathbf{x} \in S \mid x_1 = 1 \quad \text{and} \quad x_2 = 1\}$$

Each of the four subsets S_{ij}, i, $j = 0$ or 1, can be partitioned further. We can describe this procedure with a tree diagram, as shown in Figure 4.2. The numbered circles are called **nodes**; the lines connecting them are called **branches**. Node 1 represents the set S of all feasible solutions. The partitioning of S into S_0 and S_1 is represented by the branches leading to nodes 2 and 3. Node 2 represents S_0 and node 3 represents S_1.

A sequence of nodes and branches from node 1 to any other node k is called a **path** to node k. Each branch represents the imposition of one constraint on the variables (setting one variable equal to 0 or 1). Node k represents the set of all solutions to the original constraints which also satisfy the constraints imposed by the branches in the path to node k. For example, node 9 represents

$$\{x \in S \mid x_1 = 0,   x_2 = 0,   x_3 = 1\}$$

If two nodes are connected by a path, the lower node represents a subset of the solutions which are represented by the higher node.

If all possible values for the n variables in a zero-one programming problem are enumerated, we have to generate 2^n paths. The bottom node of each path would correspond to exactly one value for **x**, and this value may be infeasible depending on the nature of the constraints. Our enumeration strategy will try to eliminate as many of these paths as possible from consideration. We describe one method which might be used to perform this elimination.

Suppose our zero-one programming problem has an optimal solution for which the value of the objective function is z^*. Assume that we have

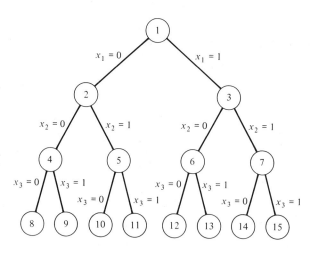

Figure 4.2

found a feasible solution to the problem which has an objective function value of z_L. Obviously, $z_L \leq z^*$. Keep in mind that z^* is unknown and is precisely what we are trying to find. At this point we know that we do not have to consider any set of solutions whose objective function values are smaller than z_L. Consider the set of solutions to the zero-one problem which is represented by node k. We can find an upper bound z_U^k for the values of the objective function at the solutions represented by node k by solving a related linear programming problem. If

$$z_U^k \leq z_L$$

then no path which includes node k will yield an improved solution; the objective function values will all be no larger than z_U^k. Consequently, we can eliminate all paths through node k without explicitly evaluating the objective function at each solution on the path. This set of solutions has been implicitly enumerated. In this case, node k is called a **terminal node**.

Node k is also called a terminal node if there are no feasible solutions satisfying the constraints imposed at node k. Besides the original constraints and those imposed by the path leading to node k, we often add the constraint

$$\mathbf{c}^T\mathbf{x} \geq z_L$$

since we are only interested in solutions which are better than the current best feasible solution.

Finally, node k is called a terminal node if it represents a single feasible solution—that is, if all the variables have been assigned a unique value.

Both branch and bound methods and methods that search in the tree of solutions generate the nodes of the tree until all paths end in terminal nodes. At that point all solutions will have been explicitly or implicitly enumerated and the best solution found is an optimal solution. The difference between the two types of methods is the manner in which the nodes are generated. Search methods follow a path until it ends at a terminal node before examining another path. Branch and bound methods examine the nodes in a less straightforward manner. They may generate, more or less simultaneously, many different paths. Branch and bound methods use more computer storage than search methods do, but they have much more flexibility in examining the nodes and thus may be faster.

We will give the details of one branch and bound method for solving an arbitrary mixed integer programming problem. This method was developed in 1966 by R. J. Dakin and is a modification of the Land–Doig method. It is widely used in the computer codes for integer programming.

Consider the mixed integer programming problem

$$\text{Maximize} \quad z = \mathbf{c}^T\mathbf{x}$$

$$\text{subject to}$$

$$\mathbf{Ax} = \mathbf{b} \qquad\qquad (1)$$

$$\mathbf{x} \geq \mathbf{0}$$

$$x_j, \ j \in I, \quad \text{integral}$$

where $\mathbf{A}$ is $m \times s$, $\mathbf{b}$ is $m \times 1$, $\mathbf{c}$ is $s \times 1$, and $\mathbf{x}$ is $s \times 1$.

Dakin's method will generate a tree very similar to the one we formed for the zero-one programming problem. For each node there will be two branches to check. If neither of the branches ends in a terminal node, the method follows the most promising branch. The other branch is called **dangling** and must be examined before the algorithm terminates. Dakin's method proceeds as follows:

Step 1. (Initial solution) Solve the problem given by (1) as a linear programming problem ignoring the integrality restrictions. If all x_j, $j \in I$, have integral values, we are done. If not, go to Step 2.

Step 2. (Branching variable selection) Choose, from among those variables x_j, $j \in I$, which do not have integral values at this node, one variable to be used to form the branching constraints. An easily implemented rule for this choice is to use the variable whose value has largest fractional part. There are other, more complicated rules which may even involve one iteration of the dual simplex method to determine which variable to choose. The x_j selected must be a basic variable; otherwise, its value would be zero. Suppose it is the ith basic variable in the final tableau for the node, so that its value is x_{Bi}. We can write

$$x_{Bi} = [x_{Bi}] + f_i$$

where $0 < f_i < 1$. Since x_j must have an integral value, it must satisfy either

$$x_j \leq [x_{Bi}] \qquad\qquad (2)$$

or

$$x_j \geq [x_{Bi}] + 1 \qquad\qquad (3)$$

Step 3. (Formation of new nodes) We create two new mixed integer problems represented by the node under consideration in Step 2. One problem is formed by adding constraint (2) and the other problem is formed by adding constraint (3). Solve each of these problems as a linear programming problem using the dual simplex method.

Step 4. (Test for terminal node) Each of the nodes formed in Step 3 may be a terminal node for one of two reasons. First, the problem represented by the node may have no feasible solutions. Or the values of x_j, $j \in I$, are all integers. In the former case, label the nodes as terminal nodes and go to Step 5. In the latter case, besides labeling the nodes, compare the values of the objective function with the current best value. If the objective function value for the new node is better, replace the current best value with it. Go to Step 5.

Step 5. (Node selection)

(a) If both nodes were terminal nodes in Step 4, the next node to be considered is the next one on the list of dangling nodes. If this dangling node has objective function value larger than the current best value, then use this node and go to Step 2. Otherwise, check the next node in the list of dangling nodes. When the list of dangling nodes is exhausted, stop. The current best value is the optimal solution.

(b) If exactly one node in Step 4 was terminal, use the nonterminal node and go to Step 2.

(c) If both nodes in Step 4 were nonterminal, we choose the more promising one. Usually the node with the largest objective function value is considered more promising. The other node is recorded in the list of dangling nodes to be considered later.

EXAMPLE 1. Consider the pure integer programming problem

$$\text{Maximize} \quad z = 7x_1 + 3x_2$$

subject to

$$2x_1 + 5x_2 \leq 30$$

$$8x_1 + 3x_2 \leq 48$$

$$x_1 \geq 0, \quad x_2 \geq 0, \quad \text{integral}$$

We solve the related linear programming problem and obtain the final tableau shown in Tableau 4.11. From this tableau we see that an optimal solution is

$$x_1 = 4\tfrac{7}{17}, \quad x_2 = 4\tfrac{4}{17}, \quad z = 43\tfrac{10}{17}$$

We choose x_1 as the branching variable, since it has the largest fractional part. The constraints to be added are

$$x_1 \leq 4, \quad x_1 \geq 5$$

Tableau 4.11

c_B		7 x_1	3 x_2	0 x_3	0 x_4	$\mathbf{x}_B$
3	x_2	0	1	$\frac{4}{17}$	$-\frac{1}{17}$	$\frac{72}{17}$
7	x_1	1	0	$-\frac{3}{34}$	$\frac{5}{34}$	$\frac{75}{17}$
		0	0	$\frac{3}{34}$	$\frac{29}{34}$	$\frac{741}{17}$

To carry out Step 3 we add each of these constraints in turn to the final tableau in Step 1. We get the tableaux shown in Tableaux 4.12 and 4.13. For Tableau 4.12 we have written

$$x_1 = \tfrac{75}{17} + \tfrac{3}{34}x_3 - \tfrac{5}{34}x_4$$

from the second row of Tableau 4.11 and then introduced a slack variable u_1 so that our new constraint is

$$x_1 + u_1 = 4$$

or

$$\tfrac{3}{34}x_3 - \tfrac{5}{34}x_4 + u_1 = 4 - \tfrac{75}{17} = -\tfrac{7}{17}$$

In the same way the new constraint for Tableau 4.13 becomes

$$-\tfrac{3}{34}x_3 + \tfrac{5}{34}x_4 + u_1 = -5 + \tfrac{75}{17} = -\tfrac{10}{17}$$

Tableau 4.12

c_B		7 x_1	3 x_2	0 x_3	0 x_4	0 u_1	$\mathbf{x}_B$
3	x_2	0	1	$\frac{4}{17}$	$-\frac{1}{17}$	0	$\frac{72}{17}$
7	x_1	1	0	$-\frac{3}{34}$	$\frac{5}{34}$	0	$\frac{75}{17}$
0	u_1	0	0	$\frac{3}{34}$	$-\frac{5}{34}$	1	$-\frac{7}{17}$
		0	0	$\frac{3}{34}$	$\frac{29}{34}$	0	$\frac{741}{17}$

Tableau 4.13

c_B		7 x_1	3 x_2	0 x_3	0 x_4	0 u_1	$\mathbf{x}_B$
3	x_2	0	1	$\frac{4}{17}$	$-\frac{1}{17}$	0	$\frac{72}{17}$
7	x_1	1	0	$-\frac{3}{34}$	$\frac{5}{34}$	0	$\frac{75}{17}$
0	u_1	0	0	$-\frac{3}{34}$	$\frac{5}{34}$	1	$-\frac{10}{17}$
		0 .	0	$\frac{3}{34}$	$\frac{29}{34}$	0	$\frac{741}{17}$

We now apply the dual simplex method to each of these tableaux to obtain Tableaux 4.14 and 4.15, respectively. At this point we have the tree in Figure 4.3. The objective function value in node 3 is larger than that in node 2. Consequently, node 2 is recorded in our list of dangling nodes and we continue from node 3. We add the constraints

$$x_2 \le 2 \quad \text{and} \quad x_2 \ge 3$$

to the problem represented by node 3 to form two new problems. As

Tableau 4.14

c_B		7 x_1	3 x_2	0 x_3	0 x_4	0 u_1	x_B
3	x_2	0	1	$\frac{1}{5}$	0	$-\frac{2}{5}$	$\frac{22}{5}$
7	x_1	1	0	0	0	1	4
0	x_4	0	0	$-\frac{3}{5}$	1	$-\frac{34}{5}$	$\frac{14}{5}$
		0	0	$\frac{3}{5}$	0	$\frac{29}{5}$	$\frac{206}{5}$

Tableau 4.15

c_B		7 x_1	3 x_2	0 x_3	0 x_4	0 u_1	x_B
3	x_2	0	1	0	$\frac{1}{3}$	$\frac{8}{3}$	$\frac{8}{3}$
7	x_1	1	0	0	0	-1	5
0	x_3	0	0	1	$-\frac{5}{3}$	$-\frac{34}{3}$	$\frac{20}{3}$
		0	0	0	1	1	43

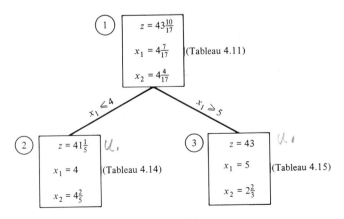

Figure 4.3

before, we write

$$x_2 = -\tfrac{1}{3}x_4 - \tfrac{8}{3}u_1 + \tfrac{8}{3}$$

from the first row of Tableau 4.15. Then by adding a slack variable to each of the constraints, we have

$$x_2 + u_2 = 2 \quad \text{and} \quad -x_2 + u_2 = -3$$

or

$$-\tfrac{1}{3}x_4 - \tfrac{8}{3}u_1 + u_2 = 2 - \tfrac{8}{3} = -\tfrac{2}{3} \tag{4}$$

and

$$\tfrac{1}{3}x_4 + \tfrac{8}{3}u_1 + u_2 = -3 + \tfrac{8}{3} = -\tfrac{1}{3} \tag{5}$$

We add each of these constraints (4) and (5) to Tableau 4.15 to form Tableaux 4.16 and 4.17. We use the dual simplex method on each of these tableaux to try to restore feasibility. Tableau 4.18 is obtained from Tableau 4.16 by this process. However, Tableau 4.17 represents an infeasible solution which satisfies the optimality criterion, since there are no negative entries in the pivotal row (labeled by u_2). Thus, Tableau 4.17 represents a terminal node.

Tableau 4.16

		7	3	0	0	0	0	
c_B		x_1	x_2	x_3	x_4	u_1	u_2	x_B
3	x_2	0	1	0	$\tfrac{1}{3}$	$\tfrac{8}{3}$	0	$\tfrac{8}{3}$
7	x_1	1	0	0	0	-1	0	5
0	x_3	0	0	1	$-\tfrac{5}{3}$	$-\tfrac{34}{3}$	0	$\tfrac{20}{3}$
0	u_2	0	0	0	$-\tfrac{1}{3}$	$\left(-\tfrac{8}{3}\right)$	1	$-\tfrac{2}{3}$
		0	0	0	1	1	0	43

Tableau 4.17

		7	3	0	0	0	0	
c_B		x_1	x_2	x_3	x_4	u_1	u_2	x_B
3	x_2	0	1	0	$\tfrac{1}{3}$	$\tfrac{8}{3}$	0	$\tfrac{8}{3}$
7	x_1	1	0	0	0	-1	0	5
0	x_3	0	0	1	$-\tfrac{5}{3}$	$-\tfrac{34}{3}$	0	$\tfrac{20}{3}$
0	u_2	0	0	0	$\tfrac{1}{3}$	$\tfrac{8}{3}$	1	$-\tfrac{1}{3}$
		0	0	0	1	1	0	43

Tableau 4.18

c_B		7 x_1	3 x_2	0 x_3	0 x_4	0 u_1	0 u_2	x_B
3	x_2	0	1	0	0	0	1	2
7	x_1	1	0	0	$\frac{1}{8}$	0	$-\frac{3}{8}$	$\frac{21}{4}$
0	x_3	0	0	1	$-\frac{1}{4}$	0	$-\frac{17}{4}$	$\frac{19}{2}$
0	u_1	0	0	0	$\frac{1}{8}$	1	$-\frac{3}{8}$	$\frac{1}{4}$
		0	0	0	$\frac{7}{8}$	0	$\frac{3}{8}$	$\frac{171}{4}$

At this point we have the tree in Figure 4.4. Note that in node 3, x_1 had an integer value, but in node 4 the value of x_1 is no longer an integer. Thus, we cannot expect a variable once it has an integer value to continue having an integer value.

We now use Step 2 on node 4, adding the constraints

$$x_1 \le 5 \quad \text{and} \quad x_1 \ge 6$$

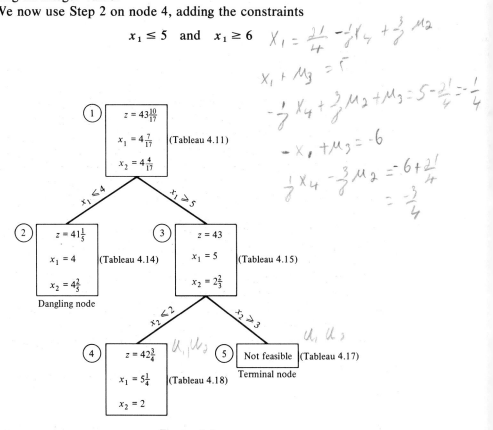

Figure 4.4

Tableau 4.19

		7	3	0	0	0	0	0	
c_B		x_1	x_2	x_3	x_4	u_1	u_2	u_3	x_B
3	x_2	0	1	0	0	0	1	0	2
7	x_1	1	0	0	$\frac{1}{8}$	0	$-\frac{3}{8}$	0	$\frac{21}{4}$
0	x_3	0	0	1	$-\frac{1}{4}$	0	$-\frac{17}{4}$	0	$\frac{19}{2}$
0	u_1	0	0	0	$\frac{1}{8}$	1	$-\frac{3}{8}$	0	$\frac{1}{4}$
0	u_3	0	0	0	$\left(-\frac{1}{8}\right)$	0	$\frac{3}{8}$	1	$-\frac{1}{4}$
		0	0	0	$\frac{7}{8}$	0	$\frac{3}{8}$	0	$\frac{171}{4}$

Tableau 4.20

		7	3	0	0	0	0	0	
c_B		x_1	x_2	x_3	x_4	u_1	u_2	u_3	x_B
3	x_2	0	1	0	0	0	1	0	2
7	x_1	1	0	0	$\frac{1}{8}$	0	$-\frac{3}{8}$	0	$\frac{21}{4}$
0	x_3	0	0	1	$-\frac{1}{4}$	0	$-\frac{17}{4}$	0	$\frac{19}{2}$
0	u_1	0	0	0	$\frac{1}{8}$	1	$-\frac{3}{8}$	0	$\frac{1}{4}$
0	u_3	0	0	0	$\frac{1}{8}$	0	$\left(-\frac{3}{8}\right)$	1	$-\frac{3}{4}$
		0	0	0	$\frac{7}{8}$	0	$\frac{3}{8}$	0	$\frac{171}{4}$

We obtain Tableaux 4.19 and 4.20 after introducing the slack variable u_3 and rewriting these constraints. We use the dual simplex algorithm on each of these tableaux to try to restore feasibility. Tableau 4.21 is obtained from Tableau 4.19 and Tableau 4.22 is obtained from Tableau 4.20 by this process. Since both Tableaux 4.21 and 4.22 give integer solutions, they both correspond to terminal nodes. The solution

$$x_1 = 6, \quad x_2 = 0, \quad z = 42$$

Tableau 4.21

c_B		7 x_1	3 x_2	0 x_3	0 x_4	0 u_1	0 u_2	0 u_3	x_B
3	x_2	0	1	0	0	0	1	0	2
7	x_1	1	0	0	0	0	0	1	5
0	x_3	0	0	1	0	0	-5	-2	10
0	u_1	0	0	0	0	1	0	1	0
0	x_4	0	0	0	1	0	-3	-8	2
		0	0	0	0	0	3	7	41

Tableau 4.22

c_B		7 x_1	3 x_2	0 x_3	0 x_4	0 u_1	0 u_2	0 u_3	x_B
3	x_2	0	1	0	$\frac{1}{3}$	0	0	$\frac{8}{3}$	0
7	x_1	1	0	0	0	0	0	-1	6
0	x_3	0	0	1	$-\frac{5}{3}$	0	0	$-\frac{34}{3}$	18
0	u_1	0	0	0	0	1	0	-1	1
0	u_2	0	0	0	$-\frac{1}{3}$	0	1	$-\frac{8}{3}$	2
		0	0	0	1	0	0	1	42

from Tableau 4.22 is the better one, since it gives a larger objective function value. We then check the list of dangling nodes to see whether other branches of the tree must be pursued. The only dangling node has objective function value $z = 41\frac{1}{5}$, which is smaller than the value obtained in Tableau 4.22. Branching at this dangling node will not increase the objective function value. Therefore, we have found an optimal solution to the original problem:

$$x_1 = 6, \quad x_2 = 0, \quad z = 42$$

The tree of problems which was constructed by the branch and bound algorithm for this example is given in Figure 4.5.

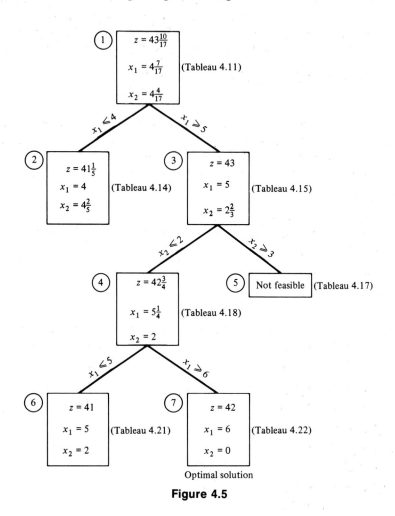

Figure 4.5

4.3   Exercises

In Exercises 1—10 solve the indicated mixed integer programming problem in Section 4.2 and draw the tree of problems used in finding the solution.

 1. Exercise 3 (page 253).
 2. Exercise 4 (page 253).
 3. Exercise 5 (page 254).
 4. Exercise 6 (page 254).
 5. Exercise 7 (page 254).
 6. Exercise 8 (page 254).
 7. Exercise 9 (page 254).
 8. Exercise 10 (page 254).
 9. Exercise 11 (page 254).
10. Exercise 12 (page 254).

In Exercises 11 through 14, use either the cutting plane method or the branch and bound method to solve the indicated mixed integer programming problem in Section 4.1.

11. Exercise 1 (page 237).
12. Exercise 2 (page 237).
13. Exercise 4 (page 237).
14. Exercise 5 (page 238). (If a computer program is available for solving linear programming problems, make use of it.)

4.3   Project

Consider the air filter manufacturer in Section 3.5, Project 1. Suppose there is an additional specification for the purposes of standardization: the diameters of the main chamber and exit duct must be integer values. Rework the problem with this additional specification and comment on its economic impact on the company.

4.4   COMPUTER ASPECTS (Optional)

The computation of a solution to a mixed integer programming problem can be an extremely difficult task. We have discussed two algorithms for finding such a solution. Experimental evidence indicates, however, that there are integer programming problems for which each of these algorithms would be very inefficient. There have been many other algorithms reported in the literature, along with several papers comparing their effectiveness. Most researchers now feel that there will never be a universally good algorithm for mixed integer programming. This is perhaps disappointing, because in the linear programming case the simplex algorithm with modifications to make it numerically stable is a universal algorithm, and one might expect that restricting the range of some of the variables to integer values would simplify the computations. In fact, just the opposite happens.

The state of the art for computational methods for integer programming is as follows. There are commercially available codes, just as for the simplex method, virtually all of which use a branch and bound technique. At various universities there are also experimental implementations of various modifications of the branch and bound algorithm and of cutting plane techniques. Several authors have proposed lists of problems against which one can test an algorithm. In running these tests, it is found that each algorithm is successful on some problems but no algorithm is successful on all problems. Worse than this is the discovery that by simply renumbering the variables or reordering the constraints of a problem, an algorithm which solved the original problem well may perform poorly on the reformulated problem.

The size of a mixed integer programming problem which can be successfully solved in a reasonable amount of computer time is much smaller than that for a linear programing problem. The larger mixed integer problems have 100–200 integer variables, maybe as many as 1000 continuous variables, and perhaps 200–300 constraints. Of course, as the computational speed of new computers increases, larger problems will become feasible.

To be a successful problem solver of mixed integer programming problems, then, one must be able to do much more than provide data to a computer program. First, one must be skilled at formulating problems and must know alternative formulations so that when the problem is given to the computer it will be in a form on which the code will work efficiently. Several factors which should be considered while constructing the model are as follows:

(a) Having a large number of variables constrained to be integers may cause the solution procedure to be very complex and lengthy. Each integer variable can cause branches in a branch and bound algorithm. A large number of these variables will cause a very large tree of problems to be constructed. It will require an extensive amount of storage for the tree and a large amount of time to process the branches. One way of limiting the number of integer variables is to agree that an almost optimum solution will be acceptable. Having a solution 10% less than optimal would not be bad if uncertainties in input parameters would cause almost that much deviation anyway. After settling for an almost optimal solution, one can argue that variables with large values need not be constrained to be integers. Simply take the noninteger values for these variables and round off. Most likely, rounding off will not change the objective function value greatly. For example, if a variable represents the number of workers but its magnitude will be at least 1000, allowing it to be noninteger and rounding off will make an error of less than .05% in the variable, and if the model is well formulated, this error should not propagate to a very large error in the objective function. In fact, some suggest that any variable whose value will be larger than 20 should not be constrained to be an integer.

(b) Tightly bounding integer variables will allow a branch and bound algorithm to find terminal nodes more quickly as the extra constraints become inconsistent with the bounds. Geometrically, the tight bounds reduce the number of lattice points within the convex set of feasible solutions to the related linear programming problem which must be considered.

(c) It helps to formulate the problem as a network problem or special-type (transportation, assignment) integer programming problem to take advantage of the special algorithms which are available. These special algorithms exist because they are more efficient than the general algorithm on a particular type of problem. Even though the combinatorial problems arising in network theory can be difficult, it is generally better to make use of the additional information or structure which the problem has.

(d) The problem formulator should avoid the use of both zero-one variables and general integer variables in the same model. There are algorithms available which work efficiently on pure zero-one programming problems. However, having this type of variable in combination with general integer variables creates unnecessary complexities. There usually are ways of reformulating a problem to avoid the need for both types of variables.

(e) It helps to organize the model so as to take advantage of the scanning procedures which are incorporated in the solution code. Perhaps the code scans the variables in order from 1 to n to check for integer values. In this case, place the integer variables first and the most tightly bounded of these at the beginning of the list. Likewise, listing the more restrictive constraints first may speed the algorithm.

Second, the user should be prepared to monitor the progress of the computer solution by dividing the problem into stages and carefully interpreting the output from each stage. If the user suspects that a particular algorithm is not proceeding quickly enough to the solution, he or she may choose to change algorithms for a few iterations. Or perhaps the structure of the problem and the output will indicate that a different selection of decision rules for the algorithm will speed the process.

Finally, the user should be familiar with the heuristic methods which are applicable to his or her type of problem. Perhaps one of these methods will determine a good starting solution or good decision rules.

Further Readings

Balinski, M. "Integer Programming: Methods, Uses, Computations." *Management Science,* 12 (1965), 253–313.

Bellmore, M., and G. L. Nemhauser. "The Traveling Salesman Problem: A Survey." *Operations Research,* 16 (1968), 538–58.

Dakin, N. J. "A Tree Search Algorithm for Mixed Integer Programming Problems." *Computer Journal,* 9 (1966), 250–55.

Garfinkel, Robert S., and George L. Nemhauser. *Integer Programming.* Wiley, New York, 1972.

Geoffrion, A. M., and R. E. Marsten. "Integer Programming Algorithms: A Framework and State-of-the-Art Survey." *Management Science,* 18 (1972), 465–91.

Gilmore, P. C., and Ralph E. Gomory. "A Linear Programming Approach to the Cutting Stock Problem." *Operations Research,* 9 (1961), 849–59.

Gilmore, P. C. and Ralph E. Gomory. "A Linear Programming Approach to the Cutting Stock Problem—Part II." *Operations Research,* 11 (1963), 863–88.

Gilmore, P. C., and Ralph E. Gomory. "Multistage Cutting Stock Problems of Two or More Dimensions." *Operations Research,* 13 (1965), 94–120.

Salkin, Harvey M. *Integer Programming.* Addison-Wesley, Reading, Mass., 1975.

Taha, Hamdy. *Integer Programming: Theory, Applications.* Academic Press, New York, 1975.

CHAPTER

SPECIAL TYPES OF
LINEAR PROGRAMMING
PROBLEMS

In this chapter we study a number of special types of linear programming problems. These problems arise in transportation systems, in communication systems, in pipeline systems, in electrical networks, in the planning and scheduling of projects and in many other applications. Each of these problems can be formulated and solved as a linear programming problem. However, because of their special structure these problems can be more effectively solved by other methods that have been specifically developed to handle them.

5.1 THE TRANSPORTATION PROBLEM

We have previously considered two examples of the transportation problem—Example 3 in Section 2.1 and Example 1 in Section 4.1. We present another example here which will be used to indicate the ways in which the simplex method can be adapted to the special structure of this problem. Our eventual goal is the transportation algorithm.

EXAMPLE 1. The plastic manufacturing company described in Example 3 of Section 2.1 has decided to build a combination plant-warehouse in San Antonio. This expansion will give it three plants and four warehouses. The following shipping costs have been determined for the new plant and warehouse:

		To			
		Los Angeles	Chicago	New York City	San Antonio
From	120 Salt Lake City	5 (1,1)	7 (1,2)	9 (1,3)	6 (1,4)
	140 Denver	6	7	10	5
	100 San Antonio	7	6	8	1
		100	60	80	120

The San Antonio plant can supply 100 tons per week, and the warehouse needs 120 tons per week to meet its demand. We want to determine how many tons of sheet polyethylene should be shipped from each plant to each warehouse to minimize the transportation cost.

Mathematical Model. The cost matrix is given by

$$
C = \begin{matrix} \text{Los Angeles} & \text{Chicago} & \text{New York} & \text{San Antonio} \\ & & \text{City} & \end{matrix}
$$

$$
C = \begin{bmatrix} 5 & 7 & 9 & 6 \\ 6 & 7 & 10 & 5 \\ 7 & 6 & 8 & 1 \end{bmatrix} \begin{matrix} \text{Salt Lake City} \\ \text{Denver} \\ \text{San Antonio} \end{matrix} \quad (1)
$$

where we have labeled the rows with the points of origin and the columns with the destinations. The supply vector is

$$
\mathbf{s} = \begin{bmatrix} 120 \\ 140 \\ 100 \end{bmatrix}
$$

and the demand vector is

$$
\mathbf{d} = \begin{bmatrix} 100 \\ 60 \\ 80 \\ 120 \end{bmatrix}
$$

We let x_{ij} denote the amount shipped from the ith plant to the jth warehouse, where the plants and warehouses are numbered as in equation (1). Our model is to minimize

c_{ij} = unit shipping cost from ith plant to jth warehouse

$$
z = \sum_{i=1}^{3} \sum_{j=1}^{4} c_{ij} x_{ij}
$$

subject to

$$
\left.
\begin{array}{ll}
\sum_{j=1}^{4} x_{ij} \leq s_i, & i = 1, 2, 3 \\[2mm]
\sum_{i=1}^{3} x_{ij} \geq d_j, & j = 1, 2, 3, 4
\end{array}
\right\}
\tag{2}
$$

demand

Supply

$$
x_{ij} \geq 0, \quad \text{integers}
$$

Clearly the company must be able to provide a supply of polyethylene at least equal to the demand. In our example, we have

$$
\sum_{i=1}^{3} s_i = 360 = \sum_{j=1}^{4} d_j
$$

Later we will indicate what to do if supply exceeds demand. We will say that the model is infeasible if demand exceeds supply. The student may show that, when demand equals supply, each of the constraints in (2) is an equality.

We will now develop an algorithm for solving the general transportation problem given by

$x_{ij} = s_i \cdot d_j / S$

$$
\text{Minimize} \quad z = \sum_{i=1}^{m} \sum_{j=1}^{n} c_{ij} x_{ij}
$$

$S = \sum_{i=1}^{m} s_i = \sum_{j=1}^{n} d_j$ subject to

$$
\left.
\begin{array}{ll}
\sum_{j=1}^{n} x_{ij} = s_i, & i = 1, 2, \ldots, m \\[2mm]
\sum_{i=1}^{m} x_{ij} = d_j, & j = 1, 2, \ldots, n
\end{array}
\right\}
\tag{3}
$$

$$
\sum_{i=1}^{m} s_i = \sum_{j=1}^{n} d_j
\tag{4}
$$

$$
x_{ij} \geq 0, \quad i = 1, 2, \ldots, m, \quad j = 1, 2, \ldots, n
$$

This form of the transportation problem has the following properties:

1. The problem always has a feasible solution (Exercise 18). /7
2. The set of feasible solutions is bounded (Exercise 19). //
3. Because of (4), one of the $m + n$ constraints in (3) is redundant (Exercise 20).

We may substitute the values for the coefficients in our model and write it as

Minimize $z = 5x_{11} + 7x_{12} + 9x_{13} + 6x_{14} + 6x_{21} + 7x_{22} + 10x_{23} + 5x_{24}$
$+ 7x_{31} + 6x_{32} + 8x_{33} + x_{34}$

subject to

$$
\begin{aligned}
x_{11} + x_{12} + x_{13} + x_{14} &= 120 \\
x_{21} + x_{22} + x_{23} + x_{24} &= 140 \\
x_{31} + x_{32} + x_{33} + x_{34} &= 100 \\
x_{11} \qquad\qquad + x_{21} \qquad\qquad + x_{31} &= 100 \\
x_{12} \qquad\qquad + x_{22} \qquad\qquad + x_{32} &= 60 \\
x_{13} \qquad\qquad + x_{23} \qquad\qquad + x_{33} &= 80 \\
x_{14} \qquad\qquad + x_{24} \qquad\qquad + x_{34} &= 120
\end{aligned}
\qquad (5)
$$

$$x_{ij} \geq 0, \quad \text{integer}, \quad i = 1, 2, 3, \quad j = 1, 2, 3, 4$$

Note that each variable appears in exactly two constraints. Since there are seven constraints in the problem, we expect that there will be seven nonzero variables in a basic feasible solution. Actually, there will only be six nonzero variables in any basic feasible solution, because one constraint is redundant. This redundancy occurs because the supply is equal to the demand.

The model can be solved by the simplex algorithm and its solution will always be an integer vector, since the constraint matrix contains only 0's and 1's. However, there is a much more efficient algorithm which uses a 3×4 tableau instead of the 6×12 tableau for the simplex algorithm.

We construct the transportation tableau by writing a 3×4 matrix to hold the values of x_{ij}. The supply vector is placed to the right of this matrix, the transpose of the demand vector is written below the matrix, and the unit costs are written in the insets. For our problem we obtain Tableau 5.1.

There are several ways of systematically filling in the tableau with

Tableau 5.1

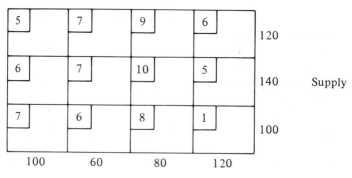

Supply

Demand

values for x_{ij}. For now, we start by allocating as much of the supply in row 1 (from plant 1) as possible to the cheapest route. This means x_{11} = 100, which satisfies the demand in the first column and leaves 20 units to be allocated to the next cheapest route in the first row. We set x_{14} = 20, which exhausts the supply from the first plant. Consequently, x_{12} = x_{13} = 0. Moving to the second row, we follow the same procedure of allocating as much as possible to the cheapest route. We set x_{24} = 100, since 20 units have already been provided in row 1. The 40 remaining units from plant 2 are shipped to warehouse 2, since that route is the cheapest remaining. The rest of the tableau is filled in using the same reasoning (Tableau 5.2). Observe that the solution represented by Tableau 5.2 is a basic feasible solution. There are six nonzero variables. Later we will develop a technique to show that the columns corresponding to these variables are linearly independent.

Tableau 5.2

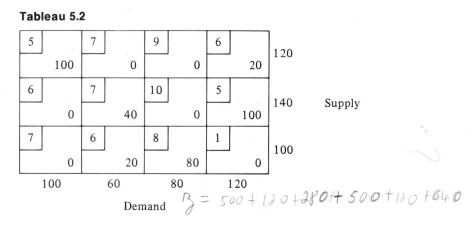

$$B = 500 + 120 + 280 + 500 + 120 + 640$$

Demand

Have we found the minimum cost solution? Our scheme for distributing the polyethylene represents \$2160 in transportation costs. It is most likely not the minimum, since no goods made in San Antonio are shipped to San Antonio. Note that our current solution has six nonzero variables and is a basic feasible solution.

We now look for a way of systematically modifying our initial solution to arrive at the minimum cost distribution scheme in a small number of steps. For each of the *unused* routes, routes along which no goods were shipped, we calculate the change in the total shipping cost due to shipping one unit of goods along that route subject to the restriction that the corresponding changes must be made in the used routes. For example, shipping one unit from plant 1 to warehouse 2 (x_{12} = 1) leads to the *modifications* in Tableau 5.2 shown in Tableau 5.3.

The change in the total shipping cost due to this one modification is

$$7 - 6 - 7 + 5 = -1$$

to determine entering variable

Tableau 5.3

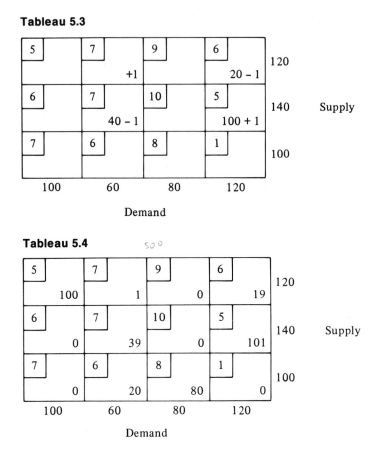

Tableau 5.4   50 0

that is, the total shipping cost for the modified tableau (Tableau 5.4) is $2159.

Thus, we have improved our solution by $1. We now compute the possible improvements for each of the unused routes and represent them in Tableau 5.5. A blank means that the route was being used. A negative value means that shipping one unit along the indicated unused route resulted in *increasing* the total shipping cost.

In doing these computations remember that all changes which are forced by the initial change must be assigned to a route which is being used. Thus, in computing the possible improvement for route (3, 4) we seem to have two choices for decreasing by 1 the amount shipped along a route in use. However, only one choice works, since the other leads to a change in an unused route which cannot be balanced with a corresponding change in a route that is being used. These two cases are illustrated in Tableaux 5.6 and 5.7.

The possible improvement tableau (Tableau 5.5) shows that the total

Tableau 5.5   Possible Improvement Tableau

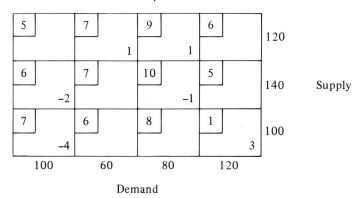

Demand

Tableau 5.6

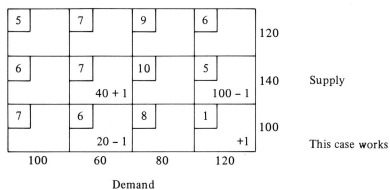

Demand

This case works

Tableau 5.7

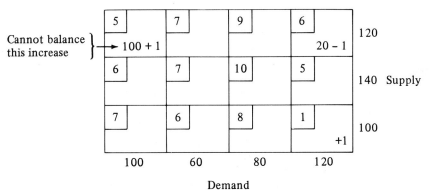

Demand

shipping cost would be substantially reduced if we allocated more to routes (2, 2) and (3, 4) and then modified routes (3, 2) and (2, 4) accordingly. If we set

$$x_{34} = 20$$

and increase x_{22} by 20, then we get Tableau 5.8, which represents a tableau whose total shipping cost is $2100. Consequently, Tableau 5.2 did *not* represent an optimal solution.

Tableau 5.8

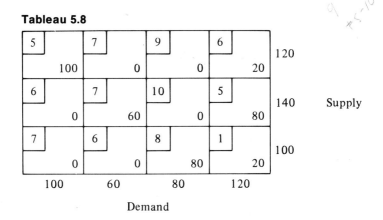

By developing a way to easily compute the possible improvements, we shall see how to reallocate shipments and how to determine when an optimal solution has been reached. We first formulate the dual problem to the transportation problem.

Let v_1, v_2, v_3 be the dual variables corresponding to the supply constraints, and let w_1, w_2, w_3, w_4 be the dual variables corresponding to the demand constraints. Then the dual problem to the transportation problem (5) is

$$\text{Maximize} \quad z' = \sum_{i=1}^{3} s_i v_i + \sum_{j=1}^{4} d_j w_j$$

subject to

$$v_i + w_j \leq c_{ij}, \qquad i = 1, 2, 3, \qquad j = 1, 2, 3, 4$$

$$v_i, w_j \quad \text{unrestricted}$$

Tableau 5.2 represents the initial basic feasible solution to the transportation problem with basic variables x_{11}, x_{14}, x_{22}, x_{24}, x_{32}, and x_{33}. From the properties of the simplex algorithm, $z_{ij} = c_{ij}$ when x_{ij} is a basic variable. As we showed in Section 3.2, there are values for the dual variables for which the left-hand side of the dual constraint

$$v_i + w_j \leq c_{ij}$$

is equal to z_{ij}, for all i and j. This means that for the pairs $(1, 1)$, $(1, 4)$, $(2, 2)$, $(2, 4)$, $(3, 2)$, and $(3, 3)$, we have

$$v_i + w_j = c_{ij} \tag{6}$$

This gives us 6 equations in 7 unknowns. The values of the dual variables can be found from (6) by giving one of the unknowns—say, the one that appears most often—an arbitrary value—say, 0—and then solving for the remaining unknowns. Our system is

$$
\begin{array}{lllll}
x_{11}: & v_1 & + w_1 & & = 5 \\
x_{14}: & v_1 & & + w_4 & = 6 \\
x_{22}: & v_2 & + w_2 & & = 7 \\
x_{24}: & v_2 & & + w_4 & = 5 \\
x_{32}: & v_3 & + w_2 & & = 6 \\
x_{33}: & v_3 & & + w_3 & = 8
\end{array}
$$

Setting $v_1 = 0$, we have

$$v_2 = -1, \quad v_3 = -2, \quad w_1 = 5, \quad w_2 = 8, \quad w_3 = 10, \quad w_4 = 6$$

Now the entries in the objective row of the simplex tableau corresponding to the solution in Tableau 5.2 can be determined. The only nonzero entries will be those for the nonbasic variables. Each entry will be of the form

$$z_{ij} - c_{ij} = v_i + w_j - c_{ij}$$

since z_{ij} is the left-hand side of the (i, j) constraint of the dual problem. The entries are

$$
\begin{array}{ll}
x_{12}: & v_1 + w_2 - c_{12} = 0 + 8 - 7 = 1 \\[4pt]
x_{13}: & v_1 + w_3 - c_{13} = 0 + 10 - 9 = 1 \\[4pt]
x_{21}: & v_2 + w_1 - c_{21} = -1 + 5 - 6 = -2 \\[4pt]
x_{23}: & v_2 + w_3 - c_{23} = -1 + 10 - 10 = -1 \\[4pt]
x_{31}: & v_3 + w_1 - c_{31} = -2 + 5 - 7 = -4 \\[4pt]
x_{34}: & v_3 + w_4 - c_{34} = -2 + 6 - 1 = 3
\end{array}
$$

Since we are dealing with a minimization problem, the largest positive value determines the entering variable. In this case it is x_{34}. To determine the departing variable, examine route $(3, 4)$. If we try to send one unit along unused route $(3, 4)$ and make the corresponding modification in the routes which are being used (basic variables), we have Tableau 5.9. The total shipping cost is changed by

$$7 - 5 - 6 + 1 = -3 \qquad (\textit{Note:} \quad v_3 + w_4 - c_{34} = 3)$$

Tableau 5.9

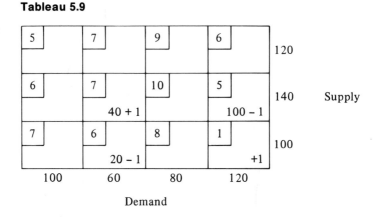

Supply

Demand

so that it has been *reduced* by \$3. We can send as many as 20 units along route (3, 4), so that if $x_{34} = 20$, we drive x_{32} to zero. Thus, x_{32} becomes the departing variable. Modifying Tableau 5.2, we obtain Tableau 5.8, as we did before. Since each unit sent along route (3, 4) reduced the total shipping cost by \$3, when 20 units are sent the cost should be reduced by \$60. Indeed, the cost represented by Tableau 5.8 is \$2100.

We now give the calculations necessary to determine the next tableau. The system of equations relating the dual variables is

$$
\begin{aligned}
x_{11}: \quad & v_1 && + w_1 && && = 5 \\
x_{14}: \quad & v_1 && && && + w_4 = 6 \\
x_{22}: \quad & v_2 && + w_2 && && = 7 \\
x_{24}: \quad & v_2 && && && + w_4 = 5 \\
x_{33}: \quad & v_3 && && + w_3 && = 8 \\
x_{34}: \quad & v_3 && && && + w_4 = 1
\end{aligned}
$$

Setting $w_4 = 0$, we obtain

$$v_1 = 6, \quad v_2 = 5, \quad v_3 = 1, \quad w_1 = -1, \quad w_2 = 2, \quad w_3 = 7$$

Consequently, the values of $z_{ij} - c_{ij}$ for the nonbasic variables are

$$
\begin{aligned}
x_{12}: \quad & v_1 + w_2 - c_{12} = 6 + 2 - 7 = 1 \\
x_{13}: \quad & v_1 + w_3 - c_{13} = 6 + 7 - 9 = 4 \\
x_{21}: \quad & v_2 + w_1 - c_{21} = 5 - 1 - 6 = -2 \\
x_{23}: \quad & v_2 + w_3 - c_{23} = 5 + 7 - 10 = 2 \\
x_{31}: \quad & v_3 + w_1 - c_{31} = 1 - 1 - 7 = -7 \\
x_{32}: \quad & v_3 + w_2 - c_{32} = 1 + 2 - 6 = -3
\end{aligned}
$$

The entering variable is x_{13}. If we send one unit along the unused route

(1, 3) and make the modifications in the routes being used (basic varia-bles), we have Tableau 5.10. The total shipping cost is changed by

$$9 - 6 - 8 + 1 = -4 \qquad (\textit{Note:} \quad v_1 + w_3 - c_{13} = 4)$$

so that it has been reduced by \$4. We can send as many as 20 units along route (1, 3), so that if $x_{13} = 20$, we drive x_{14} to 0. Thus, x_{14} becomes the departing variable. Modifying Tableau 5.8, we obtain Tableau 5.11. Since each unit sent along route (1, 3) reduced the total shipping cost by \$4, when 20 units are sent the cost should be reduced by \$80. Indeed, the cost represented by Tableau 5.11 is \$2020.

Tableau 5.10

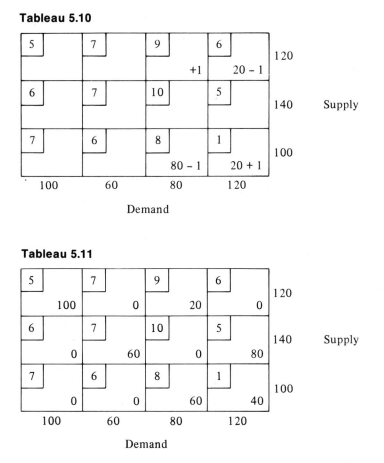

Demand

Tableau 5.11

Demand

Performing these steps several more times, we come to the point where no entering variable can be chosen (all $z_{ij} - c_{ij}$ are negative because this is a minimization problem), and the procedure stops. An optimal solution has been obtained; it is given in Tableau 5.12. The cost of this solution

Tableau 5.12

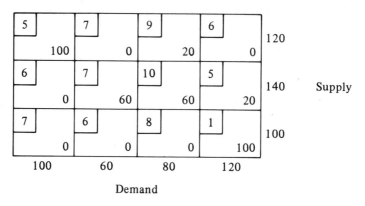

Demand

is $1900. Note that it chooses the plant in San Antonio to supply all the demand of the San Antonio warehouse.

We formalize the procedure used in this example as "the transportation algorithm."

Given the transportation problem

$$\text{Minimize} \quad z = \sum_{i=1}^{m} \sum_{j=1}^{n} c_{ij} x_{ij}$$

subject to

$$\sum_{j=1}^{n} x_{ij} = s_i, \quad i = 1, 2, \ldots, m$$

$$\sum_{i=1}^{m} x_{ij} = d_j, \quad j = 1, 2, \ldots, n$$

$$x_{ij} \geq 0, \quad \text{integers}$$

$$\sum_{i=1}^{m} s_i = \sum_{j=1}^{n} d_j$$

we choose an initial basic feasible solution using the **minimum cost rule**:

In row i assign the maximum possible amount x_{ij} of the remaining supply to the cheapest route.

The transportation algorithm then is

1. Find the entering variable:
   (a) Solve the dual constraint equations corresponding to the basic variables for the remaining $m + n - 1$ dual variables. The value of one dual variable will have to be chosen arbitrarily.

(b) Evaluate the objective row coefficients by computing $z_{ij} - c_{ij} = v_i + w_j - c_{ij}$ for each pair (i, j), where x_{ij} is a nonbasic variable.

(c) The entering variable is x_{ij}, where $z_{ij} - c_{ij}$ is the largest positive value from Step b. If all $z_{ij} - c_{ij}$ are nonpositive, stop; an optimal solution has been obtained.

2. Find the departing variable (these steps will be modified to cover certain degenerate cases):

(a) Compute which basic variables x_{pq} will decrease when x_{ij} is increased.

(b) The departing variable is the one from the list in Step a whose value is smallest.

3. Form the new tableau:

(a) Set the departing variable to 0.

(b) Set the entering variable to the previous value of the departing variable.

(c) Adjust the values of the other basic variables to make the supply and demand constraints hold.

To show that this algorithm works for all possible transportation problems, we must check the following points:

1. The minimum cost rule yields a basic feasible solution.
2. The dual constraint equations can always be solved.
3. The values of the objective row entries $z_{ij} - c_{ij}$ are independent of the choice of value for one of the dual variables.
4. The departing variable can always be computed by the given scheme. A more precise statement of this scheme will also be needed.

We need to introduce some definitions so that we can further discuss the points we raised above. We will call each block in the transportation tableau a **cell**. The value of x_{ij} is recorded in cell (i, j). In our example we saw that if we changed the entry in cell (i, j) from 0 to 1, where x_{ij} was a nonbasic variable, then this change forced changes in the values of some of the basic variables. We now systematize the recording of these changes.

A **loop** in a transportation tableau is a sequence of cells in the tableau that satisfies the following criteria:

1. The sequence consists of horizontal and vertical segments arranged so that the directions of the segments alternate.
2. Each segment joins exactly two cells.
3. The first cell of the sequence is the last, and no other cell is used twice.

Properties 1 and 2 tell us that if (i, j) is a cell in a loop and if we reached it horizontally (along row i), then the next cell in the loop must be in column j. Likewise, if we reached cell (i, j) vertically, then the next cell in the loop must be in row i. Consequently, we use the cells in each row two at a time when forming a loop. A loop must therefore have an even number of cells in it. We sketch some of the loops we found when computing possible improvements for our example in Tableau 5.13. The heavy dot in the cell indicates that the cell is included in the loop. This loop could be written as

$$(1, 2), \quad (1, 4), \quad (2, 4), \quad (2, 2)$$

if we proceeded horizontally from $(1, 2)$. The example in Tableau 5.14 shows how a loop can be more complicated. In this example the loop is

$$(3, 1), \quad (1, 1), \quad (1, 4), \quad (2, 4), \quad (2, 2), \quad (3, 2)$$

Another example is given in Tableau 5.15. This loop is

$$(1, 3), \quad (3, 3), \quad (3, 2), \quad (2, 2), \quad (2, 4), \quad (1, 4)$$

Tableau 5.13

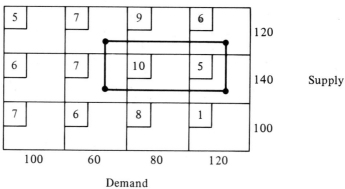

Demand

Tableau 5.14

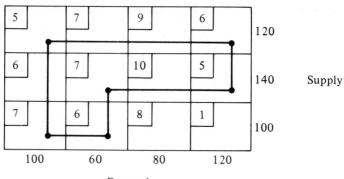

Demand

Tableau 5.15

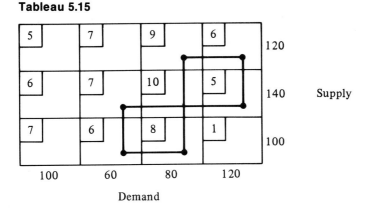

Demand

Note that cell (2, 3) is not in the loop, even though two segments cross there.

Recall that a basic feasible solution to a linear programming problem with m constraints, none of which were redundant, had at most m non-zero values, and the corresponding columns of **A** were linearly independent. Loops give us a very convenient way of determining the linear independence of the corresponding columns of **A**. We saw in (3) that each column of the constraint matrix **A** corresponded to one of the variables x_{ij} and hence to one of the cells in the transportation tableau.

Theorem 5.1.   The columns of **A** determined by

$$(i_1, j_1) \ (i_2, j_2), \ \ldots, \ (i_k, j_k)$$

are linearly dependent if and only if the corresponding cells (or some of them) can be arranged in a loop.

This theorem shows that in computing possible improvements we were discovering how to write each column of **A** that corresponds to a nonbasic variable as a linear combination of the columns that correspond to the basic variables. It also allows us to conclude that the minimum cost rule gives a basic feasible solution—one whose corresponding columns of **A** are linearly independent. For when we make the first assignment we must either satisfy the supply requirement of the first row or the demand requirement of a column in which the minimum cost cell of the first row appears. In either case there is only one cell with a nonzero entry in that column or row or both. This cell cannot belong to any loop, since a loop has an even number of cells in each row and column. Suppose the demand constraint was satisfied by the first assignment. The next assignment will either satisfy a demand constraint or a supply constraint or both. In either case the argument about an even number of cells per row or column of a loop implies that the cell which is used cannot appear in a loop. Continuing in this manner, we find that the constraints can be satisfied without forming any loops.

Theorem 5.1 also allows us to formalize the procedure for computing the departing variable. If x_{ij} has been chosen as the entering variable, then cell (i, j) must belong to a loop consisting of basic cells. That is, the column of $\mathbf{A}$ corresponding to x_{ij} must be a linear combination of the columns of $\mathbf{A}$ corresponding to the basic variables. We list the cells of the loop for x_{ij} in order, starting with (i, j). The cells which appear in the even-numbered positions in the sequence will have their values decreased when the value of x_{ij} is increased. To maintain feasibility, these values cannot be decreased to a negative number. Therefore, we may chose the departing variable x_{kl} as one which satisfies the following criteria:

(a) The variable x_{kl} is a basic variable and cell (k, l) appears in the loop for cell (i, j).
(b) Cell (k, l) appears in an even-numbered position in the loop.
(c) The value of x_{kl} is the smallest of the values of all variables in even-numbered positions in the loop.

For example, going from Tableau 5.2 to Tableau 5.8, we found that x_{34} was the entering variable and that it belonged to the loop

$$(3, 4), \quad (2, 4), \quad (2, 2), \quad (3, 2)$$

We have $x_{24} = 100$ and $x_{32} = 20$, so that x_{32} has the smaller value. Thus, x_{32} is the departing variable.

The special form of the transportation problem allows us to do the pivotal elimination in a very concise form. Suppose the value of the departing variable x_{kl} is α. Then each variable in an odd-numbered position in the loop of the entering variable has as its new value, α plus its old value. Each variable in an even-numbered position in the loop has as its new value, its old value minus α. In this scheme the entering variable automatically is set at α, and the departing variable is set at 0.

The equations which are the dual constraints can always be solved. There are more unknowns than equations, and the equations are consistent. In the following example we show a fast technique for solving these equations by hand.

EXAMPLE 2. Consider a transportation problem between three sources and five destinations where the supply vector is

$$\mathbf{s} = \begin{bmatrix} 100 \\ 160 \\ 140 \end{bmatrix}$$

and the demand vector is

$$\mathbf{d} = \begin{bmatrix} 90 \\ 60 \\ 80 \\ 100 \\ 70 \end{bmatrix}$$

The costs are shown in Tableau 5.16. We find an initial basic feasible solution by using the minimum-cost rule. It is shown in Tableau 5.17 and has a value of $z = 1990$. Note that this solution was found in the order

Tableau 5.16

					Supply
9	3	6	7	3	100
7	5	2	10	6	160
5	4	9	8	10	140
90	60	80	100	70	

Demand

Tableau 5.17

					Supply
9 — 0	3 — 60	6 — 0	7 — 0	3 — 40	100
7 — 50	5 — 0	2 — 80	10 — 0	6 — 30	160
5 — 40	4 — 0	9 — 0	8 — 100	10 — 0	140
90	60	80	100	70	

Demand

Table 5.1

Assignment	Cell	Constraint satisfied
1	(1,2)	Demand
2	(1,5)	Supply
3	(2,3)	Demand
4	(2,5)	Demand
5	(2,1)	Supply
6	(3,1)	Demand
7	(3,4)	Demand and supply

shown in Table 5.1 and that each assignment satisfied either a demand constraint or a supply constraint but not both (with the exception of the last assignment). Also note that the solution has

$$3 + 5 - 1 = 7$$

nonzero variables.

To find the entering variable, we must compute the possible improvement or $z_{ij} - c_{ij}$ for each nonbasic variable. Recall that

$$z_{ij} - c_{ij} = v_i + w_j - c_{ij}$$

For the basic variables we solve the system of equations

$$v_r + w_s = c_{rs}$$

We adjoin a row below the tableau for the values of w_s and a column to the right for the values of v_r and leave out the zeros in the nonbasic cells. We start by assuming $v_1 = 0$. Then $w_2 = 3$ and $w_5 = 3$. Using w_5, we find $v_2 = 3$. Hence, $w_3 = -1$ and $w_1 = 4$. Since $w_1 = 4$, we have $v_3 = 1$ and then $w_4 = 7$. These results are shown in Tableau 5.18. Now the values for $z_{ij} - c_{ij}$ can be filled in for the blanks in Tableau 5.18. The values of the basic variables are circled to distinguish them. We obtain Tableau 5.19. Only $z_{22} - c_{22}$ is positive. Hence, x_{22} is the entering variable. It belongs to the loop of basic variables

$$(2, 2), \quad (2, 5), \quad (1, 5), \quad (1, 2)$$

We have $x_{25} = 30$ and $x_{12} = 60$, so that the smaller x_{25} becomes the departing variable. We increase x_{22} and x_{15} by 30 and decrease x_{25} and x_{12} by 30. This gives Tableau 5.20 with objective function value $z = 1960$. Solving for v_r and w_s and then computing $z_{ij} - c_{ij}$, we obtain Tableau 5.21.

In Tableau 5.21 the entering variable is x_{14} and it belongs to the loop

$$(1, 4), \quad (3, 4), \quad (3, 1), \quad (2, 1), \quad (2, 2), \quad (1, 2)$$

Tableau 5.18

					Supply	v
9	3 (60)	6	7	3 (40)	100	0
7 (50)	5	2 (80)	10	6 (30)	160	3
5 (40)	4	9	8 (100)	10	140	1
90	60	80	100	70		

w: 4   3   −1   7   3

Demand

We have

$$x_{34} = 100, \qquad x_{21} = 50, \qquad x_{12} = 30$$

so that x_{12} is the departing variable. Decreasing x_{34}, x_{21}, and x_{12} by 30 and increasing x_{14}, x_{31}, and x_{22} by the same amount gives us Tableau

Tableau 5.19

					Supply	v
9 −5	3 (60)	6 −7	7 0	3 (40)	100	0
7 (50)	5 1	2 (80)	10 0	6 (30)	160	3
5 (40)	4 0	9 −9	8 (100)	10 −6	140	1
90	60	80	100	70		

w: 4   3   −1   7   3

Demand

Tableau 5.20

9	3 ⟨30⟩	6	7	3 ⟨70⟩	100
7 ⟨50⟩	5 ⟨30⟩	2 ⟨80⟩	10	6	160	Supply
5 ⟨40⟩	4	9	8 ⟨100⟩	10	140
90	60	80	100	70	

Demand

5.22. We have also solved for the dual variables and given the values of $z_{ij} - c_{ij}$. Since all these values are nonpositive, we have found an optimal solution. Its value is $z = 1930$.

Tableau 5.21

								v
9 −4	3 ⟨30⟩	6 −6	7 1	3 ⟨70⟩	100	0	
7 ⟨50⟩	5 ⟨30⟩	2 ⟨80⟩	10 0	6 −1	160	2	Supply
5 ⟨40⟩	4 −1	9 −9	8 ⟨100⟩	10 −7	140	0	
90	60	80	100	70		
w 5	3	0	8	3		

Demand

Tableau 5.22

						v
9 −5	3 −1	6 −7	7 (30)	3 (70)	100	0
7 (20)	5 (60)	2 (80)	10 0	6 0	160	3
5 (70)	4 −1	9 −9	8 (70)	10 −6	140	1
90	60	80	100	70		

Supply (right of second row)

w	4	2	−1	7	3

Demand

Degeneracy

Degeneracy in a transportation problem has the same meaning as it did for a general linear programming problem. That is, a basic variable has value zero. This means that we have designated a route as being used although no goods are being sent along it. In a transportation problem degeneracy can occur in two ways. It is possible that while finding an initial basic feasible solution both a supply and a demand constraint are satisfied simultaneously. On the other hand, there may be a tie for choosing a departing variable. In either case at least one basic variable will have a value of 0. As in the simplex method, degeneracy generally causes no difficulties. A transportation problem has never been known to cycle.

An algorithm to find an initial basic feasible solution to the transportation problem assigns a value to a variable to satisfy a demand constraint or a supply constraint. This variable is then an initial basic variable which corresponds to the constraints satisfied. If both a demand and a supply constraint are simultaneously satisfied by the assignment of a value to x_{rs}, *except for the final allocation,* we have degeneracy. To maintain the correspondence between basic variables and satisfied constraints, we must designate a variable other than x_{rs} as also being basic and assign to it the value zero.

EXAMPLE 3. Consider the transportation problem defined by Tableau 5.23. Using the minimum cost rule, the cells are filled in the order

$$(1, 2), \quad (1, 5), \quad (2, 3), \quad (2, 5)$$

Tableau 5.23

9	3	6	7	3	100
7	5	2	10	6	160
5	4	9	8	10	100

Supply

50	60	80	100	70

Demand

At this point the tableau looks like Tableau 5.24. The next assignment of 50 units to cell (2, 1) will complete both the first column and second row.

Thus, we have degeneracy. To systematically choose a variable as basic and having value zero, we agree in this case to say that only the second row has been completed. Moving to the third row, the smallest available cost is for cell (3, 1) and x_{31} is assigned the value 0. This makes x_{31} a basic variable. We complete the determination of an initial basic feasible solution by letting $x_{34} = 100$.

Degeneracy during the iterations of the transportation algorithm may be ignored.

Tableau 5.24

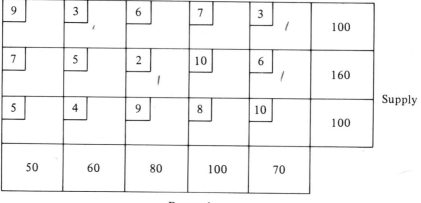

9 0	3 60	6 0	7 0	3 40	100
7 50	5 0	2 80	10 0	6 30	160
5 0	4 0	9 0	10 100	10 0	100

Supply

50	60	80	100	70

Demand

Starting Procedures

We have already described the minimum cost rule for obtaining an initial basic feasible solution to the transportation problem. A number of other methods are available. A desirable starting method is one which will provide a starting solution that is not far from the optimal one in the sense that only a small number of iterations of the transportation algorithm are required. We now describe Vogel's method, which is widely used.

Vogel's Method

Let $C = [c_{ij}]$ be the cost matrix of a transportation problem.
1. For each row and each column of C find the difference between the smallest and the next smallest entry. This difference represents the minimum penalty incurred when one fails to assign goods to the cheapest route.
2. Select the row or column with the largest difference. Ties may be broken arbitrarily.
3. Allocate as much as possible to the cell with the smallest cost in that row or column. Let us say this allocation is made to cell (r, s). Decrease the available supply in row r and the required demand in column s by the amount allocated. This allocation will satisfy either a demand constraint or a supply constraint or perhaps both. The indication of which constraint has been satisfied is the reduction of the available supply in row r or the required demand in column s to zero. Remove the constraint which is satisfied from further consideration by crossing out the corresponding row or column of the cost matrix. If both a demand and supply constraint are satisfied simultaneously, remove only one from further consideration. In this case both the available supply and required demand have been reduced to zero.
4. Repeat steps 1, 2, and 3 until *either* exactly one row or exactly one column remains. In doing step 1, do not compute differences for any row with 0 available supply or column with 0 required demand. When exactly one row or one column remains, the entries in that row or column are fully determined by the previous allocations and are filled in accordingly.

EXAMPLE 4. We apply Vogel's method to the transportation problem with the cost matrix and supply and demand vectors shown below:

$$C = \begin{bmatrix} 8 & 6 & 3 & 9 \\ 2 & 6 & 1 & 4 \\ 7 & 8 & 6 & 3 \end{bmatrix}, \quad s = \begin{bmatrix} 120 \\ 140 \\ 100 \end{bmatrix}, \quad d = \begin{bmatrix} 100 \\ 60 \\ 80 \\ 120 \end{bmatrix}$$

We compute the differences according to Step 1 of the algorithm and circle the largest.

				Differences
8	6	3	9	3
2	6	1	4	1
7	8	6	3	3

Differences: ⑤ 0 2 1

Thus, we allocate as much as possible (100 units) to the cheapest route in the first column ($x_{21} = 100$), fill in the rest of this column with zeros, and cross out the column. The allocation in x_{21} is circled to indicate that x_{21} is a basic variable. The revised supplies and demands from Step 3 of the algorithm and the new differences are shown in Tableau 5.25.

Tableau 5.25

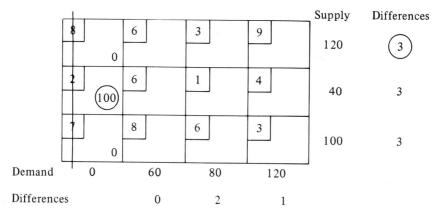

We can arbitrarily choose among rows 1, 2, and 3. Choosing the first row, we allocate 80 units to the cheapest route in that row ($x_{13} = 80$), circle the allocation, fill in the rest of the row with zeros, and cross out the third column. We revise the supplies and demands, compute the new differences, and show the result in Tableau 5.26.

Choosing row 3, we allocate 100 units to the cheapest route in that row ($x_{34} = 100$) and cross out the row. Tableau 5.27 shows the consequences of this allocation along with the new differences.

The largest difference is now in column 4, so that we allocate 20 units to the cheapest route in that column ($x_{24} = 20$), cross out the column, and obtain Tableau 5.28.

There is now exactly one column remaining. The allocations for that column are completely determined ($x_{24} = 40$ and $x_{22} = 20$). Thus, the

Tableau 5.26

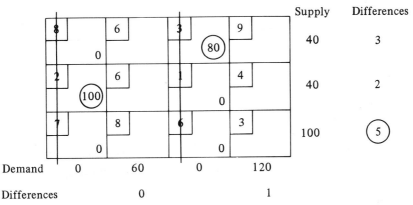

Tableau 5.27

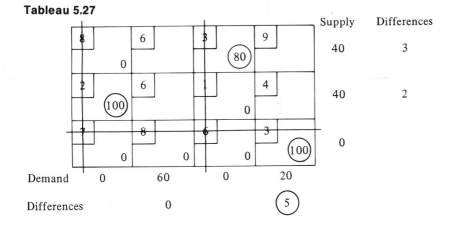

Tableau 5.28

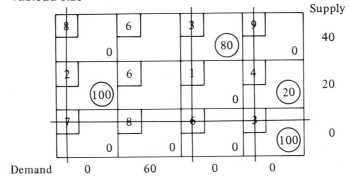

Tableau 5.29

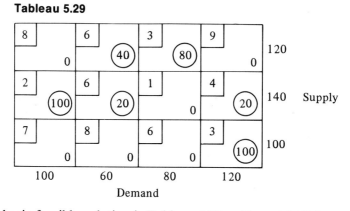

initial basic feasible solution is Tableau 5.29, with cost $1180.

The minimum cost method yields the initial basic feasible solution in Tableau 5.30, whose cost is $1240. Thus, in this case the Vogel method gives a better starting solution.

Tableau 5.30

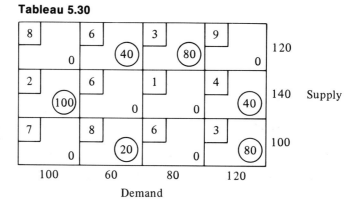

Applying the transportation algorithm to either of the two starting solutions given above, the reader may show that the minimum cost for this transportation problem is $1140.

The effectiveness of Vogel's method can be improved by using a procedure due to Roland E. Larson which is, perhaps, too complicated for hand computation but can be carried out rapidly on a computer.

Instead of using the given costs c_{ij} for Vogel's method, we use the normalized costs c'_{ij} defined by

$$c'_{ij} = c_{ij} - \frac{1}{n}\sum_{p=1}^{n} c_{pj} - \frac{1}{m}\sum_{q=1}^{m} c_{iq}$$

That is, we subtract from each c_{ij} the average of the costs of the row and column in which it appears. We then apply Vogel's method to the matrix $[c'_{ij}]$.

EXAMPLE 5. We consider the same transportation problem as in Example 4. When we compute the normalized cost matrix, we obtain

$$\mathbf{C'} = \begin{bmatrix} -\frac{25}{6} & -\frac{43}{6} & -\frac{41}{6} & -\frac{17}{6} \\ -\frac{73}{12} & -\frac{37}{12} & -\frac{67}{12} & -\frac{35}{12} \\ -\frac{14}{3} & -\frac{14}{3} & -\frac{10}{3} & -\frac{25}{3} \end{bmatrix}$$

Applying Vogel's method to $\mathbf{C'}$, we obtain the starting solution in Tableau 5.31. This solution is actually an optimal solution, and its cost is $1140.

Tableau 5.31

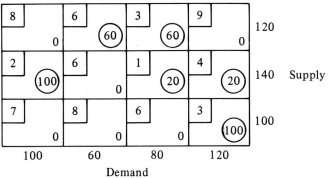

Demand

Extensions

If the total supply exceeds the total demand in the original statement of the transportation problem, a dummy destination can be set up. That is, we create a destination with demand equal to the difference between total supply and total demand. The cost of shipping to this destination from any source is 0. In fact, we have just added a slack variable to each supply constraint.

EXAMPLE 6. The problem originally given in Example 3, Section 2.1, is

Minimize $z = 5x_{11} + 7x_{12} + 9x_{13} + 6x_{21} + 7x_{22} + 10x_{23}$

subject to

$$x_{11} + x_{12} + x_{13} \le 120$$
$$x_{21} + x_{22} + x_{23} \le 140$$
$$x_{11} + x_{21} \ge 100$$
$$x_{12} + x_{22} \ge 60$$
$$x_{13} + x_{23} \ge 80$$

The difference between supply and demand is 20. We create a destination with a demand equal to this amount. The constraints then become equalities, since demand equals supply. The tableau for the problem is Tableau 5.32.

Tableau 5.32

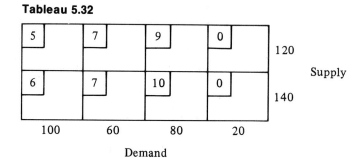

Supply

Demand

5.1 Exercises

1. Verify that Tableau 5.12 represents the optimal solution to Example 1 by performing the necessary steps to obtain Tableau 5.12 from Tableau 5.11.

In Exercises 2–4 find an initial basic feasible solution using (a) the minimum cost rule, (b) Vogel's method, (c) Larson's method if a hand calculator is available.

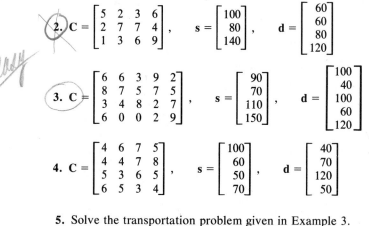

2. $C = \begin{bmatrix} 5 & 2 & 3 & 6 \\ 2 & 7 & 7 & 4 \\ 1 & 3 & 6 & 9 \end{bmatrix}$, $s = \begin{bmatrix} 100 \\ 80 \\ 140 \end{bmatrix}$, $d = \begin{bmatrix} 60 \\ 60 \\ 80 \\ 120 \end{bmatrix}$

3. $C = \begin{bmatrix} 6 & 6 & 3 & 9 & 2 \\ 8 & 7 & 5 & 7 & 5 \\ 3 & 4 & 8 & 2 & 7 \\ 6 & 0 & 0 & 2 & 9 \end{bmatrix}$, $s = \begin{bmatrix} 90 \\ 70 \\ 110 \\ 150 \end{bmatrix}$, $d = \begin{bmatrix} 100 \\ 40 \\ 100 \\ 60 \\ 120 \end{bmatrix}$

4. $C = \begin{bmatrix} 4 & 6 & 7 & 5 \\ 4 & 4 & 7 & 8 \\ 5 & 3 & 6 & 5 \\ 6 & 5 & 3 & 4 \end{bmatrix}$, $s = \begin{bmatrix} 100 \\ 60 \\ 50 \\ 70 \end{bmatrix}$, $d = \begin{bmatrix} 40 \\ 70 \\ 120 \\ 50 \end{bmatrix}$

5. Solve the transportation problem given in Example 3.

6. Solve the transportation problem given in Example 5.

7. Solve the transportation problem given in Exercise 3.

In Exercises 8–13 solve the given transportation problem.

8. $C = \begin{bmatrix} 2 & 5 & 6 & 3 \\ 9 & 6 & 2 & 1 \\ 7 & 7 & 2 & 4 \end{bmatrix}$, $\quad s = \begin{bmatrix} 100 \\ 90 \\ 130 \end{bmatrix}$, $\quad d = \begin{bmatrix} 70 \\ 50 \\ 80 \\ 120 \end{bmatrix}$

9. $C = \begin{bmatrix} 3 & 2 & 5 & 4 \\ 6 & 5 & 7 & 8 \\ 2 & 1 & 4 & 3 \\ 4 & 3 & 5 & 2 \end{bmatrix}$, $\quad s = \begin{bmatrix} 80 \\ 60 \\ 50 \\ 100 \end{bmatrix}$, $\quad d = \begin{bmatrix} 70 \\ 70 \\ 70 \\ 80 \end{bmatrix}$

10. $C = \begin{bmatrix} 4 & 3 & 2 & 5 & 6 \\ 8 & 3 & 4 & 5 & 7 \\ 6 & 8 & 6 & 7 & 5 \\ 4 & 3 & 5 & 2 & 4 \end{bmatrix}$, $\quad s = \begin{bmatrix} 70 \\ 80 \\ 60 \\ 120 \end{bmatrix}$, $\quad d = \begin{bmatrix} 60 \\ 50 \\ 50 \\ 70 \\ 100 \end{bmatrix}$

11. $C = \begin{bmatrix} 4 & 2 & 9 & 7 \\ 7 & 8 & 5 & 6 \\ 3 & 3 & 4 & 1 \\ 7 & 5 & 2 & 6 \end{bmatrix}$, $\quad s = d = \begin{bmatrix} 1 \\ 1 \\ 1 \\ 1 \end{bmatrix}$

12. $C = \begin{bmatrix} 5 & 6 & 7 & 4 \\ 2 & 9 & 7 & 5 \\ 8 & 5 & 8 & 7 \end{bmatrix}$, $\quad s = \begin{bmatrix} 75 \\ 50 \\ 60 \end{bmatrix}$, $\quad d = \begin{bmatrix} 45 \\ 50 \\ 25 \\ 50 \end{bmatrix}$

13. $C = \begin{bmatrix} 6 & 4 & 3 & 5 \\ 7 & 4 & 8 & 6 \\ 8 & 3 & 2 & 5 \end{bmatrix}$, $\quad s = \begin{bmatrix} 100 \\ 60 \\ 50 \end{bmatrix}$, $\quad d = \begin{bmatrix} 60 \\ 80 \\ 70 \\ 40 \end{bmatrix}$

14. Show that in the $m \times n$ transportation problem if

$$\sum_{i=1}^{m} s_i = \sum_{j=1}^{n} d_j$$

then the constraints are equalities.

15. Show that in the $m \times n$ transportation problem if

$$\sum_{i=1}^{m} s_i = \sum_{j=1}^{n} d_j$$

then one constraint is redundant. (*Hint:* First, sum all the supply constraints. Second, sum all but one of the demand constraints. Then show that the difference between the two sums is the other demand constraint.)

16. Write the formulation of the dual of the $m \times n$ transportation problem.

In Exercises 17 and 18 consider the general transportation problem where

$$S = \sum_{i=1}^{m} s_i = \sum_{j=1}^{n} d_j$$

17. Show that $x_{ij} = s_i d_j / S$ is a feasible solution.

18. Show that if $\mathbf{X} = [x_{ij}]$ is a feasible solution, then, for all i, j,

$$0 \le x_{ij} \le \min \{s_i, d_j\}$$

Further Readings

Berge, C., and A. Ghouila-Houri. *Programming, Games, and Transportation Networks*. Wiley, New York, 1965.

Larson, R. E. "Normalizing Vogel's Approximation Method." *Mathematics Magazine*, 45 (1972), 266–69.

Reinfeld, N. V., and W. R. Vogel. *Mathematical Programming*. Prentice-Hall, Englewood Cliffs, New Jersey, 1958.

5.2 THE ASSIGNMENT PROBLEM

We gave an example of the assignment problem in Section 4.1, Example 3. In this section we will discuss the simplest formulation of the problem. We assume that there are exactly as many persons available for assignment as there are jobs and that each person is to be assigned exactly one job. In the event that there are more persons available than jobs, we can create dummy jobs for the excess persons. Being assigned to a dummy job means in reality that the person is not assigned. We do not consider the case where are more jobs than people if all the jobs are to be completed because this violates our basic assumption that each person is assigned exactly one job.

The assignment problem, like the transportation problem, will also be considered as a minimization problem. That is, if we assign person i to job j, this assignment will cost c_{ij}. Our goal is to minimize the total cost of the assignment. Our first model for the problem is

$$\text{Minimize} \quad z = \sum_{i=1}^{n} \sum_{j=1}^{n} c_{ij} x_{ij}$$

subject to

$$\sum_{i=1}^{n} x_{ij} = 1, \qquad j = 1, 2, \ldots, n \tag{1}$$

$$\sum_{j=1}^{n} x_{ij} = 1, \qquad i = 1, 2, \ldots, n \tag{2}$$

$$x_{ij} = 0, 1, \qquad i = 1, 2, \ldots, n; \qquad j = 1, 2, \ldots, n \tag{3}$$

We can actually remove the restriction that x_{ij} take on only 0 or 1 as a value. If we simply assume that x_{ij} must be an integer, then each of the constraints (1) and (2) guarantees that x_{ij} can take on only 0 or 1 as a value. Thus, in place of (3) we write

$$x_{ij} \geq 0, \quad \text{integer}, \quad i = 1, 2, \ldots, n; \quad j = 1, 2, \ldots, n \quad (4)$$

The model defined by (1), (2), and (4) is actually a transportation problem where the demands are all equal, the supplies are all equal, the supplies equal the demands, and the number of sources is equal to the number of destinations. Consequently, this problem could be solved using the transportation algorithm of Section 4.1. With this algorithm we will find that $2n - 1$ variables are basic. However, only n of these variables will be nonzero, so that the solution is highly degenerate. We would suspect that many iterations of the transportation algorithm would simply replace a basic variable with zero value with another variable with zero value. To obtain a better algorithm for the problem it will be helpful to use a slightly different point of view.

Suppose a list of available persons and a list of the jobs are both written in some fixed order. Then one possible assignment is to give job i to person i. On the other hand, if the job list is reordered, it might be possible to reduce the total cost. Assume that person i is still assigned the job in the ith position of the new job list, but since the list was shuffled, this job is no longer job i. What we need is a shuffling or permutation of the job list that yields the minimum total cost.

One way of recording a permutation of the numbers 1, 2, ..., n is to write a list of these numbers in the desired order. For example, 4 2 1 3 is a permutation of 1 2 3 4. Another way of writing this permutation is to give a 4×4 matrix $\mathbf{M} = [m_{ij}]$ with entries which are zeros and ones. If $m_{ij} = 1$, it means that i is assigned the jth position in the new order. We see that 4 2 1 3 corresponds to the matrix

$$\begin{bmatrix} 0 & 0 & 1 & 0 \\ 0 & 1 & 0 & 0 \\ 0 & 0 & 0 & 1 \\ 1 & 0 & 0 & 0 \end{bmatrix}$$

since 1 is in third position, 2 is in second position, 3 is in fourth position, and 4 is in first position.

The set of all feasible solutions to the assignment problem is simply the set of all $n \times n$ permutation matrices. Note that each permutation matrix has the property that in each row (or column) there is precisely one entry equal to 1.

Theorem 5.2. If the cost matrix for an assignment problem has non-negatives entries and at least n zeros, then an optimal solution to the problem exists if n of the zeros lie in the positions of the ones of some $n \times n$ permutation matrix $\mathbf{P}$. The matrix $\mathbf{P}$ represents an optimal assignment.

Proof. In the situation described, the cost can never be smaller than zero, and we have found an assignment where the cost is zero.

This theorem provides a goal for our algorithm. We will show that we can modify the cost matrix without changing the optimal solution. The algorithm will then attempt to carry out this modification to reach a situation where the cost matrix has a zero in each row and in each column.

Theorem 5.3. Suppose the matrix $C = [c_{ij}]$ is the cost matrix for an $n \times n$ assignment problem. Suppose that $\hat{X} = [\hat{x}_{ij}]$ is an optimal solution to this problem. Let C' be the matrix formed by adding α to each entry in the rth row. Then $\hat{X}$ is an optimal solution to the new assignment problem defined by C'.

Proof. The objective function for the new problem is

$$z' = \sum_{i=1}^{n} \sum_{j=1}^{n} c'_{ij} x_{ij} = \sum_{\substack{i=1 \\ i \neq r}}^{n} \sum_{j=1}^{n} c_{ij} x_{ij} + \sum_{j=1}^{n} (c_{rj} + \alpha) x_{rj}$$

$$= \sum_{i=1}^{n} \sum_{j=1}^{n} c_{ij} x_{ij} + \alpha \sum_{j=1}^{n} x_{rj}$$

This shows if X minimizes
$z = \sum_{i=1}^{n} c_{ij} x_{ij}$ *then the same vector*
$$= \sum_{i=1}^{n} \sum_{j=1}^{n} c_{ij} x_{ij} + \alpha$$
X also minimizes $z = \sum_{i,j=1}^{n} c_{ij} x_{ij}$

since each row sum is 1. Therefore, the smallest value for z' will be obtained when

$$z = \sum_{i=1}^{n} \sum_{j=1}^{n} c_{ij} x_{ij}$$

is smallest; namely, it is obtained when $X = \hat{X}$.

A statement similar to Theorem 5.3 can be made if a constant is added to some column of a cost matrix. Thus, our strategy is to modify C by adding constants to rows or columns.

EXAMPLE 1. Suppose the cost matrix for an assignment problem is

$$C = \begin{bmatrix} 4 & 5 & 2 & 5 \\ 3 & 1 & 1 & 4 \\ 12 & 3 & 6 & 3 \\ 12 & 6 & 5 & 9 \end{bmatrix}$$

We can begin to introduce zeros by subtracting the smallest entry in each row from that row. We obtain

$$\begin{bmatrix} 2 & 3 & 0 & 3 \\ 2 & 0 & 0 & 3 \\ 9 & 0 & 3 & 0 \\ 7 & 1 & 0 & 4 \end{bmatrix}$$

There is no zero in column 1, but one can be forced there by subtracting 2 from each entry in that column. We have

$$\begin{bmatrix} 0 & 3 & 0 & 3 \\ 0 & 0 & 0 & 3 \\ 7 & 0 & 3 & 0 \\ 5 & 1 & 0 & 4 \end{bmatrix}$$

We now have at least one zero in each row and column. We attempt to assign the locations for the ones in a permutation matrix.

Actually, we will not produce the permutation matrix but will star the zeros in the cost matrix to indicate an assignment. We must assign the zero in position (4, 3) since it is the only zero in row 4. However, we will carry out the assignment in an order which will anticipate part of the algorithm we are developing. Starting with the first row, we assign the first 0 in each row which does not belong to a previously assigned column. We say a row or column is **assigned** when some zero in it is assigned. Thus, in row 1 we assign the zero in column 1; in row 2 we assign the zero in column 2, skipping the zero in column 1 because that column was previously assigned; in row 3 we assign the zero in column 4; and in row 4 we assign the zero in column 3. The assignment which we obtain is

$$\begin{bmatrix} 0^* & 3 & 0 & 3 \\ 0 & 0^* & 0 & 3 \\ 7 & 0 & 3 & 0^* \\ 5 & 1 & 0^* & 4 \end{bmatrix}$$

or person 1 is assigned job 1, person 2 is assigned job 2, person 3 is assigned job 4, and person 4 is assigned job 3. The cost is

$$4 + 1 + 3 + 5 = 13$$

In Example 1 we were successful in determining an assignment for each person. However, as the next example shows, we may not always be as fortunate.

EXAMPLE 2. Suppose the cost matrix for an assignment problem is

$$\mathbf{C} = \begin{bmatrix} 4 & 1 & 3 & 4 \\ 5 & 6 & 2 & 9 \\ 6 & 5 & 8 & 5 \\ 7 & 6 & 2 & 3 \end{bmatrix}$$

Subtracting the minimum entry in each row, we obtain

$$\begin{bmatrix} 3 & 0 & 2 & 3 \\ 3 & 4 & 0 & 7 \\ 1 & 0 & 3 & 0 \\ 5 & 4 & 0 & 1 \end{bmatrix}$$

Now subtracting the minimum entry in each column, we obtain

$$\begin{bmatrix} 2 & 0 & 2 & 3 \\ 2 & 4 & 0 & 7 \\ 0 & 0 & 3 & 0 \\ 4 & 4 & 0 & 1 \end{bmatrix}$$

Making the assignments row by row, we have

$$\mathbf{C}' = \begin{bmatrix} 2 & 0^* & 2 & 3 \\ 2 & 4 & 0^* & 7 \\ 0^* & 0 & 3 & 0 \\ 4 & 4 & 0 & 1 \end{bmatrix}$$

This matrix does not represent a complete assignment; person 4 has not been given a job. Two explanations are available for this situation. Either there is no possible complete assignment for the given pattern of zeros or there is a complete assignment but the algorithm failed to find it. We investigate each of these possibilities.

First notice that any pattern of zeros in an $n \times n$ matrix has the property that all the zeros can be covered by n lines. For example, choose the n lines each of which covers one column. Suppose the zeros in the $n \times n$ matrix $\mathbf{C}$ can be covered with k lines where $k < n$. Let a be the smallest of the uncovered entries of $\mathbf{C}$. We form a new matrix $\mathbf{C}'$ by subtracting a from the entries of each uncovered row and adding a to the entries of each covered column. Each uncovered entry of $\mathbf{C}$ has decreased by a, since it belongs to an uncovered row and an uncovered column. Each entry covered by one line has remained unchanged: it either belonged to a covered row and uncovered column and was not modified, or it belonged to an uncovered row and covered column and had a added to it and subtracted from it. Each entry in both a covered row and covered column has increased by a. Thus $\mathbf{C}'$ has a zero entry in a position where $\mathbf{C}$ did not have a zero and it might be possible to finish the assignment. The procedure for modifying $\mathbf{C}$ can be more simply stated as: subtract a from each uncovered entry and add a to each doubly covered entry.

For example, we can cover the last matrix as follows:

$$\begin{bmatrix} 2 & \cancel{0} & \cancel{2} & 3 \\ 2 & \cancel{4} & \cancel{0} & 7 \\ \cancel{0} & \cancel{0} & \cancel{3} & \cancel{0} \\ 4 & \cancel{4} & \cancel{0} & 1 \end{bmatrix}$$

The smallest uncovered entry of C' is 1. Subtracting 1 from each uncovered entry and adding 1 to each doubly covered entry, we obtain the matrix

$$\begin{bmatrix} 1 & 0 & 2 & 2 \\ 1 & 4 & 0 & 6 \\ 0 & 1 & 4 & 0 \\ 3 & 4 & 0 & 0 \end{bmatrix}$$

Now using the assignment algorithm on this last matrix, we have

$$\begin{bmatrix} 1 & 0^* & 2 & 2 \\ 1 & 4 & 0^* & 6 \\ 0^* & 1 & 4 & 0 \\ 3 & 4 & 0 & 0^* \end{bmatrix}$$

which is a complete assignment.

We can be guided in our use of this procedure by the following theorem, proved by the graph theorist König.

Theorem 5.4. The maximum number of zeros which can be assigned is equal to the minimum number of lines that are needed to cover all the zeros.

In Example 2, since we can cover all the zeros of matrix C' with three lines, it follows from Theorem 5.4 that at most three zeros can be assigned. We have determined such an assignment. It is impossible to assign four zeros with the given matrix C', and more zeros must be introduced using the procedure described above.

The other possibility when we do not discover a complete assignment is that the row-searching algorithm has failed.

EXAMPLE 3. Consider the cost matrix for an assignment problem given by

$$C = \begin{bmatrix} 4 & 2 & 9 & 7 \\ 7 & 8 & 5 & 6 \\ 3 & 3 & 4 & 1 \\ 7 & 5 & 2 & 6 \end{bmatrix}$$

Subtracting the minimum entry in each row from that row and then the minimum entry in each column from that column, we obtain

$$\begin{bmatrix} 0 & 0 & 7 & 5 \\ 0 & 3 & 0 & 1 \\ 0 & 2 & 3 & 0 \\ 3 & 3 & 0 & 4 \end{bmatrix}$$

Assigning the first zero entry in each row which does not lie in a previously assigned column, we get

$$\begin{bmatrix} 0^* & 0 & 7 & 5 \\ 0 & 3 & 0^* & 1 \\ 0 & 2 & 3 & 0^* \\ 3 & 3 & 0 & 4 \end{bmatrix}$$

However, there is a complete assignment for this matrix given by

$$\begin{bmatrix} 0 & 0^* & 7 & 5 \\ 0^* & 3 & 0 & 1 \\ 0 & 2 & 3 & 0^* \\ 3 & 3 & 0^* & 4 \end{bmatrix}$$

Consequently, we must develop an algorithm to search it out. Before we do this, let us summarize our method of solution for an assignment problem as we now have it.

Step 1. Assuming that the cost matrix C has nonnegative entries and the problem is a minimization problem, subtract the smallest entry in each row from that row and then subtract the smallest entry in each column from that column. The new matrix C' defines an assignment problem which has the same optimal solutions as C, and C' has at least one zero in each row and column.

Step 2. For each row assign the first zero not in any previously assigned column. If n zeros have been assigned, stop; an optimal solution has been found.

Step 3. Assuming that less than n zeros have been assigned, determine whether reassigning some zeros will give a complete assignment. If it will, stop after the reassignment.

Step 4. Find k lines ($k < n$) which cover all the zeros of C'.

Step 5. Let a be the smallest uncovered entry in C'. Rearrange the zeros in C' by subtracting a from each uncovered entry of C' and adding a to each doubly covered entry of C'. Go to Step 2 to reassign all zeros.

We now describe the details of Steps 3 and 4. The algorithm will require many searches of the rows and columns of C'. All these searches are to be done from left to right or from top to bottom (in order of increasing subscripts).

Step 3. Suppose that i_0 is the index of one of the rows for which, in Step 2, we fail to find a zero that can be assigned. However, there must be at least one zero in row i_0, since every row has a zero in it. Say one of these zeros occurs in column j_0. There must be an assigned zero in column j_0, for otherwise the zero in (i_0, j_0) could be assigned. Starting at cell (i_0, j_0), we construct a path consisting of alternating vertical and horizontal segments, joining cells which alternately contain zeros and starred zeros. Specifically, we join

$$0 \quad \text{in} \quad (i_0, j_0) \quad \text{to}$$

$$0^* \quad \text{in} \quad (i_1, j_0) \quad \text{to}$$

$$0 \quad \text{in} \quad (i_1, j_1) \quad \text{to}$$

$$0^* \quad \text{in} \quad (i_2, j_1) \quad \ldots$$

where the column indices $j_0, j_1, \ldots, j_n$ must be distinct:

$$
\begin{array}{c c c}
 & j_1 & j_0 \\
i_0 & & 0 \\
 & & | \\
i_1 & 0\!\!-\!\!0\,* \\
 & | \\
i_2 & 0\,*
\end{array}
$$

We can represent this path by the sequence of cells (i_0, j_0), (i_1, j_0), (i_1, j_1), (i_2, j_1), ... The next cell in the sequence is obtained as follows:

Case A. Suppose we are at a 0 in (i_k, j_k). We search column j_k for a 0^*. If a 0^* is found, we add its cell to the sequence. If no 0^* exists in column j_k, then we make the following reassignments in $\mathbf{C}'$: each 0 in the sequence from (i_0, j_0) to (i_k, j_k) is changed to a 0^* and each 0^* is changed to a 0. Note that we have created a 0^* for row i_0 and that a 0^* remains in every row which previously had one.

Case B. Suppose on the other hand, we are at a 0^* in cell (i_{k+1}, j_k). We search row i_{k+1} for a 0 which does not lie in a column appearing previously in the path. If a 0 is found, we add its cell to the sequence. If no 0 is found, we will not be able to modify the assignment as we did in Case A; we label column j_k as **necessary** and redirect our search. This is done by deleting the 0^* at (i_{k+1}, j_k) and the 0 at (i_k, j_k) from the sequence and returning to the 0^* at (i_k, j_{k-1}). We now repeat this process with row i_k instead of row i_{k+1}. That is, we search row i_k for a 0 other than the zero in column j_k:

$$0^*\text{------------}0\ (i_k,\ j_k)$$
$$(i_k,\ j_{k-1})\quad\Big|$$
$$0^*$$
$$(i_{k+1},\ j_k)$$

When exactly one 0 is found in a row, the column in which that 0 lies is called **necessary**. We have found that column j_k is necessary. If, however, in searching row i_{k+1} for a 0 we are successful, then this 0, say, in column j_{k+1}, is added to the sequence.

There are two ways in which the construction of this sequence can be terminated. One way is when the 0's are changed to 0^*'s and the 0^*'s are changed to 0's. In this case we have found a 0 in row i_0 which can be assigned. The other is when all the cells are deleted because they lie in necessary columns. In this case, a complete assignment has not been made and we must go to Step 4.

EXAMPLE 4.   Consider the assignment problem whose cost matrix is

$$\mathbf{C} = \begin{bmatrix} 8 & 7 & 9 & 9 \\ 5 & 2 & 7 & 8 \\ 6 & 1 & 4 & 9 \\ 2 & 3 & 2 & 6 \end{bmatrix}$$

We construct $\mathbf{C}'$ which has a zero in each row and column by performing Step 1 of the algorithm. We subtract the smallest entry in each row from that row and then subtract the smallest entry in each column from that column, obtaining

$$\mathbf{C}' = \begin{bmatrix} 1 & 0 & 2 & 0 \\ 3 & 0 & 5 & 4 \\ 5 & 0 & 3 & 6 \\ 0 & 1 & 0 & 2 \end{bmatrix}$$

We now assign zeros starting in row 1 as in Step 2. We find that row 2 is the first row which has no 0 in an unassigned column. At this point,

$$\mathbf{C}' = \begin{bmatrix} 1 & 0^* & 2 & 0 \\ 3 & 0 & 5 & 4 \\ 5 & 0 & 3 & 6 \\ 0^* & 1 & 0 & 2 \end{bmatrix}$$

We then start constructing a sequence of 0's and 0^*'s for Step 3. The first cell is $(2, 2)$, which contains 0. We search column 2 for a 0^* and find it in cell $(1, 2)$. We search row 1 for a 0 and find it in cell $(1, 4)$. We search column 4 for a 0* and find none. Consequently we are in Case A with the sequence

$$0 \quad \text{in} \quad (2, 2)$$

$$0^* \quad \text{in} \quad (1, 2)$$

$$0 \quad \text{in} \quad (1, 4)$$

Changing every 0 to a 0* and every 0* to a 0 in the sequence, we obtain the matrix

$$\mathbf{C'} = \begin{bmatrix} 1 & 0 & 2 & 0^* \\ 3 & 0^* & 5 & 4 \\ 5 & 0 & 3 & 6 \\ 0^* & 1 & 0 & 2 \end{bmatrix}$$

where we increased the number of assignments by 1. We now return to Step 2 at row 3.

There is no assignable 0 in row 3; therefore, we must construct a sequence starting at the 0 in cell $(3, 2)$. The 0* in column 2 is in row 2. But there are no other 0's in row 2. We are in Case B and column 2 is necessary. Our sequence is

$$0 \quad \text{in} \quad (3, 2)$$

$$0^* \quad \text{in} \quad (2, 2)$$

Since all the cells in our sequence lie in a necessary column, we must go to Step 4 to determine the necessary rows.

A row is **necessary** if it contains a 0* in an unnecessary column. Starting with row 1, we find that its 0* is in column 4, and consequently row 1 is necessary. Row 2 has its 0* in column 2, which is a necessary column. Therefore, row 2 is not necessary. Row 3 has no 0*. Row 4 has its 0* in column 1 and consequently is necessary.

Step 4. Covering each necessary row and column with a line provides the k lines previously described. This procedure automatically covers all zeros of $\mathbf{C'}$.

We find that the $\mathbf{C'}$ of our example is covered as follows:

$$\mathbf{C'} = \begin{bmatrix} 1 & 0 & 2 & 0^* \\ 3 & 0^* & 5 & 4 \\ 5 & 0 & 3 & 6 \\ 0^* & 1 & 0 & 2 \end{bmatrix}$$

We are now ready for Step 5. Subtract 3, the smallest uncovered entry, from each uncovered entry and add 3 to each doubly covered entry. Our new $\mathbf{C}'$, ready for Step 2, is

$$\mathbf{C}' = \begin{bmatrix} 1 & 3 & 2 & 0 \\ 0 & 0 & 2 & 1 \\ 2 & 0 & 0 & 3 \\ 0 & 4 & 0 & 2 \end{bmatrix}$$

At the completion of the assignment procedure in Step 2 we have

$$\begin{bmatrix} 1 & 3 & 2 & 0^* \\ 0^* & 0 & 2 & 1 \\ 2 & 0^* & 0 & 3 \\ 0 & 4 & 0^* & 2 \end{bmatrix}$$

which is a complete assignment.

We can now rewrite Step 3 in the algorithm as:

Step 3. Assume that no zero in row i_0 has been assigned and that there is a zero in cell (i_0, j_0). Construct a sequence of vertical and horizontal segments which alternate and join 0 to 0* to 0 to 0* and so on as follows:

(A) If we are at 0 in cell (i_k, j_k), search column j_k for 0*. If 0* is found, adjoin its cell to the sequence. If 0* is not found, change each 0 in the sequence to 0* and each 0* to 0 and search for the next row without a 0*.

(B) If we are at 0* in cell (i_{k+1}, j_k), search row i_{k+1} for 0. If 0 is found, adjoin its cell to the sequence. If 0 is not found, label column j_k as necessary and delete cells (i_k, j_k) and (i_{k+1}, j_k) from the sequence. If there are no more cells in the sequence, go to Step 4. If there are more cells, we are at 0* in cell (i_k, j_{k-1}).

The algorithm which we have developed is called the **Hungarian method** in honor of the mathematicians König and Egervary on whose work it is based.

The Hungarian method as we have described it assumes that the given assignment problem is a minimization problem. However, a maximization problem can be modified in a way that will enable the Hungarian method to produce an optimal solution. Suppose we are given the problem

$$\text{Maximize} \quad z = \sum_{i=1}^{n} \sum_{j=1}^{n} c_{ij} x_{ij}$$

subject to

$$\left.\begin{array}{ll}
\sum_{j=1}^{n} x_{ij} = 1, & i = 1, 2, \ldots, n \\[2ex]
\sum_{i=1}^{n} x_{ij} = 1, & j = 1, 2, \ldots, n
\end{array}\right\} \tag{5}$$

$$x_{ij} \geq 0, \quad \text{integer}$$

We can convert this problem to the minimization problem

$$\text{Minimize} \quad z = \sum_{i=1}^{n} \sum_{j=1}^{n} (-c_{ij}) x_{ij}$$

subject to the constraints in (5)

If some of the entries in the matrix $[c_{ij}]$ were positive, we now have negative entries in $[-c_{ij}]$, and the optimality criterion in Theorem 5.2 does not apply. However, we can use Theorem 5.3 and add to each entry in $[-c_{ij}]$ the negative of the smallest (most negative) entry. This new matrix will have nonnegative entries, and the assignment problem for it can be solved using the Hungarian method.

EXAMPLE 5.   Suppose an assignment problem asks to maximize the total value of the assignment where the individual values are given by

$$[c_{ij}] = \begin{bmatrix} 3 & 7 & 4 & 6 \\ 5 & 2 & 8 & 5 \\ 1 & 3 & 4 & 7 \\ 6 & 5 & 2 & 6 \end{bmatrix}$$

The corresponding minimization problem is

$$\text{Minimize} \quad z = \sum_{i=1}^{n} \sum_{j=1}^{n} (-c_{ij}) x_{ij}$$

subject to the constraints in (5)

The smallest entry in $[-c_{ij}]$ is -8; thus, we add 8 to each entry in $[-c_{ij}]$

to obtain

$$\begin{bmatrix} 5 & 1 & 4 & 2 \\ 3 & 6 & 0 & 3 \\ 7 & 5 & 4 & 2 \\ 2 & 3 & 6 & 2 \end{bmatrix}$$

This matrix is now used as the cost matrix for the Hungarian method.

5.2 Exercises

In Exercises 1–6 solve the assignment problem with given cost matrix.

1. $\begin{bmatrix} 4 & 2 & 3 & 5 \\ 2 & 3 & 4 & 6 \\ 3 & 2 & 5 & 2 \\ 2 & 5 & 3 & 4 \end{bmatrix}$

2. $\begin{bmatrix} 3 & 2 & 5 & 8 & 9 \\ 6 & 7 & 4 & 2 & 3 \\ 5 & 3 & 5 & 4 & 2 \\ 4 & 7 & 3 & 2 & 4 \\ 2 & 6 & 5 & 5 & 3 \end{bmatrix}$

3. $\begin{bmatrix} 3 & 4 & 0 & 2 & 6 & 7 \\ 4 & 6 & 4 & 5 & 3 & 6 \\ 5 & 7 & 7 & 8 & 2 & 8 \\ 0 & 8 & 8 & 4 & 6 & 4 \\ 6 & 4 & 3 & 7 & 4 & 9 \\ 7 & 5 & 5 & 0 & 6 & 7 \end{bmatrix}$

4. $\begin{bmatrix} 3 & 2 & 7 & 4 & 8 \\ 5 & 4 & 3 & 8 & 5 \\ 3 & 7 & 9 & 1 & 2 \\ 4 & 2 & 6 & 5 & 7 \\ 2 & 8 & 4 & 6 & 6 \end{bmatrix}$

5. $\begin{bmatrix} 3 & 5 & 4 & 2 & 8 & 1 \\ 8 & 3 & 6 & 6 & 4 & 3 \\ 4 & 4 & 8 & 8 & 3 & 5 \\ 3 & 8 & 7 & 4 & 9 & 7 \\ 7 & 7 & 9 & 2 & 3 & 5 \\ 9 & 7 & 2 & 7 & 5 & 8 \end{bmatrix}$

6. $\begin{bmatrix} 9 & 7 & 4 & 7 & 3 \\ 0 & 8 & 3 & 5 & 8 \\ 6 & 3 & 2 & 8 & 9 \\ 5 & 6 & 7 & 0 & 4 \\ 2 & 9 & 5 & 7 & 3 \end{bmatrix}$

7. A company leases offices along one side of a hall in a large office building. Suppose there are 15 offices all the same size leased by the company, and they are numbered 701 through 715. Because of a recent series of job reassignments the desks in some of the offices must be moved. Specifically, the desks in 702, 705, 708, 709, and 713 must be moved to 706, 707, 712, 714, and 715, but it does not matter which desk goes to which office. Consequently, the facilities supervisor has decided to assign desks in such a way as to minimize the total distance which they would need to be moved. What assignment should the supervisor use?

5.2 Projects

1. Consider the problem of scheduling the full-time registered nurses on a surgical floor of a hospital. Adequate patient care requires that there be 4 RN's during the day, 2 RN's in the evening, and 1 RN at night. Assume that the scheduling

is done two weeks at a time. In the two-week planning horizon each nurse must work 10 shifts.

(a) How many nurses are required?

(b) A schedule for a nurse is a list indicating which of the 42 shifts (why 42?) are to be worked by that nurse. How many possible schedules are there? Which of these schedules would be less desirable? What guidelines might be used for making up acceptable schedules for the nurses?

(c) Assume the decisions about schedules for each nurse are made in the following manner: Each nurse is given a list of 10 possible schedules following the guidelines from part b and asked to rank them in order of desirability. Each nurse gets a different list of schedules. Each nurse's first choice is to be assigned a weight of 10, the second a weight of 9, and so on. How might ties be handled? The choice of one schedule for each nurse from among those presented is determined by maximizing the sum of the weights of the chosen schedules. Set up a model for this situation.

(d) Estimate the human time and the computer time involved in determining the schedule for staffing the floor in this way. (Computer time estimates can come from the size of the assignment problem.)

2. Consider the following generalizations of the desk-moving problem* in Exercise 7:

(a) Assume there are N offices from which desks must be moved. Let $\mathbf{D} = [d_{ij}]$ be the matrix of distances from office i to office j. Suppose the current locations are $s_1 < s_2 < \cdots < s_N$, where "<" means ordered from near to far from the elevator along the hall. Suppose the future locations are $t_1 < t_2 < \cdots < t_N$. Show that an optimal assignment is $s_k \rightarrow t_k$, $k = 1, 2, \ldots, N$.

(b) Suppose instead of being located along a hall, the offices are located around the entire perimeter of a floor of the office building. Suppose that the distance to walk completely around the floor is L and that $\mathbf{D} = [d_{ij}]$ is the matrix of shortest distances from location i to location j. What is the minimal-distance assignment?

Further Readings

Kuhn, H. W. "The Hungarian Method for the Assignment Problem." *Naval Research Logistics Quarterly*, 2 (1955), 83–97.

Kuhn, H. W. "Variants of the Hungarian Method for the Assignment Problem." *Naval Research Logistics Quarterly*, 3 (1956), 253–58.

---

* R. M. Karp and S-Y. Li, "Two Special Cases of the Assignment Problem." *Discrete Mathematics*, 13 (1975), 129–42.

5.3   GRAPHS AND NETWORKS. BASIC DEFINITIONS

A **graph** is a collection of points, called **nodes** (also called **vertices**), some of which (possibly all) are joined by lines, called **arcs**, **edges**, or **branches**. Every graph can be conveniently represented by a figure in the plane. Thus, Figure 5.1 shows a graph with seven nodes and nine arcs. We may note that the arc joining nodes 1 and 3 and the arc joining nodes 2 and 4 intersect at a point which is not a node in this representation of the graph. The arc joining nodes i and j will be denoted by (i, j).

An arc of the form (a, a) is called a **loop**. We will not allow loops in our graphs. If the nodes of a graph are numbered consecutively from 1 to n, the graph can be represented by a matrix. The **incidence matrix** of a graph G is the matrix $\mathbf{M} = [m_{ij}]$, where

$$m_{ij} = \begin{cases} 1 & \text{if node } i \text{ is connected to node } j \\ 0 & \text{otherwise} \end{cases}$$

For an arbitrary graph, the incidence matrix is symmetric $(\mathbf{M} = \mathbf{M}^{\mathsf{T}})$. Why? The incidence matrix for the graph in Figure 5.1 is

$$\mathbf{M} = \begin{bmatrix} 0 & 1 & 1 & 1 & 1 & 0 & 0 \\ 1 & 0 & 1 & 1 & 0 & 0 & 0 \\ 1 & 1 & 0 & 0 & 0 & 0 & 0 \\ 1 & 1 & 0 & 0 & 1 & 0 & 1 \\ 1 & 0 & 0 & 1 & 0 & 1 & 0 \\ 0 & 0 & 0 & 0 & 1 & 0 & 0 \\ 0 & 0 & 0 & 1 & 0 & 0 & 0 \end{bmatrix}$$

A **path** between the nodes i and j of a graph is an ordered set of arcs

$$(i, a_1), (a_1, a_2), (a_2, a_3), \ldots, (a_r, j)$$

joining i and j. Thus, a path between nodes 2 and 6 in Figure 5.1 is

$$(2, 4), (4, 5), (5, 6)$$

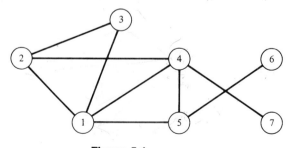

Figure 5.1

Other paths between nodes 2 and 6 are

$$(2, 1), (1, 4), (4, 5), (5, 6)$$
$$(2, 1), (1, 5), (5, 6)$$

and

$$(2, 3), (3, 1), (1, 4), (4, 5), (5, 6)$$

A **cycle** is a path joining a node to itself. Examples of cycles in Figure 5.1 are

$$(2, 3), (3, 1), (1, 2)$$

and

$$(2, 4), (4, 5), (5, 1), (1, 2)$$

A graph is said to be **connected** if there is a path joining any two nodes of the graph. The graph shown in Figure 5.1 is connected, whereas the graph shown in Figure 5.2 is not connected.

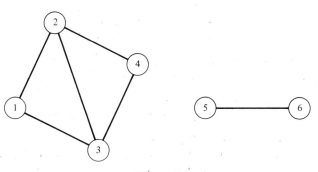

Figure 5.2

An arc of a graph is called **directed** or **oriented** if there is a sense of direction so that one node is considered the point of origin and the other node is the point of termination. The directed arc from node i to node j will be denoted by $\overline{(i, j)}$. A graph in which every arc is directed is called a **directed graph**, a **digraph**, or an **oriented graph**. An example of a directed graph is shown in Figure 5.3.

An incidence matrix can also be used to represent a directed graph. In this case $m_{ij} = 1$ if a directed arc connects i to j. Consequently, the incidence matrix of a digraph is usually not symmetric. The incidence

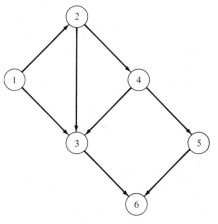

Figure 5.3

matrix of the digraph in Figure 5.3 is

$$
\mathbf{M} =
\begin{bmatrix}
0 & 1 & 1 & 0 & 0 & 0 \\
0 & 0 & 1 & 1 & 0 & 0 \\
0 & 0 & 0 & 0 & 1 & 1 \\
0 & 0 & 1 & 0 & 1 & 0 \\
0 & 0 & 0 & 0 & 0 & 1 \\
0 & 0 & 0 & 0 & 0 & 0
\end{bmatrix}
$$

For certain applications of linear programming we are interested in the idea of a network. Intuitively, a network is an interconnection of several terminals by routes between certain pairs of these terminals. Each route has a capacity, and we are interested in studying the movement or flow of material along these routes. We can study pipeline networks, highway networks, electrical power distribution networks, as well as many other kinds.

More formally, a **network** is a connected, directed graph for which a nonnegative number has been assigned to each ordered pair of nodes. This number is thought of as the **capacity** of the directed arc joining the two nodes. If node i is not connected to node j by an arc, the capacity c_{ij} is set to zero. The capacities represent maximum amounts which may pass along arcs of networks; specifically, they may be tons of oil/hour in a transcontinental oil pipeline network, cubic meters of water/minute in a city water system, pulses/second in a communications networks, or number of vehicles/hour on a regional highway system. The nodes may represent shipping depots, relay stations, highway interchanges, or pumping stations. A network may be represented by its **capacity matrix**, which is a generalization of the incidence matrix of a graph and consists of the

matrix of capacities of the arcs. The directed graph in Figure 5.3 becomes
a network when we specify its capacity matrix as

$$
\mathbf{C} =
\begin{bmatrix}
0 & 7 & 5 & 0 & 0 & 0 \\
0 & 0 & 2 & 3 & 0 & 0 \\
0 & 0 & 0 & 0 & 4 & 8 \\
0 & 0 & 8 & 0 & 2 & 0 \\
0 & 0 & 0 & 0 & 6 & 0 \\
0 & 0 & 0 & 0 & 0 & 0
\end{bmatrix}
$$

If there is no limit on the amount passing from node i to node j in a
network, the capacity c_{ij} is set equal to a very large number M.

If flow is permitted in both directions between a pair of nodes (as in
a two-way street system), the nodes are connected with two directed
arcs, one going in each direction, each with its own capacity. That is, a
two-way street is thought of as two one-way streets.

A **flow** in a network is an assignment to each ordered pair of nodes
(i, j) in the network of a nonnegative number x_{ij} which represents the
amount of material moving in that directed arc. If node i is not connected
to node j, then $x_{ij} = 0$. By definition, the flow may not exceed the
capacity, so that we have for each i and j, $i, j = 1, 2, \ldots, n$,

$$
0 \leq x_{ij} \leq c_{ij}
$$

In many networks we single out two types of nodes for special consideration. A node S is called a **source** if every arc joining S to another node
is oriented so that the flow is away from S. A node T is called a **sink** if
every arc joining T to another node is directed so that the flow is toward
T. That is, a flow is produced at a source and it is absorbed at a sink.

We also specify that any flow cannot cause material to accumulate at
any node other than a source or a sink. That is, with the exception of a
source or a sink, the flow into a node is equal to the flow out of the node.
If we have a network with exactly one source and exactly one sink,
number the source as node 1 and the sink as node n. Then the flow must
satisfy, for each node k, $k = 2, \ldots, n - 1$,

$$
\sum_{i=1}^{n} x_{ik} = \sum_{j=1}^{n} x_{kj}
$$

In the next few sections we will consider certain questions about flows
in networks and certain situations that can be modeled with such a
mathematical structure.

5.3 Exercises

1. For the graph

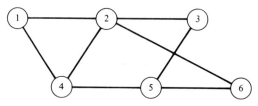

(a) find its incidence matrix;

(b) find three paths joining nodes 1 and 4;

(c) find two cycles from node 2.

2. Follow the instructions in Exercise 1 for the graph

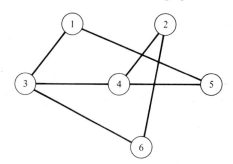

3. Sketch the graph whose incidence matrix is

(a) $\begin{bmatrix} 0 & 1 & 1 & 1 & 0 \\ 1 & 0 & 0 & 1 & 1 \\ 1 & 0 & 0 & 1 & 0 \\ 1 & 1 & 1 & 0 & 1 \\ 0 & 1 & 0 & 1 & 0 \end{bmatrix}$
(b) $\begin{bmatrix} 0 & 0 & 1 & 0 & 1 & 1 \\ 0 & 0 & 0 & 1 & 0 & 1 \\ 1 & 0 & 0 & 1 & 1 & 0 \\ 0 & 1 & 1 & 0 & 0 & 0 \\ 1 & 0 & 1 & 0 & 0 & 1 \\ 1 & 1 & 0 & 0 & 1 & 0 \end{bmatrix}$

4. Find the incidence matrix for the graph in Figure 5.2.

5. For the network

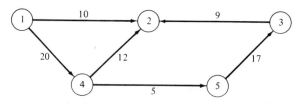

where the numbers on the edges are the capacities

(a) find the capacity matrix;

(b) find the sources;

(c) find the sinks.

6. Follow the instructions for Exercise 5 for the network

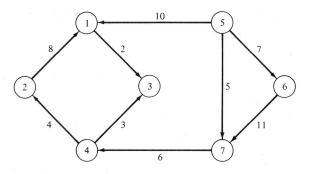

7. Show that a node i is a source in a network if and only if the ith column of the capacity matrix of the network is zero.

8. Show that a node j is a sink in a network if and only if the jth row of the capacity matrix of the network is zero.

Further Reading

Wilson, Robin J. *Introduction to Graph Theory*. Academic Press, New York, 1972.

5.4 THE MAXIMAL FLOW PROBLEM

Consider a network with n nodes which includes a single source and a single sink. For convenience we label the source as node 1 and the sink as node n. Let c_{ij} denote the capacity of arc (i, j). Consider a flow in this network, letting x_{ij} denote the amount of material flowing from node i to node j along arc (i, j). As we discussed in Section 5.3, the x_{ij} must satisfy

$$0 \le x_{ij} \le c_{ij}, \qquad i, j = 1, 2, \ldots, n \tag{1}$$

and

$$\sum_{i=1}^{n} x_{ik} = \sum_{j=1}^{n} x_{kj}, \qquad k = 2, 3, \ldots, n-1 \tag{2}$$

We may also write (2) as

$$\sum_{i=1}^{n} x_{ik} - \sum_{j=1}^{n} x_{kj} = 0, \qquad k = 2, 3, \ldots, n-1 \tag{2a}$$

The total flow starting from the source, which is to be maximized, is

$$f = \sum_{k=1}^{n} x_{1k} \tag{3}$$

The total flow into the sink is

$$\sum_{k=1}^{n} x_{kn}$$

which by the conservation of flow is precisely the expression on the right side of (3). (Verify.) Thus, a mathematical formulation of the **maximal flow problem** is

$$\text{Maximize} \quad f = \sum_{k=1}^{n} x_{1k}$$

subject to

$$\sum_{i=1}^{n} x_{ik} - \sum_{j=1}^{n} x_{kj} = 0, \qquad k = 2, 3, \ldots, n-1 \tag{4}$$

$$0 \le x_{ij} \le c_{ij}, \qquad i, j = 1, 2, \ldots, n$$

The mathematical formulation of the maximal flow problem shows that we have a linear programming problem, which could be solved by the simplex method. However, this approach is quite inefficient and in this section we present several better procedures.

Before turning to the computational procedures, we note that the maximal flow problem occurs in many applications. As a typical applied problem, consider an electrical power distribution system represented by a network which has one source and many sinks. When a brownout appears imminent at a particular location (node), that location (node) is made a sink and the goal becomes to send as much electrical power as possible to this endangered location. Thus, we have a maximal flow problem.

The following intuitive method appears to offer some hope for solving the maximal flow problem. We start at the source, and by proceeding along arcs with positive capacity we find a path from source to sink. (Why can we find this path?) The maximum amount f_1 of material that can be sent along this path is the minimum of the capacities of the arcs in the path. We now subtract f_1 from the capacity of each arc in the path just used. The capacity of at least one of the arcs in this path is now reduced to zero. We next return to the source and proceed along arcs with positive capacity to find another path to the sink. The maximum amount f_2 of material that can be sent along this path is the minimum of the capacities of the arcs in the path. The capacity of at least one of the arcs in this path is now reduced to zero by subtracting f_2 from each capacity in the path. We continue choosing possible paths from source to sink until there are no more paths all of whose arcs have positive capacity. The total flow f is the sum of the flows

$$f = f_1 + f_2 + \cdots f_k$$

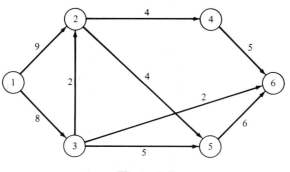

Figure 5.4

To illustrate this intuitive method, consider the network with source 1 and sink 6 (Figure 5.4).

We first choose the path

$$1 \rightarrow 3 \rightarrow 2 \rightarrow 4 \rightarrow 6$$

which has flow $f_1 = 2$. Subtracting f_1 from the capacity of each arc in this path, we obtain the network in Figure 5.5.

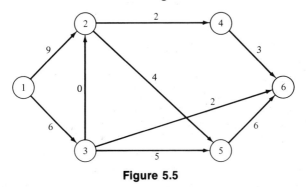

Figure 5.5

Next, choose the path

$$1 \rightarrow 3 \rightarrow 5 \rightarrow 6$$

whose flow $f_2 = 5$ yields the network in Figure 5.6.

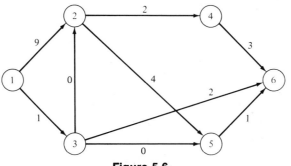

Figure 5.6

Now choose the path

$$1 \rightarrow 2 \rightarrow 5 \rightarrow 6$$

with flow $f_3 = 1$, obtaining the network in Figure 5.7.

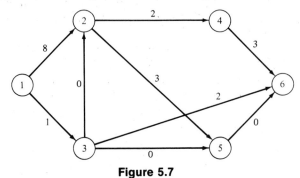

Figure 5.7

For the next path we have

$$1 \rightarrow 2 \rightarrow 4 \rightarrow 6 \quad \text{with flow} \quad f_4 = 2 \qquad \text{(Figure 5.8)}$$

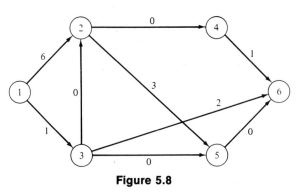

Figure 5.8

We now choose the path

$$1 \rightarrow 3 \rightarrow 6 \quad \text{with flow} \quad f_5 = 1 \qquad \text{(Figure 5.9)}$$

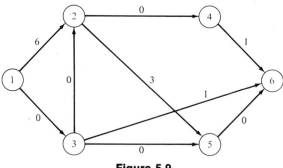

Figure 5.9

Since we cannot find any other paths from source to sink, all of whose arcs have positive capacities, we are finished. The total flow is

$$f = f_1 + f_2 + f_3 + f_4 + f_5 = 2 + 5 + 1 + 2 + 1 = 11$$

However, suppose that instead of choosing the above sequence of paths from source to sink, we choose the sequence of paths indicated below:

$1 \to 2 \to 4 \to 6,$ Flow $f_1 = 4$ (Figure 5.10)

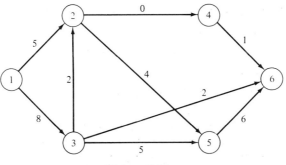

Figure 5.10

$1 \to 3 \to 5 \to 6,$ Flow $f_2 = 5$ (Figure 5.11)

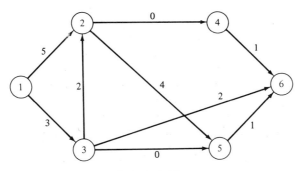

Figure 5.11

$1 \rightarrow 3 \rightarrow 6,$ Flow $f_3 = 2$ (Figure 5.12)

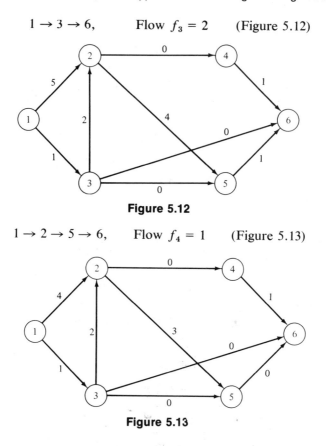

Figure 5.12

$1 \rightarrow 2 \rightarrow 5 \rightarrow 6,$ Flow $f_4 = 1$ (Figure 5.13)

Figure 5.13

Since we cannot find any other paths from source to sink, all of whose arcs have positive capacities, we are finished. The total flow is

$$f = f_1 + f_2 + f_3 + f_4 = 4 + 5 + 2 + 1 = 12$$

Thus, the intuitive method does not always yield the maximal flow. It can be modified to a correct algorithm by allowing fictitious flows in the opposite direction, so that an alternate path can be chosen. However, for sizable problems this procedure becomes too clumsy, so we shall not pursue this method any further. Instead, we describe the labeling procedure of Ford and Fulkerson. In this algorithm we do not successively start at the source and proceed along a path to the sink. Instead, we branch out from the source along many different paths at one time.

The Labeling Method

We continue to number the nodes so that node 1 is the source and node n is the sink. The directed arc joining node i to j has capacity c_{ij}. If flow cannot take place from i to j, then $c_{ij} = 0$. For each arc we now define the **excess capacity**

$$d_{ij} = c_{ij} - x_{ij} + x_{ji}$$

We begin with all flows set at value zero. Next label every directed arc with its excess capacity d_{ij}.

Step 1. Starting at the source, we form the set N_1 of all nodes which are connected to the source by an arc with positive excess capacity. We use the index k for nodes in N_1. Now label each node in N_1 with the ordered pair of numbers (e_k, p_k), where

$e_k = d_{1k} =$ Excess capacity of the arc from the source to node k
$p_k =$ Node which led to node k

Here $p_k = 1$ for each node in N_1, since we got to this node from the source.
  If we have labeled node n, the sink of the network, we then proceed directly to Step 5, where we increase the flow.

Step 2. Choose the node in N_1 with smallest index; say it is node k. Let N_2 denote the set of all unlabeled nodes which are joined to node k by an arc with positive excess capacity. From now on we must assume that the source is a labeled node. If there are no such unlabeled nodes, we pick the node in N_1 with the next smallest index and again form N_2. We use the index m for each node in N_2. Label each unlabeled node in N_2 with the ordered pair (e_m, p_m), where

$$e_m = \min \{d_{km}, e_k\}$$

$$p_m = k$$

Observe that e_m is the minimum of the excess capacities of the arcs from the source to node k and from node k to node m. Also, p_m denotes the node which led to node m. We repeat for each node in N_1.

Step 3. Repeat Step 2 with N_r replacing N_{r-1}. After a finite number of steps, we arrive at one of two possibilities:
  (i) The sink has not been labeled and no other nodes can be labeled.
  (ii) The sink has been labeled.

Step 4. If we are in case (i), then the current flow can be shown to be maximal and we stop.

Step 5. If we are in case (ii), we now increase the flow as follows. Suppose that the sink has the label (e_r, p_r). The first number in the label, e_r, indicates the amount by which we can increase the flow. The second number in the label, p_r, gives the node that led to the sink, making it possible to move backwards along this path to the source. Let d_{st} denote the excess capacities of the arcs in the path P. To increase the flow by e_r we now calculate the excess capacities as

$$d'_{st} = d_{st} - e_r$$

$$d'_{ts} = d_{ts} + e_r$$

$$d'_{ij} = d_{ij} \qquad \text{for arcs not in } P$$

Step 6. Return to Step 1.

Assuming that a maximal flow exists, the algorithm terminates after a finite number of iterations. That is, after a finite number of iterations we reach case (i).

We now calculate the net flows in each arc as follows. If $c_{ji} = 0$, so that flow cannot take place from node j to node i, then the flow in the arc from i to j is

$$x_{ij} = c_{ij} - d_{ij}$$

where d_{ij} is the most recent excess capacity calculated as described in Step 5. If both c_{ij} and c_{ji} are positive, so that flow can take place from i to j as well as from j to i, observe that

$$c_{ij} - d_{ij} = x_{ij} - x_{ji}$$

$$c_{ji} - d_{ji} = x_{ji} - x_{ij} = -(c_{ij} - d_{ij})$$

Hence, $c_{ij} - d_{ij}$ and $c_{ji} - d_{ji}$ cannot both be positive. We let

$$\left. \begin{array}{l} x_{ij} = c_{ij} - d_{ij} \\ x_{ji} = 0 \end{array} \right\} \quad \text{if} \quad c_{ij} - d_{ij} \geq 0$$

$$\left. \begin{array}{l} x_{ji} = c_{ji} - d_{ji} \\ x_{ij} = 0 \end{array} \right\} \quad \text{if} \quad c_{ji} - d_{ji} \geq 0$$

We illustrate the labeling method with the network considered in the intuitive method. Start with all flows set at value zero. Figure 5.14 shows the given network with each arc labeled with its excess capacity.

Step 1. Starting at the source, node 1, we find all nodes which are connected to it by an arc with positive excess capacity. These are nodes 2 and 3. Thus,

$$C_{12} - X_{12} + X_{21} \qquad C_{13} - X_{13} + X_{31}$$

$$e_2 = d_{12} = 9, \qquad e_3 = d_{13} = 8$$

$$p_2 = 1, \qquad p_3 = 1$$

We now label nodes 2 and 3 with the respective ordered pairs $(9, 1)$ and $(8, 1)$, shown in Figure 5.15.

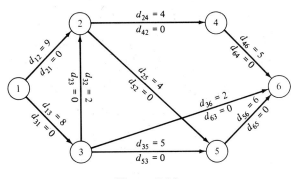

Figure 5.14

Step 2. Starting from node 2, we find all unlabeled nodes which are joined to node 2 by an arc with positive excess capacity. These are nodes 4 and 5. The label on node 4 is (e_4, p_4), where

$$e_4 = \min \{d_{24}, e_2\}$$
$$= \min \{4, 9\} = 4$$
$$p_4 = 2$$

Similarly, the label on node 5 is (e_5, p_5), where

$$e_5 = \min \{d_{25}, e_2\}$$
$$= \min \{4, 9\} = 4$$
$$p_5 = 2$$

We proceed in the same manner from node 3. The only unlabeled node that can be reached from node 3 is node 6, the sink. This node is labeled

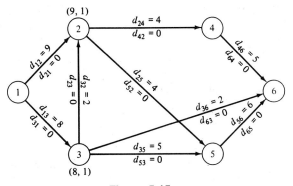

Figure 5.15

(e_6, p_6), where

$$e_6 = \min \{d_{36}, e_3\}$$
$$= \min \{2, 8\} = 2$$
$$p_6 = 3$$

The network in Figure 5.16 shows the labels obtained in Step 2. Observe that we have labeled the sink.

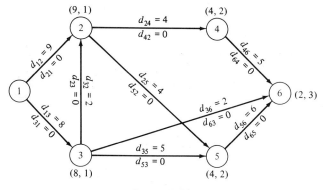

Figure 5.16

Since we have labeled the sink, we proceed to Step 5. We can increase the flow by $e_6 = 2$ units. We now move backwards to the source and calculate new excess capacities as follows. Since $p_6 = 3$, we go back to node 3, and let

$$d'_{36} = d_{36} - e_6 = 2 - 2 = 0$$
$$d'_{63} = d_{63} + e_6 = 0 + 2 = 2$$

Since $p_3 = 1$, we go back to node 1, the source, and let

$$d'_{13} = d_{13} - e_6 = 8 - 2 = 6$$
$$d'_{31} = d_{31} + e_6 = 0 + 2 = 2$$

The network in Figure 5.17 shows the new excess capacities. We now return to Step 1 and relabel the nodes. In Figure 5.17 we have also indicated these new labels.

The sink has again been labeled. We can increase the flow by 4 units. We move backwards along the path

$$1 \rightarrow 2 \rightarrow 4 \rightarrow 6$$

and calculate new excess capacities which are indicated in the network in Figure 5.18. In this network we have also indicated the newest labels on the nodes.

The sink has been labeled and evidently we can increase the flow by

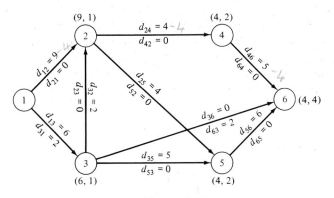

Figure 5.17

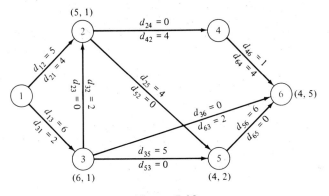

Figure 5.18

4 units. We move backwards along the path

$$1 \rightarrow 2 \rightarrow 5 \rightarrow 6$$

and calculate new excess capacities, which are indicated in the network in Figure 5.19. In this network we have also indicated the newest labels on the nodes.

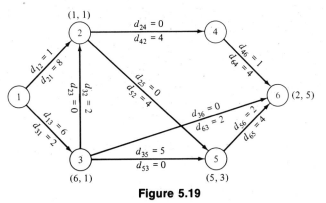

Figure 5.19

We have labeled the sink and can now increase the flow by 2 units. We move backwards along the path

$$1 \rightarrow 3 \rightarrow 5 \rightarrow 6$$

and calculate new excess capacities, which are indicated in the network in Figure 5.20. In this network we have also indicated the newest labels on the nodes.

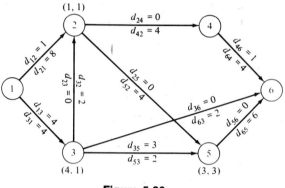

Figure 5.20

At this point, the sink has not been labeled and no other nodes can be labeled. The current flow is maximal and we stop.

The maximal flow is the sum of all that has been sent to the sink:

$$2 + 4 + 4 + 2 = 12$$

The net flow in each arc is shown in the network in Figure 5.21.

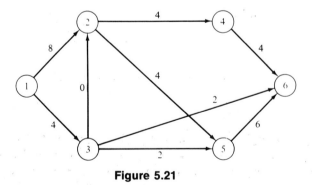

Figure 5.21

We shall now show that if the above procedure yields case (i), then the current flow is optimal. We must start with another definition. A **cut** in a network is a set of directed arcs with the property that every path from the source to the sink contains at least one arc from the set. Since the number of directed arcs in a network is finite, the number of cuts is also finite. From this finite set of cuts we shall soon single out one cut. We define the **capacity** of a cut as the sum of the capacities of its directed arcs.

EXAMPLE 1.   Consider the network in Figure 5.22.

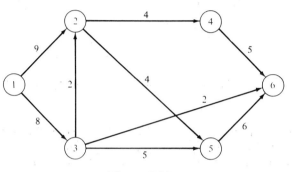

Figure 5.22

The set of directed arcs

$$\{(\overline{4, 6}), (\overline{2, 5}), (\overline{3, 2}), (\overline{1, 3})\}$$

is a cut with capacity 19. The set of directed arcs

$$\{(\overline{4, 6}), (\overline{3, 6}), (\overline{3, 5})\}$$

is not a cut because the path

$$1 \rightarrow 2 \rightarrow 5 \rightarrow 6$$

contains no directed arc from the set.

For a given path, the flow along this path cannot exceed the smallest of the capacities of its arcs. Hence, for any cut the flow along this path cannot exceed the capacity of whichever of its arcs it is that belongs to the cut. Now a maximal flow consists of a sum of flows along various paths from the source to the sink. Therefore, for any cut a maximal flow cannot exceed the sum of the capacities of the arcs of the various paths which belong to the cut. Hence, a maximal flow cannot exceed the capacity of the cut.

Suppose that we are now in case (i) of the labeling algorithm: the sink has *not* been labeled and no other nodes can be labeled. The set N of all nodes of the network can be partitioned into two disjoint subsets N_{L}, the labeled nodes, and N_{U}, the unlabeled nodes.

Let A be the set of directed arcs which join nodes in N_L to nodes in N_U. We first show that A is a cut in the network. Suppose it is not a cut, so that there is a path from source to sink which does not contain any directed arc from A. It then follows that all the nodes in this path belong either to N_L or to N_U (why?). Since by definition the source is labeled and the sink is unlabeled, we have a contradiction to our assumption that A is not a cut. Therefore, it must be one.

We next show that the capacity of the cut A is equal to the maximal flow in the network. From Equation (2a) and the definition of the excess capacities, we can write

$$\sum_{j=1}^{n} (c_{ij} - d_{ij}) = 0, \qquad i = 2, \ldots, n-1 \tag{5}$$

When $i = 1$, we obtain from the definition of excess capacity

$$\sum_{j=2}^{n} (c_{1j} - d_{1j}) = \sum_{j=2}^{n} x_{1j} \tag{6}$$

since $x_{j1} = 0$ for $j = 1, 2, \ldots, n$. (Why?) Thus, the sum in (6) gives the total flow. We now combine (5) and (6) and obtain

$$\sum_{i \in N_L} \sum_{j=1}^{n} (c_{ij} - d_{ij}) = \sum_{j=2}^{n} x_{1j} = \text{Total flow} \tag{7}$$

since the source, node 1, belongs to N_L. Consider the left side of (7). If i and j both belong to N_L, then $c_{ij} - d_{ij}$ and $c_{ji} - d_{ji}$ both occur on the left side of (7) and cancel each other out. Thus, only the terms for which j belongs to N_U are left in the sum. If j is in N_U, then $d_{ij} = 0$. Hence, the left side of (7) becomes

$$\sum_{i \in N_L} \sum_{j \in N_U} c_{ij}$$

and thus (7) can be written as

$$\sum_{i \in N_L} \sum_{j \in N_U} c_{ij} = \text{Total flow} \tag{8}$$

Now the left side of (8) is the capacity of the cut A, since A consisted of exactly those arcs joining nodes in N_L to nodes in N_U. Thus, for this particular cut, the total flow is exactly equal to its capacity.

Since we have already shown that the maximal flow of a network cannot exceed the capacity of any cut, we conclude that for the cut A, just constructed, the flow is maximal. Thus, the labeling algorithm described earlier does yield a maximal flow. Moreover, similar reasoning shows that the cut A has the minimum capacity among all cuts. The results established here can also be stated as the max flow–min cut theorem.

Theorem 5.5. (Max Flow–Min Cut Theorem) The maximum flow in a network is the minimum of the capacities of all cuts in the network.

EXAMPLE 2. The cut A defined by the network in Figure 5.20 is $\{(\overline{2, 4}), (\overline{3, 6}), (\overline{5, 6})\}$. It takes the name *cut* from the graphical representation in Figure 5.23. The cut line is drawn through precisely those arcs which belong to the cut A.

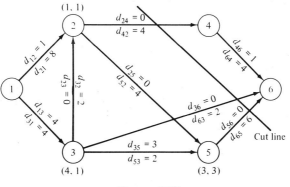

Figure 5.23

5.4 Exercises

In Exercises 1 and 2, a network is given which has been labeled with the labeling algorithm. In each case the sink has been labeled. Following the appropriate steps of the labeling algorithm, adjust the excess capacities and relabel the network.

1.

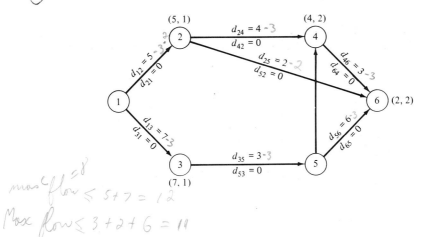

masc flow ≤ 15
≤ 8

2.

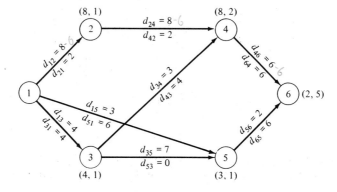

In Exercises 3–6, find the maximal flow in the given network using the labeling algorithm.

3.

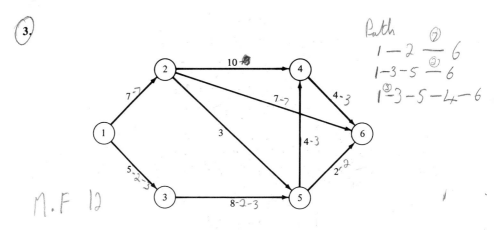

Path
$1-2 \overset{⑦}{—} 6$
$1-3-5 \overset{②}{—} 6$
$1-3-5-4-6$

M.F 12

4.

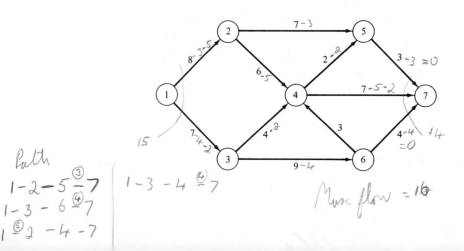

Path
$1-2-5 \overset{③}{—} 7$
$1-3-6 \overset{④}{—} 7$
$1 \overset{⑤}{—} 2-4-7$

$1-3-4 \overset{④}{—} 7$

Masc flow = 18

5.

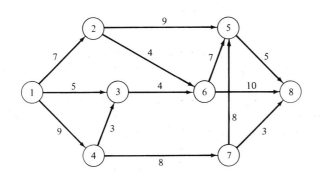

6.

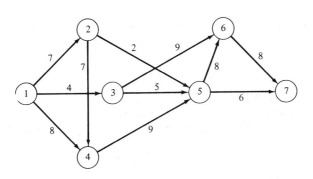

7. (a) Model the following situation as a maximal flow problem. There are an
   equal number, say n, of boys and girls at a school dance. It is also known
   for each boy–girl pair whether they are friends (This friendship relation
   could be given by an $n \times n$ matrix.) Find the maximum number of friendly
   pairs which can be formed.

   (b) Solve the above problem for the friendship relation matrix

$$\begin{bmatrix} 1 & 1 & 0 & 0 & 0 \\ 0 & 0 & 0 & 1 & 1 \\ 1 & 0 & 1 & 0 & 0 \\ 0 & 0 & 1 & 0 & 1 \\ 0 & 1 & 0 & 1 & 0 \end{bmatrix}$$

   (c) Solve the problem in part a for the friendship relation matrix

$$\begin{bmatrix} 0 & 1 & 0 & 1 & 0 \\ 1 & 0 & 0 & 1 & 1 \\ 0 & 0 & 1 & 0 & 1 \\ 0 & 0 & 0 & 1 & 1 \\ 0 & 1 & 0 & 0 & 0 \end{bmatrix}$$

(d) Find a condition on the friendship relation matrix which will guarantee that each boy and girl is dancing with a friend.

8. (a) Model the following situation as a maximal flow problem. Each day at the CPR Clothing Factory there are a large number of tasks to be assigned to the various machines. Assume that any task when assigned to a machine will occupy the machine for the entire day. Of course, only some tasks can be assigned to each machine—some machines are cutting machines, others are pleaters, others can be set for basting, straight-stitching, or top-stitching. For a particular day suppose the list of tasks and possible machines is as given in the accompanying table.

Task	Machine					
	1	2	3	4	5	6
1		✓				
2	✓					
3	✓	✓				
4	✓	✓				
5			✓	✓		
6				✓	✓	
7				✓		✓
8			✓		✓	✓
9					✓	
10			✓			✓

(b) How many tasks can be completed?

5.4 Project

There are many situations in which one is concerned about flow along one route. If there are no delays, the maximum flow is easily determined as the time available divided by time per unit flow. The interesting cases occur when there are delays at certain points and limited waiting facilities at these delay points. Examples of such a situation are trains moving on a single track with sidings at certain points and a production line with certain slowly executed operations which cause bottlenecks. Of course, a railroad siding can only hold a certain number of trains. Likewise, we assume that the holding areas on the production line can contain only a limited number of items.

These situations can be modeled as maximal flow problems. To construct the model, we must determine the network and the meanings of its nodes, arcs, and capacities. As a first step, let P be the duration of the period in which we want to maximize the flow. Divide P into k equal units. Then we may examine the state of the system at any time 0, 1, ..., k. For example, a train schedule would probably have $P = 24$ hours and $k = 240$, meaning that we know the status of each train every 0.1 hour.

Assume there are r delay points. Let n_i $(i = 1, 2, \ldots, r)$ be the capacity of the ith delay point. Let t_0 be the time needed to go from the beginning of the route to the first delay point. Let t_i $(i = 1, 2, \ldots, r - 1)$ be the time needed to go from the ith delay point to the $(i + 1)$st delay point. Finally, let t_r be the time needed to go from the rth delay point to the end of the route.

(a) Assuming there are no delays, what is the maximal flow in the route during a period of length P?
(b) What is the earliest arrival time at the end of the route, assuming that a unit of flow starts at the beginning of the time period?
(c) What is the latest departure time from the beginning of the route which will allow a unit of flow to arrive at the end of the route before the time period is up?

To construct the maximal flow model we will need a network which has a different set of nodes for each of the $k + 1$ specified times in the period under consideration. We will also need a decision function which governs whether a unit of flow can move from one delay point to the next. In the case of the train example, this function will be based in part on information about express train schedules and the need for a local train to be able to move from one siding to the next without being overtaken by an express. Let this function be

$$\delta_{ij} = \begin{cases} 1 & \text{if at time } j \text{ the unit of flow can safely move} \\ & \text{from delay point } i \text{ to delay point } i + 1 \\ 0 & \text{otherwise} \end{cases}$$

(d) Formulate the maximal flow model of this situation.
(e) Assume

$$k = 13, \quad r = 3$$

$$t_0 = 2, \quad t_1 = 3, \quad t_2 = 1, \quad t_3 = 2$$

$$n_1 = 2, \quad n_2 = 1, \quad n_3 = 3$$

$$\delta = \begin{bmatrix} 1 & 1 & 1 & 0 & 1 & 1 & 0 & 0 & 1 & 0 & 1 & 1 & 1 \\ 1 & 0 & 0 & 1 & 1 & 0 & 1 & 1 & 0 & 1 & 0 & 1 & 0 \\ 1 & 0 & 1 & 1 & 0 & 1 & 1 & 1 & 0 & 1 & 0 & 0 & 1 \end{bmatrix}$$

Find the maximal flow along this route for the given period of time.

Further Readings

Busacker, Robert G., and Thomas L. Saaty. *Finite Graphs and Networks: An Introduction with Applications.* McGraw-Hill, New York, 1965.

Elmaghraby, S. E. *Some Network Models in Management Science.* Springer-Verlag, New York, 1970.

Ford, L. R., and D. R. Fulkerson. *Flows in Networks.* Princeton University Press, Princeton, New Jersey, 1962.

5.5 THE SHORTEST ROUTE PROBLEM

In many applied problems it is necessary to find the shortest path from a given node in a network to another node in the network. These include problems in the optimal distribution of goods (along highways or railroads), routing of communication signals, and transporting people in metropolitan areas. Another very interesting application of the **shortest route problem** is the problem of optimally replacing equipment which deteriorates with age. A number of these applications will be considered in the exercises.

Many algorithms have been developed for the solution of the shortest route problem. The algorithm which we now present is *not* the most efficient one; however, it does have the advantage of being one of the simplest to describe and understand. Other algorithms are described in materials listed in the Further Readings.

Once we have designated our origin, the algorithm sweeps out to find the shortest route to each node in the network; in particular, it gives the shortest route from the origin to the designated destination node. The nodes are examined in order of distance from the origin.

The set of nodes of the network is divided into two disjoint subsets, N_r and N_u, as follows:

> N_r consists of the origin and all nodes which have been *reached* from the origin by a shortest route.
>
> N_u consists of all nodes which have not yet been reached from the origin by a shortest route.

Initially, N_r only contains the origin; we transfer nodes from N_u to N_r in a systematic fashion to be described below. The algorithm terminates when N_u is empty. Of course, from a practical point of view, the algorithm can be stopped as soon as the destination node is in N_r.

To start the algorithm we find the node closest to the origin and place this node in N_r. In case of ties we place all nodes which are closest to the origin in N_r. This completes the first stage of the algorithm.

At all subsequent stages of the algorithm we will need to find the node (or nodes) in N_u which is (are) closest to the origin and move it (them) to N_r. We now describe how to find such a node. Let j be a node in N_u and assume j is at least as close to the origin as any other node in N_u. We claim j is connected to a node in N_r by a directed arc (a path which does not go through any other nodes). Certainly j is connected to a node (the origin) in N_r by a path. (Why?) If this path contains a node in N_u other than j, then that node is closer to the origin, contradicting the choice of j. Thus, in this path the arc from j must go to a node in N_r. Hence, the candidates for node j are all those nodes in N_u which are connected to some node in N_r by a directed arc.

Thus, we select node j in N_u as follows:

For each node i in N_r (including the origin) which is connected to a node k in N_u by a directed arc, form $s' = s + d$

where $s =$ Length of the shortest route from the origin to i (We keep track of this number for each node we add to N_r)

$d =$ Length of the directed arc from i to k

Select j as the node for which s' is minimal, put j in N_r, and record its distance to the origin. In case of ties, choose all the tying nodes and place them in N_r. This completes the algorithm. This second step of selecting node j is repeated until the destination node is placed in N_r.

We shall illustrate the algorithm and describe a way of keeping track of the necessary information by using the network in Figure 5.24. Its origin is node 1, and its destination is node 8. The number on each arc represents the length of the arc rather than the capacity as for a flow problem.

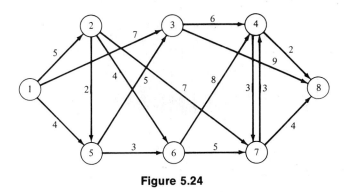

Figure 5.24

Starting with the origin, we list each node across the top of the page. Under each node we list all the nodes which are connected to this node by a directed arc which *leaves* this node. Next to each node in the list we indicate the length of the directed arc which joins the node at the top to that node in the list. For example, the list shows that node 3 is connected to nodes 4 and 8 by directed arcs of lengths 6 and 9, respectively (Table 5.2).

Table 5.2

1	2	3	4	5	6	7	8
5–4	5–2	4–6	8–2	6–3	7–5	4–3	
2–5	6–4	8–9	7–3	3–5	4–8	8–4	
3–7	7–7						

Initially, $N_r = \{1\}$. We now find the node closest to the origin. The candidates are nodes 5, 2, and 3. The closest one is node 5. In the list, we circle the entry 5–4 under node 1 and mark the distance 4 next to node 5, indicating that the distance from the origin to node 5 is 4. We also cross out all node 5's in every column, since we have now found the shortest route to node 5 and the other directed arcs leading to node 5 will not be used. At this point, $N_r = \{1, 5\}$, and our modified list is shown in Table 5.3.

Table 5.3

1	2	3	4	5–4	6	7	8
⑤–4	5̶–2̶	4–6	8–2	6–3	7–5	4–3	
2–5	6–4	8–9	7–3	3–5	4–8	8–4	
3–7	7–7						

The nodes in N_u which are connected to nodes in N_r by directed arcs can be found by reading the uncircled and not-crossed-off entries in the columns labeled by nodes in N_r. Thus, nodes 2, 3, and 6 are candidates for the next node to be added to N_r. The distance from the origin to node 6 via node 5 can be obtained by adding the number, 4, next to node 5 (the distance from the origin to node 5) to the length of the directed arc (5,6) which is 3. The distances from the origin to nodes 2 and 3 come from the table. Since the smallest of these distances is 5, we choose node 2 to be put in N_r. We circle 2–5 under 1 in the list, mark the distance 5 next to node 2, indicating that the shortest distance to node 2 is 5, and cross out all arcs leading *into* node 2 (there are none). At this point, $N_r = \{1, 2, 5\}$, and our modified list is shown in Table 5.4.

Table 5.4

1	2–5	3	4	5–4	6	7	8
⑤–4	5̶–2̶	4–6	8–2	6–3	7–5	4–3	
②–5	6–4	8–9	7–3	3–5	4–8	8–4	
3–7	7–7						

Using the technique described before, we find that nodes 3, 6, and 7 are the next candidates for the node to be put in N_r. The length of the routes associated with these nodes are given in Table 5.5.

Table 5.5

Route	Length
$1 \rightarrow 3$	7
$1 \rightarrow 2 \rightarrow 6$	$5 + 4 = 9$
$1 \rightarrow 2 \rightarrow 7$	$5 + 7 = 12$
$1 \rightarrow 5 \rightarrow 6$	$4 + 3 = 7$
$1 \rightarrow 5 \rightarrow 3$	$4 + 5 = 9$

We select both nodes 3 and 6 to be reached along routes with length 7. We circle 3–7 under node 1 and 6–3 under node 5, mark the distance 7 next to nodes 3 and 6 at the top of the columns, and cross out all node 3's and node 6's in every column. At this point, $N_r = \{1, 2, 3, 5, 6\}$, and our modified list is shown in Table 5.6. Since the first and fifth columns have no other available directed arcs, we can ignore them, so we place a check mark over these columns.

Table 5.6

✓ 1	2–5	3–7	4	✓ 5–4	6–7	7	8
(5–4)	~~3–2~~	4–6	8–2	(6–3)	7–5	4–3	
(2–5)	~~6–4~~	8–9	7–3	~~6–5~~	4–8	8–4	
(3–7)	7–7						

The next candidates for inclusion in N_r are nodes 7, 4, and 8. The lengths of the routes associated with these nodes are given in Table 5.7.

Table 5.7

Route	Length
$1 \rightarrow 2 \rightarrow 7$	$5 + 7 = 12$
$1 \rightarrow 3 \rightarrow 4$	$7 + 6 = 13$
$1 \rightarrow 3 \rightarrow 8$	$7 + 9 = 16$
$1 \rightarrow 5 \rightarrow 6 \rightarrow 7$	$7 + 5 = 12$
$1 \rightarrow 5 \rightarrow 6 \rightarrow 4$	$7 + 8 = 15$

We must choose node 7, mark its distance as 12 (along two different paths), and circle 7–7 in column 2 and 7–5 in column 6. We also cross out all other occurrences of node 7 in any of the columns. Our list is now as shown in Table 5.8.

Table 5.8

√ 1	√ 2–5	3–7	4	√ 5–4	6–7	7–12	8
(5–4)	2̶–5̶	4–6	8–2	(6–3)	(7–5)	4–3	
(2–5)	6̶–4̶	8–9	8̶–2̶	5̶–3̶	4–8	8–4	
(3–7)	(7–7)						

After the next stage the list becomes as shown in Table 5.9. (Verify.)

Table 5.9

√ 1	√ 2–5	3–7	4–13	√ 5–4	√ 6–7	7–12	8
(5–4)	2̶–5̶	(4–6)	8–2	(6–3)	(7–5)	4̶–8̶	
(2–5)	6̶–4̶	8–9	8̶–2̶	5̶–3̶	4̶–8̶	8–4	
(3–7)	(7–7)						

Verify that the final list is Table 5.10.

Table 5.10

√ 1	√ 2–5	√ 3–7	√ 4–13	√ 5–4	√ 6–7	√ 7–12	8–15
(5–4)	2̶–5̶	(4–6)	(8–2)	(6–3)	(7–5)	4̶–8̶	
(2–5)	6̶–4̶	8̶–9̶	8̶–2̶	5̶–3̶	4̶–8̶	8̶–4̶	
(3–7)	(7–7)						

From this list we see that the shortest route to the destination (node 8) has length 15. The route can be found by looking in the circled numbers and working back from node 8. Among the circles find 8; it occurs in column 4. Thus, the last arc in the route is $4 \rightarrow 8$. Now find a 4 among the circles; it is in column 3. Thus, the route goes $3 \rightarrow 4 \rightarrow 8$. After another step, we find the shortest route is

$$1 \rightarrow 3 \rightarrow 4 \rightarrow 8$$

The list also tells us that to get to node 6, for example, the shortest route has length 7. It is

$$1 \to 5 \to 6$$

Equipment Replacement Problem

Another situation which can be modeled as a shortest route problem is the question of when equipment which deteriorates with age should be replaced. Over a period of several years we would expect the price of the equipment to increase gradually and the cost of maintenance for the equipment to increase rapidly as it ages. Certainly this situation is familiar to every car owner.

Consider a factory which must occasionally replace a piece of equipment which deteriorates with age. Assume that they are using a planning horizon of 5 years. At the beginning of year 1 they will purchase a new piece of equipment which will be replaced after every j years. Their problem is to determine j so that the combined maintenance costs and purchase prices will be a minimum. The purchase prices and maintenance costs are given in Table 5.11.

Table 5.11 Purchase price

Beginning of year:	1	2	3	4	5
Price (in $10,000's):	17	19	21	25	30

Maintenance costs					
Age (in years):	0–1	1–2	2–3	3–4	4–5
Cost (in $10,000's):	3.8	5.0	9.7	18.2	30.4

We now construct a network which will model the equipment replacement problem. The shortest path in the network will represent the optimal replacement scheme. Each node of the network will represent the beginning of a year in the planning horizon. We must also include a node to represent the end of the last year in the planning horizon. Each node is connected to *every* subsequent node by an arc. The arcs represent replacement strategies. For example, the arc connecting node 1 to node 4 represents the strategy of purchasing a new machine at the beginning of year 1 and then replacing it at the beginning of year 4. The length of each arc represents the total cost (purchase price plus maintenance costs) of the corresponding replacement strategy. For example, the length of arc $(\overline{1, 4})$ is $17 + (3.8 + 5.0 + 9.7) = 20.5$. Notice that this arc represents owning a piece of equipment for 3 years, so that there is a maintenance cost for each of the years.

All the arcs for this network and the computations for their lengths are given in Table 5.12.

Table 5.12

$\overline{(1,2)}$	$17 + 3.8$	$=$	20.8
$\overline{(1,3)}$	$17 + (3.8 + 5.0)$	$=$	25.8
$\overline{(1,4)}$	$17 + (3.8 + 5.0 + 9.7)$	$=$	35.5
$\overline{(1,5)}$	$17 + (3.8 + 5.0 + 9.7 + 18.2)$	$=$	53.7
$\overline{(1,6)}$	$17 + (3.8 + 5.0 + 9.7 + 18.2 + 30.4)$	$=$	84.1
$\overline{(2,3)}$	$19 + 3.8$	$=$	22.8
$\overline{(2,4)}$	$19 + (3.8 + 5.0)$	$=$	27.8
$\overline{(2,5)}$	$19 + (3.8 + 5.0 + 9.7)$	$=$	37.5
$\overline{(2,6)}$	$19 + (3.8 + 5.0 + 9.7 + 18.2)$	$=$	55.7
$\overline{(3,4)}$	$21 + 3.8$	$=$	24.8
$\overline{(3,5)}$	$21 + (3.8 + 5.0)$	$=$	29.8
$\overline{(3,6)}$	$21 + (3.8 + 5.0 + 9.7)$	$=$	39.5
$\overline{(4,5)}$	$25 + 3.8$	$=$	28.8
$\overline{(4,6)}$	$25 + (3.8 + 5.0)$	$=$	33.8
$\overline{(5,6)}$	$30 + 3.8$	$=$	33.8

The network which models this equipment replacement problem is shown in Figure 5.25.

Any route through the network will represent a five-year plan for equipment replacement. For example, the route $1 \rightarrow 3 \rightarrow 4 \rightarrow 6$ represents purchasing a new piece of equipment at the beginning of year 1, of year

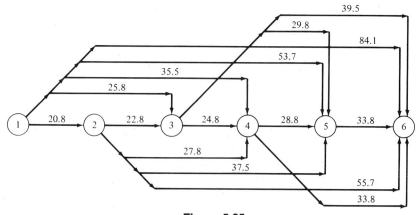

Figure 5.25

3, and of year 4. The cost of this strategy is 25.8 + 24.8 + 33.8 = 84.4 or $844,000. The reader may use the shortest route algorithm to show that the optimal replacement strategy is to purchase a new piece of equipment at the beginning of years 1 and 3. The cost of this strategy is $653,000.

In general, if the planning horizon encompasses n time units, the network will have $n + 1$ nodes. The length of the arc joining node i to node j $(i < j)$ will be the sum of the purchase price at the beginning of year i and the maintenance costs for the first $j - i$ time periods.

5.5 Exercises

In Exercises 1–4 find the shortest path between the indicated origin and destination.

1.

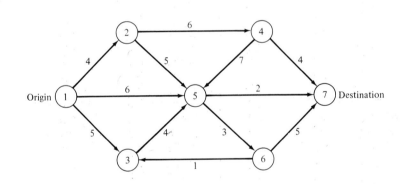

2.

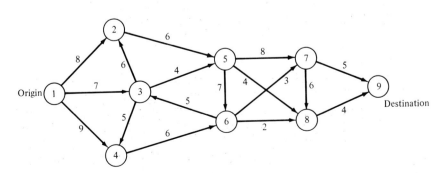

3.

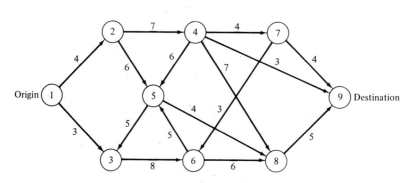

4.

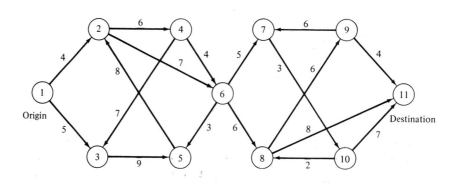

In Exercises 5 and 6 we give tables of purchase prices and maintenance costs for pieces of equipment. Develop a replacement policy for the equipment.

5.

		Year				
	1	2	3	4	5	6
Purchase price (in $10,000's):	5	6	8	9	11	12
Age (in years):	0–1	1–2	2–3	3–4	4–5	5–6
Maintenance cost (in $10,000's):	3.0	3.5	5.5	8.5	12.0	18.0

6.

	Year									
	1	2	3	4	5	6	7	8	9	10
Purchase price (in $10,000's):	3.0	3.5	4.1	4.9	6.0	6.5	ɔ.7	6.7	7.5	8.0
Age (in years):	0–1	1–2	2–3	3–4	4–5	5–6	6–7	7–8	8–9	9–10
Maintenance cost (in $10,000's):	1.0	1.5	2.1	2.8	4.0	7.0	20.0	7.0	8.0	9.5

5.5   Projects

1. Charlie Anderson lives in the town of Hatboro in eastern Montgomery County. Frequently he travels to northeast Philadelphia to visit his parents. The network of roads connecting the two points is shown in Figure 5.26. Node 1 is Charlie's home and node 45 is his destination. This network is a combination

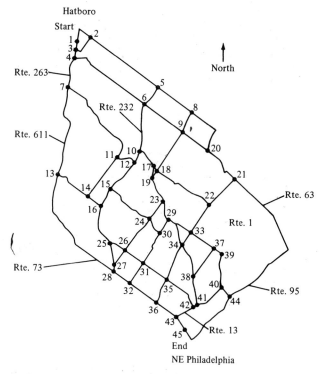

Figure 5.26

of rural roads, urban streets, and expressways. The rural roads are in the area from Hatboro to the outskirts of Philadelphia (nodes 1 through 20). These roads have long stretches without intersections, few traffic signals and stop signs, and generally light traffic. The urban streets begin near the city limits

Table 5.13   Distances and Times between Nodes

Nodes	Distance (miles)	Time (sec)	Nodes	Distance (miles)	Time (sec)
1–2	1.5	380	22–33	1.2	220
1–3	0.7	190	23–24	0.8	120
2–5	3.1	360	23–29	0.8	140
3–2	1.4	210	24–26	1.6	220
3–4	0.7	200	24–30	0.6	175
4–6	3.6	380	25–26	0.6	170
4–7	1.1	230	25–27	0.8	170
5–6	0.9	120	26–27	0.6	160
5–8	1.6	140	26–31	0.8	190
6–9	1.8	225	27–28	0.3	135
6–10	1.8	220	28–32	0.7	180
7–11	3.5	450	29–30	0.6	120
7–13	3.1	420	29–33	0.8	150
8–9	0.9	150	29–34	0.6	180
8–20	2.0	275	30–31	1.4	225
9–18	1.7	255	31–32	0.9	180
9–20	1.2	230	31–35	1.3	320
10–12	0.4	85	32–36	1.3	320
10–17	0.8	170	33–34	0.5	100
11–12	0.8	110	33–37	1.0	140
11–14	1.8	235	34–35	1.4	240
12–15	1.5	220	34–38	1.3	220
13–14	1.5	280	35–36	1.3	230
13–28	4.8	750	35–42	1.5	240
14–16	0.6	110	36–43	0.9	190
15–16	0.8	140	37–38	1.4	210
15–24	2.2	340	37–39	0.4	80
16–25	1.6	230	38–41	1.0	140
17–18	0.2	60	39–40	1.3	240
17–19	0.4	95	40–41	1.2	305
18–19	0.3	80	40–44	0.5	55
18–22	2.6	400	41–42	0.2	70
19–23	1.6	160	42–43	0.7	240
20–21	1.3	260	43–45	0.5	100
21–22	1.6	250	44–45	3.3	520
21–44	6.5	480			

of Philadelphia (nodes 21 through 45). They are generally more congested and have many traffic signals. The expressway joins nodes 21, 44, and 45.

Charlie has computed average times for traveling over each arc of the network. These times along with the length of each arc are given in Table 5.13.

(a) Compute the route Charlie should use to travel the shortest distance when going from Hatboro to northeast Philadelphia.

(b) Compute the route Charlie should use to get to northeast Philadelphia most quickly.

(c) Compare your answers in parts a and b to the route $1 \to 3 \to 4 \to 6 \to 10 \to 17 \to 19 \to 23 \to 29 \to 34 \to 38 \to 41 \to 42 \to 43 \to 45$. Discuss what penalties Charlie incurs by not using either the minimum distance or minimum time routes.

2. On the basis of published figures for new car prices and costs for maintenance, develop an optimal replacement policy for the time period 10 years ago until now.

Further Readings

Djiskstra, E. W. "A Note on Two Problems in Connection with Graphs." *Numerische Mathematik,* 1 (1959), 269–71.

Dreyfus, S. E. "An Appraisal of Some Shortest-Path Algorithms." Reprinted in Arthur M. Geoffrion, ed., *Perspectives on Optimization: A Collection of Expository Articles.* Addison-Wesley, Reading, Mass., 1972. Pp. 197–238.

5.6 THE CRITICAL PATH METHOD

Scheduling models are an important area of study in operations research. We have examined some of these models in the examples and exercises in previous sections. The assignment problem gave one form of a scheduling model as we showed in Project 1 of Section 5.2. This project discussed an idealized method of scheduling nurses on a floor of a hospital. The traveling salesman problem occurred as a scheduling model for a job shop. In this situation we were minimizing total setup time. The literature has extensive discussions of the airline flight crew problem. Some references to this problem and other scheduling problems are given in the Further Readings.

In this section we discuss yet another model for scheduling. This model, the **critical path method** or **CPM**, was originally conceived by researchers at E. I. duPont de Nemours Company and Remington Rand in a collaborative effort. John W. Mauchly, James E. Kelley, and Morgan Walker had the lead roles in its development. It was tested in early 1958 on scheduling the construction of a $10 million chemical plant in Louisville, Kentucky. Its advantage in this situation was the small effort needed to incorporate design changes into the schedule.

Modifications of the original model have been made, but the basic concepts remain the same. The model is based on a network which represents the activities and completion times of a process. CPM is used in a variety of settings, including larger construction projects, complicated maintenance procedures, installation of computer systems, and production of motion pictures. A related planning process called PERT (Program Evaluation and Review Technique) is widely used for government activities, especially in the Department of Defense. We will limit our discussion to the network model for CPM.

It is typical of construction projects that many unrelated activities can be performed simultaneously with the only constraint being that certain activities must precede others. For example, the schedule for building a house cannot call for tarring the roof before the supporting beams have been erected and covered with sheeting. Aside from these mechanical constraints, the other important constraint is meeting the scheduled completion date. CPM models the schedule of the activities of a project and indicates an order in which the individual activities should be performed.

The nodes in the CPM network represents *times*, but not measured in days from the beginning of the project. Rather the times are completions of activities such as *excavation completed* or *plumbing roughed in*. In some of the literature these times are called **events**. The arcs of the network represent **activities** such as *install floor* or *paint walls*. The source node is the beginning of the project and the sink node is the end of the project, and these can be the only source and sink, respectively, for the network.

Instead of the usual specification giving a criterion for connecting two nodes with an arc, it is easier for CPM networks to give the dual specification: a criterion for having two arcs join at a node with one arc entering the node and one leaving. Two such arcs representing activities A_i and A_j are joined at a node with A_i entering and A_j leaving if activity A_i must be completed before activity A_j can begin. For example, part of a CPM network might be

Erect roof support beams   Install sheeting   Tar roof

In joining arcs one must be careful not to produce sequences in the network which do not fit with reality. If activities B and C must follow activity A, but are independent, they should be diagrammed as

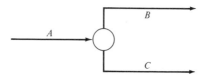

and not as

A slightly more complicated situation occurs when there are four activities, *A*, *B*, *C*, and *D* where *A* and *B* both precede *C* but only *B* precedes *D*. None of the following correctly diagrams this situation. (Why?)

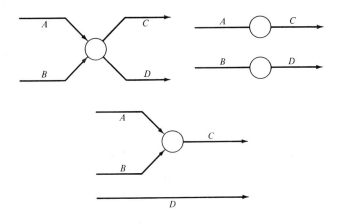

This dilemma is solved by introducing a **dummy activity** or **logical restraint**. It is an arc of the network, usually shown as a dashed rather than a solid line, which represents no work. Using this device we can correctly represent the above situation as

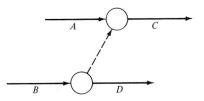

It is also important that no loops be introduced in a CPM network. A loop would mean that an activity must both precede and follow another—a logical impossibility. Loops are easily detected in a small network. In large networks with thousands of nodes, loops may be introduced by errors in the input data. Consequently, a first step in a CPM algorithm is to check the network for loops.

Once a correct precedence network has been established for a problem, the key step in the CPM process must take place. An estimate for the

time necessary to complete each activity must be given. The time estimates should be as good as possible. However, after the critical path is
found, the planner can reexamine his or her estimates for the path to see
if any are unrealistic.

Typically the events of the projects are numbered but not necessarily
in any particular order. In this way more events can be added to a
network without having to renumber all previous events. The activities
are specified by giving the pairs of nodes which are joined by their
representing arcs. To insure that each activity has a unique designation,
dummy activities are inserted in case two activities connect the same
two nodes. Thus,

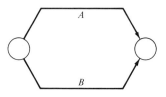

becomes

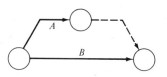

EXAMPLE 1. Consider the several-weekend project faced by a subur-
ban homeowner who wants to replace his small metal storage shed with
a larger, more solidly constructed one, building the newer one on the
same site as the old shed. Since he has no other place to store tools, he
plans to build the new shed around the old one and then to demolish the
old shed. He has drawn the precedence network shown in Figure 5.27.
The number below each activity represents the time estimate for that
activity.

After the network representing the project activities has been determined and checked for its accuracy, the critical path algorithm can be
started. The first step of the algorithm is to compute the early event time
for each node. The **early event time** $T_E(j)$ for node j represents the
soonest that node j can be reached once all activities on all paths leading
to the node have been completed. Mathematically it can be computed by
modifying the shortest path algorithm to compute the longest path.

In this example, we give the details of computing the early event times
for the first six nodes. All the early event times are shown in square
boxes in Figure 5.28. For the source node we have $T_E(1) = 0$, repre-

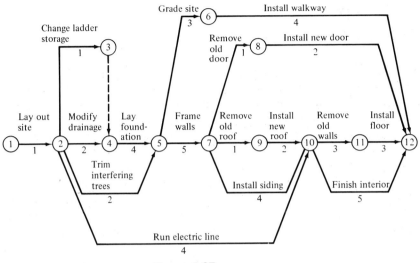

Figure 5.27

senting the beginning of the project. Node 2 has only one arc leading into it, so that $T_E(2) = T_E(1) + 1 = 1$, that is, $T_E(2)$ is the sum of the activity time and the previous event time. Likewise, $T_E(3) = T_E(2) + 1 = 2$.

Since node 4 has two arcs leading to it, we must choose the *longer* of the arcs to compute $T_E(4)$. Remember that $T_E(j)$ represents the earliest that the event can occur after *all* activities leading to it are completed.

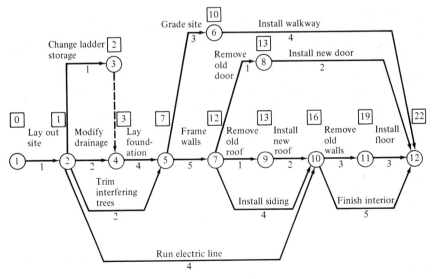

Figure 5.28

Thus,

$$T_E(4) = \max \{T_E(2) + 2, T_E(3) + 0\}$$
$$= \max \{3, 2\}$$
$$= 3$$

Likewise, node 5 has two arcs coming into it. Therefore,

$$T_E(5) = \max \{T_E(2) + 2, T_E(4) + 4\}$$
$$= \max \{3, 7\}$$
$$= 7$$

We also find $T_E(6) = T_E(5) + 3 = 10$.
   Thus, in general,

$$T_E(j) = \max_i \{T_E(i) + \text{Length } \overline{(i, j)}\}$$

where the maximum is taken over all arcs leading into node j.

The early event time computed for the node which represents project completion (the sink) is the total amount of time necessary to complete the project. If this time is unacceptable, the project manager can at this point investigate ways of speeding up some of the activities. He or she can choose the activities to concentrate on after the critical path (or several of them) is (are) known for the project.

The second step of the algorithm is to compute the late event time for each node. The **late event time** $T_L(j)$ for node j is the latest time by which node j can be reached to still have the project completed in the shortest possible time. Again in this example, we give the details of computing the late event times for nodes 7 through 12. All late event times are shown in diamonds below the early event times in Figure 5.29.

For the sink node we have $T_L(12) = 22$, the total project time. For node 11, since the activity of laying the floor takes 3 units, the start of floor laying must occur at 19 units. That is,

$$T_L(11) = T_L(12) - 3 = 22 - 3 = 19$$

There are two paths from node 10 to node 12. We must use the longer of the two to compute $T_L(10)$. We have

$$T_L(10) = \min \{T_L(11) - 3, T_L(12) - 5\}$$
$$= \min \{16, 17\}$$
$$= 16$$

For nodes 8 and 7, $T_L(9) = T_L(10) - 2 = 14$ and $T_L(8) = T_L(12) - 2$

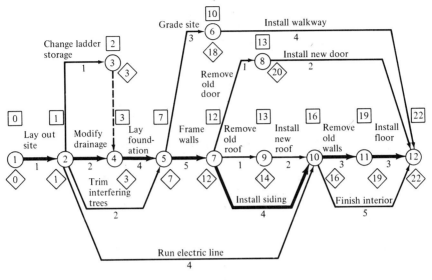

Figure 5.29

= 20, since only one arc leaves each of these nodes. For node 7, we find

$$T_L(7) = \min \{T_L(8) - 1, \; T_L(9) - 1, \; T_L(10) - 4\}$$

$$= \min \{19, 13, 12\}$$

$$= 12$$

In general, we have

$$T_L(j) = \min_{i} \{T_L(i) - \text{Length} \; (\overline{j, i})\}$$

where i runs through all nodes for which there is an arc from node j to node i. The fact that $T_L(1)$ must be 0 is a good check on the accuracy of the event time calculations.

Recall that for any node j, $T_L(j)$ is the *latest* time by which all the activities leading from node j must be started to have the project finished on schedule. Also, $T_E(j)$ is the *soonest* time by which all the activities leading from node j can be started. These activities could not be started earlier because at least one activity leading to node j had not been completed. The difference $T_L(j) - T_E(j)$ is called the **float time** or **slack time**. Any event which has zero float time is called a **critical event**. The next activity must be started without delay. A **critical path** is a path in the network which only passes through critical events and whose length is the value of T_E at the project-completion node. There may be more than one critical path in a network. The activities along a critical path are the ones whose lengths are determining the total length of the project.

In our example there is only one critical path, namely, $1 \to 2 \to 4 \to 5 \to 7 \to 10 \to 11 \to 12$. This path is indicated by the heavy lines in Figure 5.29.

5.6 Exercises

For the precedence networks given in Exercises 1–4,
(a) find the early event times;
(b) find the late event times;
(c) find a critical path.

1.

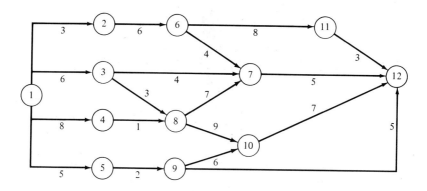

2.

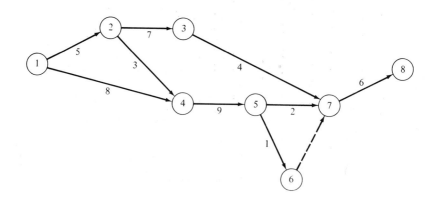

3.

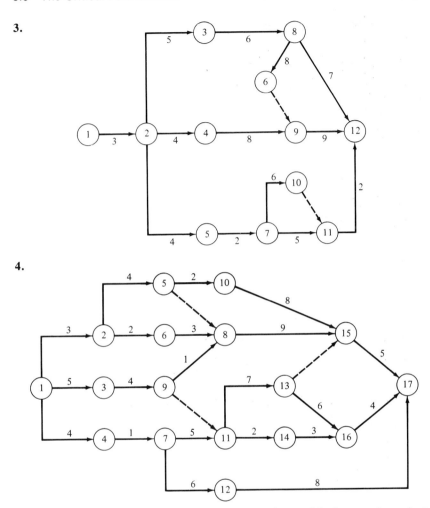

4.

The networks given in Exercises 5 and 6 are to be models for certain projects. However, they are incorrectly constructed. Find and correct, if possible, the errors in the networks.

5.

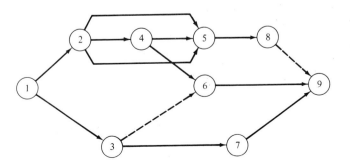

6.

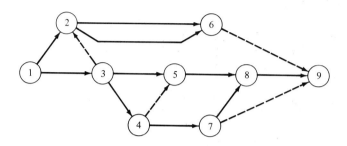

In Exercises 7 and 8 lists of activities and activity times are given. Also, we give the logical constraints on each activity. Construct a precedence network for each list of activities and find a critical path in the network.

7. A_1 precedes A_2, A_3, A_4, A_5, and A_6.
   A_{10} follows A_2.
   A_{11} follows A_3.
   A_7 follows A_5.
   A_8 follows A_4 and precedes A_{12}.
   A_9 follows A_4 and A_7.
   A_6 and A_9 precede A_{13}.
   A_{12} follows A_{10} and A_{11}.
   A_{12} and A_{13} terminate at the same time.

Activity	Time	Activity	Time
A_1	7	A_7	8
A_2	5	A_8	6
A_3	8	A_9	11
A_4	3	A_{10}	2
A_5	1	A_{11}	5
A_6	2	A_{12}	7
		A_{13}	9

8. A_1, A_2, A_3, and A_{17} start at the same time.
   A_1 precedes A_4 and A_5.
   A_6 follows A_2.
   A_3 precedes A_7.
   A_{10} follows A_5.
   A_{13} follows A_6, A_7, and A_{10}.
   A_4 precedes A_8 and A_9.
   A_{12} follows A_8.
   A_9 precedes A_{15} and A_{16}.
   A_{11} follows A_{17}.

A_{14} follows A_{12}.

A_{11} precedes A_{18}.

A_{19} follows A_{15}.

A_{13}, A_{14}, A_{16}, A_{18}, and A_{19} terminate at the same time.

Activity	Time	Activity	Time
A_1	6	A_{11}	4
A_2	7	A_{12}	3
A_3	5	A_{13}	6
A_4	4	A_{14}	8
A_5	8	A_{15}	7
A_6	9	A_{16}	5
A_7	3	A_{17}	4
A_8	7	A_{18}	6
A_9	6	A_{19}	7
A_{10}	9		

Further Readings

Antell, J. M., and R. W. Woodhead. *Critical Path Methods in Construction Practice*. 2nd ed. Wiley, New York, 1970.

Moder, J. J., and Cecil R. Phillips. *Project Management with CPM and PERT*. 2nd ed. Van Nostrand, New York, 1970.

Shaffer, L. R., J. B. Ritter, and W. L. Meyer. *The Critical Path Method*. McGraw-Hill, New York, 1965.

5.7 COMPUTER ASPECTS (Optional)

In this chapter we have described several linear programming problems which because of their specialized structure are more efficiently solved by their own algorithms than by the simplex algorithm. In fact, all these models can be framed in terms of networks. The algorithm which is most widely available for general network problems is the **out-of-kilter algorithm**, which solves the minimum cost flow problem in a capacitated network.

The **minimum cost flow problem** can be mathematically formulated as follows. Consider a network with n nodes. With each node we associate a number b_i, indicating the availability of a single commodity at node i. If $b_i > 0$, there is b_i of the commodity available at node i, and node i is called a **source**; if $b_i < 0$, there is a demand for b_i of the commodity at node i, and node i is called a **sink**; and if $b_i = 0$, node i is called an **intermediate** or **transshipment** node. Let u_{ij} and l_{ij} denote the upper and lower capacities of arc $\overline{(i, j)}$. Let c_{ij} denote the cost of shipping one unit from node i to node j along arc $\overline{(i, j)}$. Since flow can neither be created nor destroyed at any node, we obtain the conservation of flow equations

$$\sum_{j=1}^{n} x_{ij} - \sum_{k=1}^{n} x_{ki} = b_i, \qquad i = 1, 2, \ldots, n$$

Of course, we are interested in minimizing the total cost

$$\sum_{i=1}^{n} \sum_{j=1}^{n} c_{ij} x_{ij}$$

Thus, a mathematical formulation of the minimum cost flow problem is

$$\text{Minimize} \quad \sum_{i=1}^{n} \sum_{j=1}^{n} c_{ij} x_{ij}$$

subject to

$$\sum_{j=1}^{n} x_{ij} - \sum_{k=1}^{n} x_{ki} = b_i, \qquad i = 1, 2, \ldots, n$$

$$0 \le l_{ij} \le x_{ij} \le u_{ij}, \qquad \begin{cases} i = 1, 2, \ldots, n \\ j = 1, 2, \ldots, n \end{cases}$$

When this problem is represented by a network diagram, it is convenient to label each arc with the triple numbers $[c_{ij}, u_{ij}, l_{ij}]$. If an arc has no upper bound on its capacity, only the cost of shipping one unit of goods along the arc is given.

The out-of-kilter algorithm finds, for a given amount of flow, the path which is cheapest. We examined a special case of this problem as we discussed the shortest route algorithm. There the amount of flow was one unit and each arc had upper capacity of one unit and lower capacity of zero units.

All the models which we have discussed in this chapter can be transformed to minimum cost flow models. Thus, the out-of-kilter algorithm could be used to solve any of them. As an example of the types of network computer codes available, we discuss the features of a typical code for the out-of-kilter algorithm. However, a description of the algorithm itself is beyond the scope of this book.

Input

The network is specified by giving a list of arcs. Each arc is specified by giving the names of the nodes it joins in the order "from—to." For each arc the user must also specify the upper bound on the capacity (using a large number if the capacity is unbounded), the lower bound on the capacity, the cost, and the initial flow. In many cases the initial flow is chosen to be zero. The computer code may be designed to handle only integer flow problems, so that the cost, initial flow, and capacities must be restricted to integer variables.

Since some codes do not have built-in routines for input verification, the user may have to check the inputs to make sure that:

1. there are no dangling nodes in the network;
2. there is only one source and one sink;
3. the initial flow is conserved at each interior node, namely, at node j:

$$\sum_p x_{pj} = \sum_q x_{jq}$$

4. the upper bound on each arc is greater than or equal to the lower bound for that arc.

Output

The code will have output options which allow the user to save the problem and restart after making modifications to the network. The printed output will include a listing of the active arcs at an optimal solution along with the flow for each of these arcs. The output may also include a list of the $z_{ij} - c_{ij}$, which are the marginal costs for increasing the flow one unit along the arcs (i, j).

EXAMPLE 1. Consider the transportation problem (Example 1, Section 5.1) with cost matrix and demand and supply vectors as follow:

$$C = \begin{bmatrix} 5 & 7 & 9 & 6 \\ 6 & 7 & 10 & 5 \\ 7 & 6 & 8 & 1 \end{bmatrix}, \quad s = \begin{bmatrix} 120 \\ 140 \\ 100 \end{bmatrix}, \quad d = \begin{bmatrix} 100 \\ 60 \\ 80 \\ 120 \end{bmatrix}$$

A minimum cost flow network which models this problem is given in Figure 5.30. Note that each source is a node and each destination is a node. Each source node is connected to every destination node by an arc. The cost of each arc is that specified by the matrix C. Also included are two additional nodes, a **supersource** and a **supersink**. These nodes serve to place the supply and demand constraints on the sources and destinations. The supersource is connected to each source by an arc which has cost zero, lower bound zero, and upper bound equaling the supply capacity of the source. A flow from the supersource to a source can be thought of as the process of manufacturing the goods at the source. The conservation of flow at each source guarantees that no more than the available supply may be shipped from that source.

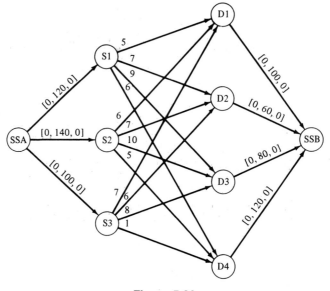

Figure 5.30

Likewise, each destination is connected to the supersink by an arc with upper bound equaling the demand at the destination, lower bound zero, and cost zero.

We show in Figure 5.31 the input necessary for solving this problem using the Sperry Univac UKILT 1100 Out-of-Kilter Network Optimization System. There is one additional arc required for this system connecting the supersink to the supersource. The system expects all nodes to have arcs leading both in and out of them. Such a network is called a **circularization network**. The cost of the new arc is zero, and the lower and upper bounds are equal to the total supply. This insures that the total supply is used and that the total demand is met. The nodes are labeled by numbers and the arcs by the pair of nodes which they join. The initial flows are all chosen to be zero.

Figure 5.32 shows the output of UKILT 1100 for this problem.

Besides the out-of-kilter algorithm, there are implementations of the various specialized algorithms for particular settings. For example, the critical path method is typically used outside operations research settings, so that easily used codes for this algorithm have been developed.

CPM Codes

The CPM codes will produce charts and tables which can be used in the field by the project supervisors and foremen. Much of the effort in writing

	From	To	Cost	Upper bound	Lower bound	Initial flow
1	BEGIN					
2	TRANSPORTATION					
3	ARCS					
4	S1	D1	5	1000	0	0
5	S1	D2	7	1000	0	0
6	S1	D3	9	1000	0	0
7	S1	D4	6	1000	0	0
8	S2	D1	6	1000	0	0
9	S2	D2	7	1000	0	0
10	S2	D3	10	1000	0	0
11	S2	D4	5	1000	0	0
12	S3	D1	7	1000	0	0
13	S3	D2	7	1000	0	0
14	S3	D3	8	1000	0	0
15	S3	D4	1	1000	0	0
16	SSA	S1	0	120	0	0
17	SSA	S2	0	140	0	0
18	SSA	S3	0	100	0	0
19	D1	SSB	0	100	0	0
20	D2	SSB	0	60	0	0
21	D3	SSB	0	80	0	0
22	D4	SSB	0	120	0	0
23	SSB	SSA	0	360	360	0
24	END					
25	SOLVE					
26	OUTPUT 11					
27	REPORT 11					
28	STOP					

Figure 5.31

```
                    TITLE  TRANSPORTATION

          NUMBER OF NODES:      9

          NUMBER OF ARCS:      20

             TOTAL COST:     1900

     **********************************
     *                                *
     *   THE SOLUTION IS OPTIMAL      *
     *                                *
     **********************************

          TRANSPORTATION
```

ARC NUMBER	FROM NODE	TO NODE	COST	MARG COST	UPPER BOUND	LOWER BOUND	FLOW
1	S1	D1	5	0	1000	0	100
2	S1	D2	7	1	1000	0	0
3	S1	D3	9	0	1000	0	20
4	S1	D4	6	2	1000	0	0
5	S2	D1	6	0	1000	0	0
6	S2	D2	7	0	1000	0	60
7	S2	D3	10	0	1000	0	60
8	S2	D4	5	0	1000	0	20
9	S3	D1	7	5	1000	0	0
10	S3	D2	7	4	1000	0	0
11	S3	D3	8	2	1000	0	0
12	S3	D4	1	0	1000	0	100
13	SSA	S1	0	-1	120	0	120
14	SSA	S2	0	0	140	0	140
15	SSA	S3	0	-4	100	0	100
16	D1	SSB	0	-4	100	0	100
17	D2	SSB	0	-3	60	0	60
18	D3	SSB	0	0	80	0	80
19	D4	SSB	0	-5	120	0	120
20	SSB	SSA	0	10	360	360	360

Figure 5.32

a CPM code goes into making this output meaningful to those who must use it.

There are about 100 commercially available CPM codes. These are either owned by larger corporations who do many complicated scheduling tasks or by consulting firms. Experience has shown that networks representing more than 250 activities should certainly be processed on a computer. In fact, the break-even point between hand and computer processing may drop to as low as 100 activities if the network is extremely complicated. The larger CPM codes can handle problems with as many as 10,000 activities.

Besides using the CPM algorithm on a given network, a typical code does a substantial amount of input verification. It will check for loops, dangling activities, and duplicate activities. Most codes also have provisions for saving the network in computer-readable form (e.g., on disk or tape) so that changes in the network can be easily made.

5.7 Exercises

1. Why is a supersink necessary in modeling a transportation problem as minimum cost flow problem? What aspect of the problem could it represent?

2. Model the assignment problem as a minimum cost flow problem. Assume that there are n persons and n jobs.

3. Model the shortest route problem as a minimum cost flow problem.

4. Model the critical path method as a minimum cost flow problem.

5.7 Projects

1. (**The transshipment problem**) In many applications of the transportation problem we encounter a situation where a node serves merely as a transshipment point. That is, there is neither supply nor demand at the point, but goods are shipped through the point. Consider the network in Figure 5.33, where the numbers beside the nodes are the b_i described in this section. The nodes can

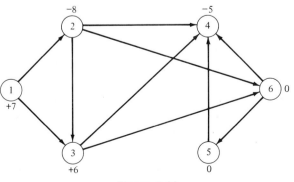

Figure 5.33

be divided into five classes:

Pure source—all arcs lead away from node, and $b_i > 0$
Pure sink—all arcs lead into node, and $b_i < 0$
Transshipment source—arcs leading in and out of node, and $b_i > 0$
Transshipment sink—arcs leading in and out of node, and $b_i < 0$
Pure transshipment point—arcs leading in and out of node, and $b_i = 0$.

(a) Classify the nodes in the network in Figure 5.33.

Every transshipment problem can be formulated as a minimum cost flow problem. This formulation may require the introduction of a supersource or supersink. A supersource will be necessary when there is more than one pure source in the network. This supersource is connected to each pure source with an arc whose capacity has an upper bound equal to the supply at the pure source. Likewise, a supersink is necessary when there is more than one pure sink. A minimum cost flow problem specifies a particular value of flow which must be sent from the supersource to the supersink. Since there must be conservation of flow at each node, the sum of the amounts available at the sources of a transshipment problem must equal the sum of the amounts required at the sinks before the problem can be formulated as a minimum cost flow problem. If these sums are not equal, an additional pure source or pure sink must be added as a dummy to the network to compensate for this difference.

(b) Why does a dummy sink node have to be connected only to the source node and not to the transshipment nodes or other sink nodes?
(c) Classify the nodes of the network in Figure 5.34.
(d) Formulate the network in Figure 5.34 as a minimum cost flow problem.

2. If your computer center has a network code, formulate and run the following:
   (a) Figure 5.4, Section 5.4 (page 321).
   (b) Example 2, Section 5.1 (page 286).
   (c) Example 1, Section 5.1 (page 271).
   (d) The shortest route example in Section 5.5.

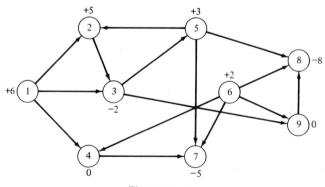

Figure 5.34

SOLUTIONS TO ODD-NUMBERED EXERCISES

CHAPTER 0

Section 0.1, page 14

1. $a = 4$, $b = 2$, $c = 9$, $d = -1$

3. (a) Not possible   (b) Not possible

(c) $\begin{bmatrix} 2 & 3 \\ 3 & 1 \\ 1 & 2 \end{bmatrix}$ (d) $\begin{bmatrix} 4 & 4 & 4 \\ 2 & 4 & 4 \\ 1 & 11 & 4 \end{bmatrix}$

5. $AB = \begin{bmatrix} 9 & 11 \\ 10 & 2 \end{bmatrix}$, $BA = \begin{bmatrix} -5 & 1 \\ 12 & 16 \end{bmatrix}$

7. $AC = BC = \begin{bmatrix} -3 & -12 \\ -8 & -32 \end{bmatrix}$

11. $\begin{bmatrix} 16 & 17 & 12 & 26 & 20 \\ 6 & 19 & -4 & 18 & 17 \\ 26 & 33 & 18 & 52 & 46 \\ 5 & 16 & -5 & 15 & 9 \\ 3 & 15 & -9 & 17 & 11 \\ 18 & 29 & 12 & 38 & 33 \end{bmatrix}$

Section 0.2, page 22

1. $\begin{bmatrix} 1 & 0 & 0 & 11 \\ 0 & 1 & 0 & 0 \\ 0 & 0 & 1 & -4 \end{bmatrix}$

3. $\begin{bmatrix} 1 & 0 & -\frac{7}{3} & 2 & -\frac{13}{3} \\ 0 & 1 & 2 & -1 & 3 \\ 0 & 0 & 0 & 0 & 0 \\ 0 & 0 & 0 & 0 & 0 \end{bmatrix}$

367

5. (a) $x = 1$, $y = 2$, $z = -2$
   (b) No solution

7. (a) No solution
   (b) $x = 3$, $y = -2$, $z = 0$, $w = 1$

9. (a) No solution
   (b) $x = \frac{7}{3}$, $y = \frac{4}{3}$, $z = 0$

11. (1) $a = -3$
   (2) a = Any real number except 3 or -3
   (3) $a = 3$

Section 0.3, page 28

1. $\begin{bmatrix} \frac{4}{5} & -\frac{1}{5} \\ -\frac{3}{5} & \frac{2}{5} \end{bmatrix}$

3. (a) $\begin{bmatrix} -\frac{5}{7} & \frac{3}{7} \\ \frac{4}{7} & -\frac{1}{7} \end{bmatrix}$ (b) No inverse   (c) $\begin{bmatrix} \frac{14}{9} & -\frac{1}{3} & -\frac{1}{9} \\ -\frac{2}{3} & 0 & \frac{1}{3} \\ -\frac{5}{9} & \frac{1}{3} & \frac{1}{9} \end{bmatrix}$

5. (a) $\begin{bmatrix} \frac{2}{5} & -\frac{3}{10} \\ \frac{1}{5} & \frac{1}{10} \end{bmatrix}$ (b) $\begin{bmatrix} \frac{1}{2} & 0 & -\frac{1}{2} \\ \frac{7}{4} & -\frac{3}{2} & -\frac{5}{4} \\ -1 & 1 & 1 \end{bmatrix}$ (c) No inverse

7. (a) Does not exist

   (b) $\begin{bmatrix} 4 & -2 & -\frac{3}{2} \\ -\frac{13}{7} & 1 & \frac{9}{14} \\ -\frac{12}{7} & 1 & \frac{11}{14} \end{bmatrix}$

   (c) Does not exist

Section 0.4, page 33

3. (a) No   (b) Yes   (c) Yes
5. (a) Yes   (b) Yes   (c) No

Section 0.5, page 41
1. a and b

3. a and b

5. (c) $\begin{bmatrix} 1 \\ 1 \\ 2 \end{bmatrix} + \begin{bmatrix} 1 \\ 2 \\ -1 \end{bmatrix} + \begin{bmatrix} 3 \\ 4 \\ -2 \end{bmatrix} = \begin{bmatrix} 5 \\ 7 \\ -1 \end{bmatrix}$

   (d) $\begin{bmatrix} 2 \\ 1 \\ 3 \end{bmatrix} + 2\begin{bmatrix} 1 \\ 2 \\ -1 \end{bmatrix} = \begin{bmatrix} 4 \\ 5 \\ 1 \end{bmatrix}$

7. a

9. a

11. (c) $\begin{bmatrix} -3 \\ 0 \\ 1 \end{bmatrix} = \begin{bmatrix} 0 \\ 1 \\ 0 \end{bmatrix} - 2\begin{bmatrix} 3 \\ 2 \\ 1 \end{bmatrix} + 3\begin{bmatrix} 1 \\ 1 \\ 1 \end{bmatrix}$

13. (a) $\begin{bmatrix} 1 \\ 0 \\ 0 \end{bmatrix}$ (b) $\begin{bmatrix} 1 \\ 2 \\ -2 \end{bmatrix}$ (c) $\begin{bmatrix} 0 \\ 1 \\ 2 \end{bmatrix}$ (d) $\begin{bmatrix} -1 \\ 2 \\ 4 \end{bmatrix}$

CHAPTER 1

Section 1.2, page 57

9. (a)

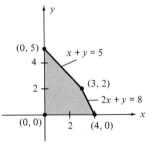

(b)

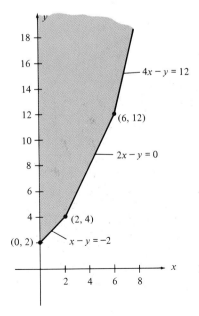

(c)

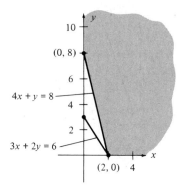

(d)

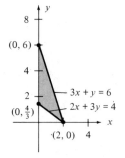

11. (a)

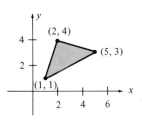

(b)

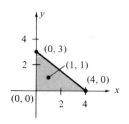

(c)

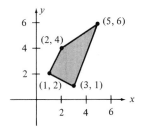

CHAPTER 2

Section 2.1, page 71

1. Let x = Amount of PEST (in kg)

   y = Amount of BUG (in kg)

   Minimize $z = 3x + 2.5y$

   subject to

   $$30x + 40y \geq 120$$
   $$40x + 20y \leq 80$$
   $$x \geq 0, \quad y \geq 0$$

   To change to standard form, change the objective function and the first constraint.

3. Let x = Number of Palium pills prescribed per day

   y = Number of Timade pills prescribed per day

   Minimize $z = .4x + .3y$

   subject to

   $$4x + 2y \geq 10$$
   $$.5x + .5y \leq 2$$
   $$x \geq 0, \quad y \geq 0$$

   To change to standard form, change the objective function and the first constraint.

5. Let x = Number of kilograms of Super

   y = Number of kilograms of Deluxe

Maximize $z = 20x + 30y$

subject to

$$.5x + .25y \leq 120$$
$$.5x + .75y \leq 160$$
$$x \geq 0, \quad y \geq 0$$

This model is in standard form.

7. Let x_1 = Number of bags of Regular Lawn (in thousands)

   x_2 = Number of bags of Super Lawn (in thousands)

   x_3 = Number of bags of Garden (in thousands)

Maximize $z = 300x_1 + 500x_2 + 400x_3$

subject to

$$4x_1 + 4x_2 + 2x_3 \leq 80$$
$$2x_1 + 3x_2 + 2x_3 \leq 50$$
$$x_1 \geq 0, \quad x_2 \geq 0, \quad x_3 \geq 0$$

This model is in standard form.

9. Let x_1 = Number of books in paperback binding

   x_2 = Number of books in bookclub binding

   x_3 = Number of books in library binding

Maximize $z = .5x_1 + .8x_2 + 1.2x_3$

subject to

$$2x_1 + 2x_2 + 3x_3 \leq 420$$
$$4x_1 + 6x_2 + 10x_3 \leq 600$$
$$x_1 \geq 0, \quad x_2 \geq 0, \quad x_3 \geq 0$$

This model is in standard form.

11. Let x_{ij} = amount of ith ingredient in jth mixture (in kg),

   where   Ingredient 1 = Sunflower seeds
           Ingredient 2 = Raisins
           Ingredient 3 = Peanuts
             Mixture 1 = Chewy
             Mixture 2 = Crunchy
             Mixture 3 = Nutty

$$\text{Maximize} \quad 2 \sum_{i=1}^{3} x_{i1} + 1.6 \sum_{i=1}^{3} x_{i2} + 1.2 \sum_{i=1}^{3} x_{i3}$$

$$- \sum_{j=1}^{3} x_{1j} - 1.5 \sum_{j=1}^{3} x_{2j} - .8 \sum_{j=1}^{3} x_{3j}$$

subject to

$$\sum_{j=1}^{3} x_{1j} \leq 100$$

$$\sum_{j=1}^{3} x_{2j} \leq 80$$

$$\sum_{j=1}^{3} x_{3j} \leq 60$$

$$.6x_{11} - .4x_{21} + .6x_{31} \leq 0$$

$$-.2x_{11} - .2x_{21} + .8x_{31} \leq 0$$

$$-.4x_{12} + .6x_{22} + .6x_{32} \leq 0$$

$$.8x_{13} - .2x_{23} - .2x_{33} \leq 0$$

$$.6x_{13} + .6x_{23} - .4x_{33} \leq 0$$

$$x_{ij} \geq 0, \quad i = 1, 2, 3; \quad j = 1, 2, 3$$

Section 2.2, page 96

1. *Example 4*

$$\text{Maximize} \quad z = [20 \quad 30] \begin{bmatrix} x \\ y \end{bmatrix}$$

subject to

$$\begin{bmatrix} .4 & .3 \\ .2 & .4 \end{bmatrix} \begin{bmatrix} x \\ y \end{bmatrix} \leq \begin{bmatrix} 18 \\ 14 \end{bmatrix}, \qquad \begin{bmatrix} x \\ y \end{bmatrix} \geq \begin{bmatrix} 0 \\ 0 \end{bmatrix}$$

Example 9

$$\text{Maximize} \quad z = [2 \quad 3 \quad 4] \begin{bmatrix} x_1 \\ x_2 \\ x_3 \end{bmatrix}$$

subject to

$$\begin{bmatrix} 3 & 2 & -3 \\ 2 & 3 & 2 \\ -3 & 1 & -2 \end{bmatrix} \begin{bmatrix} x_1 \\ x_2 \\ x_3 \end{bmatrix} \leq \begin{bmatrix} 4 \\ 6 \\ 8 \end{bmatrix}, \qquad \begin{bmatrix} x_1 \\ x_2 \\ x_3 \end{bmatrix} \geq \begin{bmatrix} 0 \\ 0 \\ 0 \end{bmatrix}$$

Example 10

Maximize $z = \begin{bmatrix} 3 & 2 \end{bmatrix} \begin{bmatrix} x \\ y \end{bmatrix}$

subject to

$$\begin{bmatrix} 2 & -6 \\ 3 & 2 \\ -5 & 4 \\ 6 & 7 \end{bmatrix} \begin{bmatrix} x \\ y \end{bmatrix} \le \begin{bmatrix} 7 \\ 8 \\ -15 \\ 4 \end{bmatrix}, \qquad \begin{bmatrix} x \\ y \end{bmatrix} \ge \begin{bmatrix} 0 \\ 0 \end{bmatrix}$$

3. (a) $\mathbf{x}_2 = \begin{bmatrix} 2 \\ 1 \end{bmatrix} \ge \mathbf{0}, \quad \begin{bmatrix} 2 & 2 \\ 5 & 3 \end{bmatrix} \begin{bmatrix} 2 \\ 1 \end{bmatrix} = \begin{bmatrix} 6 \\ 13 \end{bmatrix} \le \begin{bmatrix} 8 \\ 15 \end{bmatrix}$

$\mathbf{x}_3 = \begin{bmatrix} 1 \\ 3 \end{bmatrix} \ge \mathbf{0}, \quad \begin{bmatrix} 2 & 2 \\ 5 & 3 \end{bmatrix} \begin{bmatrix} 1 \\ 3 \end{bmatrix} = \begin{bmatrix} 8 \\ 14 \end{bmatrix} \le \begin{bmatrix} 8 \\ 15 \end{bmatrix}$

(b) $\mathbf{x}_4$: $\begin{bmatrix} 2 & 2 \\ 5 & 3 \end{bmatrix} \begin{bmatrix} 3 \\ 1 \end{bmatrix} = \begin{bmatrix} 8 \\ 18 \end{bmatrix} \nleq \begin{bmatrix} 8 \\ 15 \end{bmatrix}$

$\mathbf{x}_5$: $\begin{bmatrix} 2 & 2 \\ 5 & 3 \end{bmatrix} \begin{bmatrix} 2 \\ 2 \end{bmatrix} = \begin{bmatrix} 8 \\ 16 \end{bmatrix} \nleq \begin{bmatrix} 8 \\ 15 \end{bmatrix}$

$\mathbf{x}_6 = \begin{bmatrix} -2 \\ 3 \end{bmatrix} \ngeq \mathbf{0}$

5. (a)

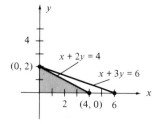

(b) $\begin{bmatrix} 0 \\ 2 \end{bmatrix}$ **(c)** $\begin{bmatrix} 0 \\ 2 \end{bmatrix}$; $z = -6$

7. (a)

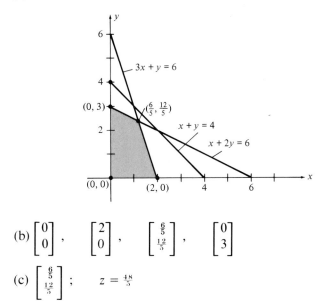

(b) $\begin{bmatrix} 0 \\ 0 \end{bmatrix}$, $\begin{bmatrix} 2 \\ 0 \end{bmatrix}$, $\begin{bmatrix} \frac{6}{5} \\ \frac{12}{5} \end{bmatrix}$, $\begin{bmatrix} 0 \\ 3 \end{bmatrix}$

(c) $\begin{bmatrix} \frac{6}{5} \\ \frac{12}{5} \end{bmatrix}$; $\quad z = \frac{48}{5}$

9. (a)

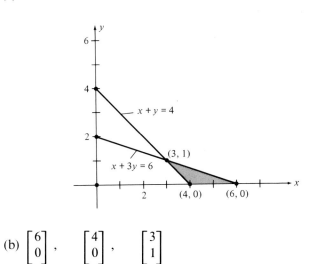

(b) $\begin{bmatrix} 6 \\ 0 \end{bmatrix}$, $\begin{bmatrix} 4 \\ 0 \end{bmatrix}$, $\begin{bmatrix} 3 \\ 1 \end{bmatrix}$

(c) Any point on the line segment joining

$$\begin{bmatrix} 6 \\ 0 \end{bmatrix} \quad \text{and} \quad \begin{bmatrix} 3 \\ 1 \end{bmatrix} ; \quad z = 3$$

11. (a)

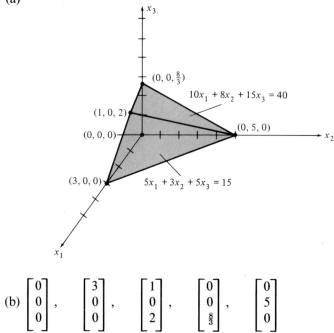

(b) $\begin{bmatrix} 0 \\ 0 \\ 0 \end{bmatrix}$, $\begin{bmatrix} 3 \\ 0 \\ 0 \end{bmatrix}$, $\begin{bmatrix} 1 \\ 0 \\ 2 \end{bmatrix}$, $\begin{bmatrix} 0 \\ 0 \\ \frac{8}{3} \end{bmatrix}$, $\begin{bmatrix} 0 \\ 5 \\ 0 \end{bmatrix}$

(c) $\begin{bmatrix} 0 \\ 5 \\ 0 \end{bmatrix}$; $z = 20$

13. (a)

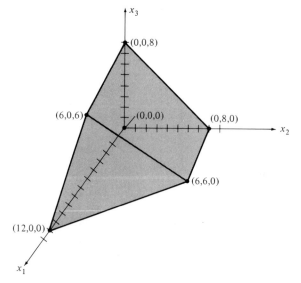

(b) $\begin{bmatrix} 0 \\ 0 \\ 0 \end{bmatrix}$, $\begin{bmatrix} 12 \\ 0 \\ 0 \end{bmatrix}$, $\begin{bmatrix} 6 \\ 0 \\ 6 \end{bmatrix}$, $\begin{bmatrix} 0 \\ 0 \\ 8 \end{bmatrix}$, $\begin{bmatrix} 0 \\ 8 \\ 0 \end{bmatrix}$, $\begin{bmatrix} 6 \\ 6 \\ 0 \end{bmatrix}$

(c) $\begin{bmatrix} 0 \\ 0 \\ 0 \end{bmatrix}$; $z = 0$

15. No feasible solution

17. 80 kg pizza-flavored potato chips
   0 kg chili-flavored potato chips
   Net profit = \$9.60

19. Let x_1 = Amount of pizza-flavored potato chips (in kg)

   x_2 = Amount of chili-flavored potato chips (in kg)

   Maximize $z = .12x_1 + .10x_2$
   subject to

$$3x_1 + 3x_2 + x_3 \qquad\qquad = 240$$
$$5x_1 + 4x_2 \qquad + x_4 \qquad = 480$$
$$2x_1 + 3x_2 \qquad\qquad + x_5 = 360$$
$$x_j \geq 0, \qquad j = 1, 2, \ldots, 5$$
$$x_3 = 0, \quad x_4 = 80, \quad x_5 = 200$$

The values of x_3, x_4, and x_5 represent the unused time (in hours) of the fryer, flavorer, and packer, respectively.

25. d

Section 2.3, page 112

1.

	x	y	u	v	
u	3	5	1	0	8
v	2	7	0	1	12
	-2	-5	0	0	0

3. (a) x_2 (b) x_1 (c) No finite optimal solution

5.

	x_1	x_2	x_3	x_4	
x_2	$\frac{1}{2}$	1	0	$\frac{1}{2}$	$\frac{3}{2}$
x_3	$-\frac{1}{4}$	0	1	$-\frac{5}{4}$	$\frac{3}{4}$
	2	0	0	-2	$\frac{23}{2}$

7.

	x_1	x_2	x_3	x_4	
x_1	1	1	5	0	4
x_4	0	1	7	1	10
	0	3	13	0	19

9. $[3 \quad 3 \quad 0 \quad 0 \quad 0 \quad 9 \quad 0]^T$

11. $\begin{bmatrix} 0 \\ 0 \end{bmatrix}$; $\quad z = 0$

13. $x = \frac{38}{23}$, $\quad y^+ = \frac{12}{23}$, $\quad y^- = 0$

15. Make $\quad$ 0 kg Super blend

$\qquad$ 640 kg Deluxe blend

Profit = \$64.00

17. Make 0 books in either paperback or library bindings, and 100 books in bookclub binding. $\quad$ Profit = \$80.00

19. $[0 \quad 4 \quad \frac{4}{3} \quad 0]^T$; $\quad z = \frac{28}{3}$

Section 2.4, page 120

1. $[0 \quad \frac{15}{2}]^T$; $\quad z = \frac{75}{2}$

3. $[0 \quad 0 \quad 3]^T$; $\quad z = 15$

5. $[\frac{22}{7} \quad \frac{15}{7}]^T$; $\quad z = \frac{207}{7}$

Section 2.5, page 140

1. (a)

	x_1	x_2	x_3	y_1	y_2	
y_1	1	2	7	1	0	4
y_2	1	3	1	0	1	5
	-2	-5	-8	0	0	-9

(b)

	x_1	x_2	x_3	y_1	y_2	
y_1	1	2	7	1	0	4
y_2	1	3	1	0	1	5
	$-1-2M$	$-5M$	$-3-8M$	0	0	$-9M$

3. (a)

	x_1	x_2	x_3	x_4	x_5	y_1	y_2	
y_1	1	3	2	-1	0	1	0	7
y_2	2	1	1	0	-1	0	1	4
	-3	-4	-3	1	1	0	0	-11

(b)

	x_1	x_2	x_3	x_4	x_5	y_1	y_2	
y_1	1	3	2	-1	0	1	0	7
y_2	2	1	1	0	-1	0	1	4
	$3-3M$	$-2-4M$	$-3M$	M	M	0	0	$-11M$

5.

	x_1	x_2	x_3	x_4	x_5	
x_2	$-\frac{1}{4}$	1	0	$-\frac{1}{4}$	$\frac{3}{4}$	$\frac{1}{2}$
x_3	$\frac{3}{4}$	0	1	$-\frac{1}{4}$	$-\frac{1}{4}$	$\frac{3}{2}$
	0	0	0	0	0	0

7. (a)

	x_1	x_2	x_3	x_4	x_5	
x_2	-1	1	$\frac{3}{10}$	0	$-\frac{1}{2}$	$\frac{3}{5}$
x_4	$\frac{1}{2}$	0	$\frac{1}{5}$	1	$-\frac{1}{10}$	$\frac{7}{10}$
	-3	0	$-\frac{7}{10}$	0	$-\frac{1}{2}$	$\frac{3}{5}$

(b) No finite optimal solution

9. An artificial variable y_1 has a nonzero value at the end of Phase 1. There are no feasible solutions.

11. Invest $70,000 in bond AAA and $30,000 in stock BB. Return = $7600

13. No feasible solutions

15. Make no PEST and 3 kg BUG.   Profit = $7.50

17. No feasible solutions

19. Invest $40,000 in the utilities stock, $0 in the electronic stock, and $160,000 in the bond.   Return = $116,000

21. No finite optimal solution

CHAPTER 3

Section 3.1, page 162

1. Maximize $z' = 8w_1 + 12w_2 + 6w_3$

subject to

$$w_1 + 2w_2 + 2w_3 \leq 3$$

$$4w_1 + 3w_2 + w_3 \leq 4$$

$$w_1 \geq 0, \ w_2 \geq 0, \ w_3 \geq 0$$

3. Minimize $z' = 8w_1 + 7w_2 + 12w_3$

subject to

$$3w_1 + 5w_2 + 4w_3 \geq 3$$

$$2w_1 + w_2 \qquad \geq 2$$

$$w_1 + 2w_2 + w_3 \geq 5$$

$$4w_2 - 2w_3 \geq 7$$

$$w_1 \geq 0, \ w_3 \geq 0$$

5. Minimize $z' = 18w_1 + 12w_2$

subject to

$$3w_1 + 2w_2 \geq 3$$

$$3w_1 + 2w_2 = 1$$

$$w_1 + 4w_2 \geq 4$$

$$w_1 \geq 0$$

7. Maximize $z' = 320w_1 - 12w_2$

subject to

$$30w_1 - w_2 \leq 15,000$$

$$50w_1 - 2w_2 \leq 20,000$$

$$w_1 \geq 0, \ w_2 \geq 0$$

9. Use $\frac{18}{11}$ ounces of walnuts, $\frac{48}{11}$ ounces of pecans, and no almonds.   Cost = 58.9 cents

13. Minimize $z' = \mathbf{b}^T\mathbf{w}$

subject to

$$\mathbf{A}^T\mathbf{w} \geq \mathbf{c}$$

$$\mathbf{B}^T\mathbf{w} = \mathbf{d}$$

$$\mathbf{w} \geq \mathbf{0}$$

Section 3.2, page 187

1.

$\mathbf{c_B}$		-1 x_1	2 x_2	-6 x_3	0 x_4	0 x_5	5 x_6	$\mathbf{x_B}$
2	x_2	-3	1	-1	-2	0	0	2
5	x_6	2	0	0	-1	0	1	3
0	x_5	6	0	7	6	1	0	1
		5	0	4	-9	0	0	19

3.

c_B		2 x_1	3 x_2	5 x_3	1 x_4	0 x_5	0 x_6	0 x_7	0 y_1	0 y_2	x_B
2	x_1	1	$\frac{1}{2}$	1	0	3	0	-1	0	0	$\frac{1}{2}$
1	x_4	0	1	$-\frac{1}{2}$	1	-2	0	-1	0	0	4
0	x_6	0	2	$-\frac{2}{3}$	0	$-\frac{1}{2}$	1	0	0	0	$\frac{3}{2}$
0	y_1	0	$\frac{3}{2}$	2	0	-3	0	0	1	0	0
0	y_2	0	4	-1	0	2	0	2	0	1	0
		0	-1	$-\frac{7}{2}$	0	4	0	-3	0	0	5

$$[\tfrac{1}{2} \quad 0 \quad 0 \quad 4 \quad 0 \quad \tfrac{3}{2} \quad 0]^T; \quad z = 5$$

7. $[5 \quad 0 \quad 3 \quad 0]^T$ or $[0 \quad 0 \quad 8 \quad 0]^T; \quad z = 8$

9. No feasible solution

11. $[0 \quad 2 \quad 0 \quad 4]^T; \quad z = 6$

13. *Exercise 6*

1st: $\quad \mathbf{B} = \begin{bmatrix} 1 & 0 & 0 \\ 0 & 1 & 0 \\ 0 & 0 & 1 \end{bmatrix}, \quad \mathbf{B}^{-1} = \begin{bmatrix} 1 & 0 & 0 \\ 0 & 1 & 0 \\ 0 & 0 & 1 \end{bmatrix}$

2nd: $\quad \mathbf{B} = \begin{bmatrix} 3 & 0 & 0 \\ 5 & 1 & 0 \\ 1 & 0 & 1 \end{bmatrix}, \quad \mathbf{B}^{-1} = \begin{bmatrix} \frac{1}{3} & 0 & 0 \\ -\frac{5}{3} & 1 & 0 \\ -\frac{1}{3} & 0 & 1 \end{bmatrix}$

3rd: $\quad \mathbf{B} = \begin{bmatrix} 3 & -1 & 0 \\ 5 & 3 & 0 \\ 1 & 0 & 1 \end{bmatrix}, \quad \mathbf{B}^{-1} = \begin{bmatrix} \frac{3}{14} & \frac{1}{14} & 0 \\ -\frac{5}{14} & \frac{3}{14} & 0 \\ -\frac{3}{14} & -\frac{1}{14} & 1 \end{bmatrix}$

Final: $\quad \mathbf{B} = \begin{bmatrix} 3 & -1 & 2 \\ 5 & 3 & 1 \\ 1 & 0 & 2 \end{bmatrix}, \quad \mathbf{B}^{-1} = \begin{bmatrix} \frac{2}{7} & \frac{2}{21} & -\frac{1}{3} \\ -\frac{3}{7} & \frac{4}{21} & \frac{1}{3} \\ -\frac{1}{7} & -\frac{1}{21} & \frac{2}{3} \end{bmatrix}$

Exercise 9

1st: $\quad \mathbf{B} = \begin{bmatrix} 1 & 0 & 0 & 0 \\ 0 & 1 & 0 & 0 \\ 0 & 0 & 1 & 0 \\ 0 & 0 & 0 & 1 \end{bmatrix}, \quad \mathbf{B}^{-1} = \begin{bmatrix} 1 & 0 & 0 & 0 \\ 0 & 1 & 0 & 0 \\ 0 & 0 & 1 & 0 \\ 0 & 0 & 0 & 1 \end{bmatrix}$

2nd: $\quad \mathbf{B} = \begin{bmatrix} 1 & 3 & 0 & 0 \\ 0 & 4 & 0 & 0 \\ 0 & 2 & 1 & 0 \\ 0 & 3 & 0 & 1 \end{bmatrix}$, $\quad \mathbf{B}^{-1} = \begin{bmatrix} 1 & -\frac{3}{4} & 0 & 0 \\ 0 & \frac{1}{4} & 0 & 0 \\ 0 & -\frac{1}{2} & 1 & 0 \\ 0 & -\frac{3}{4} & 0 & 1 \end{bmatrix}$

3rd: $\quad \mathbf{B} = \begin{bmatrix} 1 & 3 & 1 & 0 \\ 0 & 4 & 0 & 0 \\ 0 & 2 & 3 & 0 \\ 0 & 3 & 1 & 1 \end{bmatrix}$, $\quad \mathbf{B}^{-1} = \begin{bmatrix} 1 & -\frac{7}{12} & -\frac{1}{3} & 0 \\ 0 & \frac{1}{4} & 0 & 0 \\ 0 & -\frac{1}{6} & \frac{1}{3} & 0 \\ 0 & -\frac{7}{12} & -\frac{1}{3} & 1 \end{bmatrix}$

Final: $\quad \mathbf{B} = \begin{bmatrix} 1 & 1 & 1 & 0 \\ 0 & 2 & 0 & 0 \\ 0 & -1 & 3 & 0 \\ 0 & 4 & 1 & 1 \end{bmatrix}$, $\quad \mathbf{B}^{-1} = \begin{bmatrix} 1 & -\frac{2}{3} & -\frac{1}{3} & 0 \\ 0 & \frac{1}{2} & 0 & 0 \\ 0 & \frac{1}{6} & \frac{1}{3} & 0 \\ 0 & -\frac{13}{6} & -\frac{1}{3} & 1 \end{bmatrix}$

15. Invest \$70,000 in bond AAA and \$30,000 in stock BB. Return = \$7600

17. Use only large backhoe for $6\frac{2}{3}$ hours. Cost = \$266.67

Section 3.3, page 195

1. $[4 \quad \frac{10}{3} \quad 2 \quad 0 \quad 0]^T;$ $\quad z = 40$
3. $[0 \quad \frac{5}{2} \quad 2 \quad 2 \quad \frac{5}{2} \quad 0 \quad 0];$ $\quad z = \frac{37}{2}$
5. No feasible solutions

Section 3.4, page 203

1. $\begin{bmatrix} 1 & 0 & 0 & -\frac{1}{3} \\ 0 & 1 & 0 & \frac{2}{3} \\ 0 & 0 & 1 & 0 \\ 0 & 0 & 0 & \frac{1}{3} \end{bmatrix}$

3. $\begin{bmatrix} 1 & -1 & \frac{3}{2} \\ -1 & -\frac{1}{2} & \frac{9}{4} \\ 0 & 1 & \frac{1}{2} \end{bmatrix}$

5. $[\frac{10}{3} \quad \frac{5}{3} \quad \frac{1}{3}]^T;$ $\quad z = \frac{28}{3}$

7. Optimal solution is unbounded

9. $[0 \quad 0 \quad 10 \quad 0 \quad 0 \quad 0 \quad 18 \quad 0]^T;$ $\quad z = 66$

Section 3.5, page 213

1. (a) $-\infty < \Delta c_1 \leq \frac{7}{9}$ (b) $-4 \leq \Delta b_1 \leq 17$

     $-\frac{7}{6} \leq \Delta c_2 < \infty$ $-12 \leq \Delta b_2 \leq 12$

     $-1 \leq \Delta c_3 \leq 5$ $-\frac{34}{3} \leq \Delta b_3 < \infty$

     $-\infty < \Delta c_4 \leq \frac{14}{9}$

3. (a) $-2 \leq \Delta c_1 < \infty$ (b) $-4 \leq \Delta b_1 \leq 12$

     $-\infty < \Delta c_2 \leq 4$ $9 \leq \Delta b_2 < \infty$

     $-1 \leq \Delta c_3 < \infty$ $-12 \leq \Delta b_3 \leq 12$

     $-\infty < \Delta c_4 \leq \frac{1}{2}$

     $-\infty < \Delta c_5 \leq 4$

     $-1 \leq \Delta c_6 < \infty$

5. (a) Plant 9 acres of corn, 3 acres of oats, no soybeans. Profit = \$420
   (b) Plant 7 acres of corn, 3 acres of oats, no soybeans. Profit = \$340
   (c) Plant 6 acres of corn, 6 acres of oats, no soybeans. Profit = \$366
   (d) At least \$10/acre

CHAPTER 4

Section 4.1, page 236

1. Let x_1 = Number of type A

    x_2 = Number of type B

Minimize $z = 22{,}000x_1 + 48{,}000x_2$

subject to

$$100x_1 + 200x_2 \geq 600$$

$$50x_1 + 140x_2 \leq 350$$

$$x_1 \geq 0, \; x_2 \geq 0, \quad \text{integers}$$

3. Let $x_i = \begin{cases} 1 & \text{if } i\text{th record is purchased} \\ 0 & \text{otherwise} \end{cases}$

     $a_{ij} = \begin{cases} 1 & \text{if } j\text{th song is on } i\text{th record} \\ 0 & \text{otherwise} \end{cases}$

$$\text{Minimize} \quad z = \sum_{i=1}^{10} c_i x_i$$

subject to

$$\sum_{i=1}^{10} a_{ij} x_i \geq 1$$

$$j = 1, 2, \ldots, 6$$

5. Let w_i = Person-weeks necessary for project i
   c_i = Cost of project i (in thousands of dollars)
   v_i = Value of completion of project i
   $$x_i = \begin{cases} 1 & \text{if project } i \text{ is to be completed} \\ 0 & \text{otherwise} \end{cases}$$

$$\text{Maximize} \quad z = \sum_{i=1}^{10} v_i x_i$$

subject to

$$\sum_{i=1}^{10} w_i x_i \leq 1000$$

$$\sum_{i=1}^{10} c_i x_i \leq 1500$$

Section 4.2, page 252

1. $-\frac{1}{2}x_3 + u_1 = -\frac{7}{8}$
3. $x = 2, \quad y = 2; \qquad z = 4$

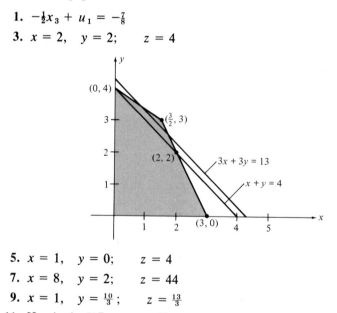

5. $x = 1, \quad y = 0; \qquad z = 4$
7. $x = 8, \quad y = 2; \qquad z = 44$
9. $x = 1, \quad y = \frac{10}{3}; \qquad z = \frac{13}{3}$
11. $[0 \quad 4 \quad \frac{4}{3} \quad 0]^{\mathrm{T}}; \qquad z = \frac{28}{3}$

Section 4.3, page 266

1.

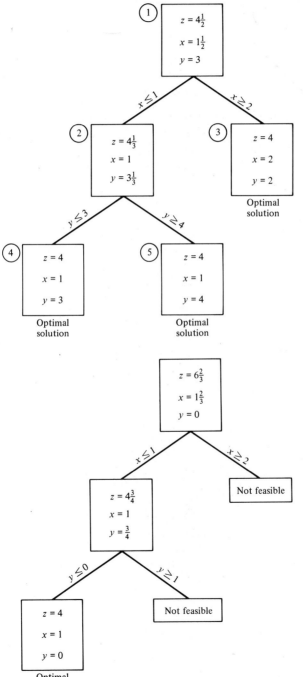

3.

5.

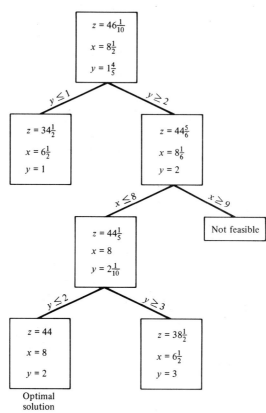

7.

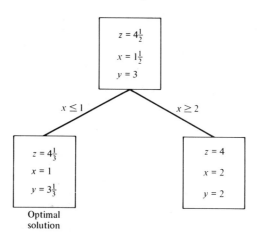

9.

$$\boxed{\begin{array}{c} z = 9\frac{1}{3} \\ x_1 = x_4 = 0 \\ x_2 = 4 \\ x_3 = 1\frac{1}{3} \end{array}}$$

Optimal
solution

11. Buy 2 machines of type A and 2 machines of type B.
Cost = $140,000

13. Buy 2 bottles each of cola and ginger ale, 5 bottles of root beer, and
3 bottles of grape.　　Cost = $6.59

CHAPTER 5

Section 5.1, page 298

3. (a)

				90
		70		
50			60	
50	40	30		30

(b)

				90
			40	30
100			10	
	40	100	10	

(c)

				90
			40	30
100			10	
	40	100	10	

5.

			30	70
20	60	80		
30			70	

$z = \$1730$

7.

				90
			40	30
100			10	
	40	100	10	

$z = \$950$

9.

10	70		
10		50	
50			
		20	80

$z = \$940$

11.

	1		
1			
			1
		1	

$z = 12$

13.

20	20	20	40	
	60			
		50		
40				

← Dummy supply

$z = \$800$

Section 5.2, page 312

1.
$$\begin{bmatrix} 0 & 1 & 0 & 0 \\ 1 & 0 & 0 & 0 \\ 0 & 0 & 0 & 1 \\ 0 & 0 & 1 & 0 \end{bmatrix}; \qquad z = 9$$

3.
$$\begin{bmatrix} 0 & 0 & 1 & 0 & 0 & 0 \\ 0 & 0 & 0 & 0 & 0 & 1 \\ 0 & 0 & 0 & 0 & 1 & 0 \\ 1 & 0 & 0 & 0 & 0 & 0 \\ 0 & 1 & 0 & 0 & 0 & 0 \\ 0 & 0 & 0 & 1 & 0 & 0 \end{bmatrix}; \qquad z = 12$$

5.
$$\begin{bmatrix} 0 & 0 & 0 & 0 & 0 & 1 \\ 0 & 1 & 0 & 0 & 0 & 0 \\ 0 & 0 & 0 & 0 & 1 & 0 \\ 1 & 0 & 0 & 0 & 0 & 0 \\ 0 & 0 & 0 & 1 & 0 & 0 \\ 0 & 0 & 1 & 0 & 0 & 0 \end{bmatrix}; \qquad z = 14$$

7. $702 \rightarrow 706$
$705 \rightarrow 707$
$708 \rightarrow 714$
$709 \rightarrow 712$
$713 \rightarrow 715$

Section 5.3, page 318

1. (a)
$$\begin{bmatrix} 0 & 1 & 0 & 1 & 0 & 0 \\ 1 & 0 & 1 & 1 & 0 & 1 \\ 0 & 1 & 0 & 0 & 1 & 0 \\ 1 & 1 & 0 & 0 & 1 & 0 \\ 0 & 0 & 1 & 1 & 0 & 1 \\ 0 & 1 & 0 & 0 & 1 & 0 \end{bmatrix}$$

(b) (i) $(1, 4)$
(ii) $(1, 2)$, $(2, 4)$
(iii) $(1, 2)$, $(2, 6)$, $(6, 5)$, $(5, 3)$, $(3, 2)$, $(2, 4)$
(c) (i) $(2, 3)$, $(3, 2)$
(ii) $(2, 1)$, $(1, 4)$, $(4, 2)$

3. (a)

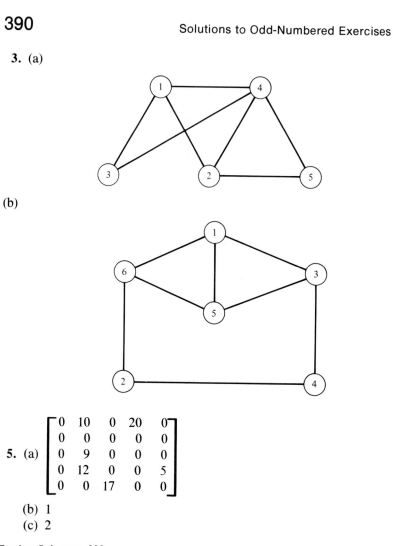

(b)

$$\begin{bmatrix} 0 & 10 & 0 & 20 & 0 \\ 0 & 0 & 0 & 0 & 0 \\ 0 & 9 & 0 & 0 & 0 \\ 0 & 12 & 0 & 0 & 5 \\ 0 & 0 & 17 & 0 & 0 \end{bmatrix}$$

5. (a)

(b) 1

(c) 2

Section 5.4, page 333

1.

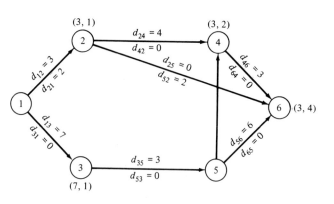

3.

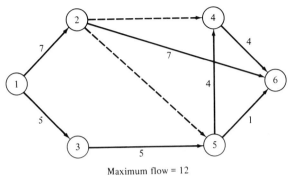

Maximum flow = 12

5.

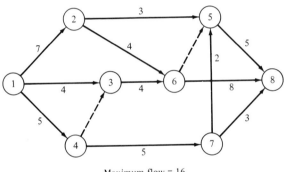

Maximum flow = 16

7. (a)

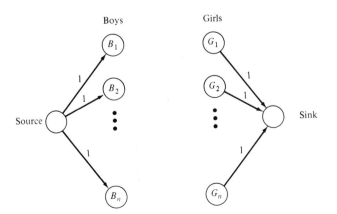

Connect B_i to G_j with an arc of capacity 1 if boy i and girl j are friends.

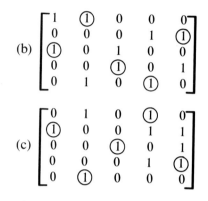

(d) *n* lines are necessary to cover the 1's.

Section 5.5, page 345

1. Path: $1 \to 5 \to 7$;   Length = 8
3. Path: $1 \to 2 \to 4 \to 9$;   Length = 14
5. Replace equipment at beginning of fourth year.
   Total equipment cost = \$380,000

Section 5.6, page 356

1. (a) and (b)

Node	Early event times	Late event times
1	0	0
2	3	8
3	6	6
4	8	8
5	5	10
6	9	14
7	16	20
8	9	9
9	7	12
10	18	18
11	17	22
12	25	25

(c) $1 \to 4 \to 8 \to 10 \to 12$

3. (a) and (b)

Node	Early event times	Late event times
1	0	0
2	3	3
3	8	8
4	7	14
5	7	21
6	22	22
7	9	23
8	14	14
9	22	22
10	15	29
11	15	29
12	31	31

(c) $1 \rightarrow 2 \rightarrow 3 \rightarrow 8 \rightarrow 6 \rightarrow 9 \rightarrow 12$

5. Insert dummy activity between nodes 2 and 5.

7.

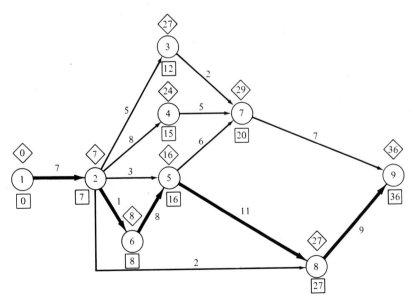

Critical path: $1 \rightarrow 2 \rightarrow 6 \rightarrow 5 \rightarrow 8 \rightarrow 9$

Section 5.7, page 365

3.

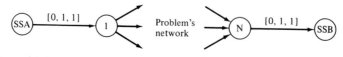

All remaining arcs should be labeled $[c, 1, 0]$, where c is the length of the arc.

INDEX

A

Accounting prices, 156
Activity
  in critical path method, 350
  dummy, 351
Adjacent extreme points, 100
Algorithm
  cutting plane, 240
  out-of-kilter, 359
  transportation, 282
Arc, 314
  directed, 315
Artificial variable, 125
Assigned column, 303
Assigned row, 303
Assignment problem, 229, 300
Augmented matrix, 11

B

Basic feasible solution, 93, 125
  initial, 101, 125
Basic solution, 93
Basic variable, 93
Basis, 38
  reinverting the, 224
Big M method, 136

B (continued — right column)

Bounded convex set, 55
Branch, 256, 314
  dangling, 258
Branch and bound method, 254

C

Canonical form for linear programming
    problem, 66
Capacity
  of a cut, 331
  of a directed arc, 316
  excess, 325
Capacity matrix, 316
Cell, 283
Circularization network, 362
Closed convex set, 55
Closed half-space, 46
Coefficient matrix, 11
Column
  assigned, 303
  of a matrix, 7
  necessary, 307, 308
  pivotal, 18, 105
Combination
  convex, 52
  linear, 34

Complementary slackness, 158
Components of a vector, 29
Computer codes, 216−217, 219, 362
Connected graph, 315
Constraint, 4, 65
    cutting plane, 242
Convex combination, 52
Convex function, 56
Convex hull, 54
Convex polyhedron, 55
Convex polytope, 54
Convex set, 51
    bounded, 54
    closed, 55
    unbounded, 55
Coordinate vector, 40
CPM (critical path method), 349
Critical event, 355
Critical path, 355
Critical path method, 349
    activity in, 350
    dummy activity in, 351
    event in, 350
Cut, 330
Cutting plane algorithm, 240
Cutting plane constraint, 242
Cycle, 315
Cycling, 116

D

Dangling branch, 258
Decision variable, 4
Degeneracy, 115, 291
Degenerate solution, 116
Departing variable, 105, 191, 282
Diagonal, main, 8
Diet problem, 60, 157
Digraph, 315
Dimension of a subspace, 39
Directed arc, 315
Directed graph, 315
Dual problem, 144
    economic interpretation of, 155
    solution to, 178
Dual simplex method, 190

Duality theorem, 152
Dummy activity, 351

E

Early event time, 352
Edge, 314
Element of a matrix, 8
Elementary row operation, 17
Entering variable, 104, 192, 282
Entry
    leading, 16
    of a matrix, 8
Equipment replacement problem, 343
Event, 350
    critical, 355
Event time
    early, 352
    late, 354
Excess capacity, 325
Extreme point, 53
    adjacent, 100

F

Feasible solution, 79
Fictitious price, 156
Fixed charge problem, 233
Float time, 355
Flow, 316
FMPS, 216, 217
Function
    convex, 56
    linear, 56
    objective, 4, 65

G

Gauss−Jordan reduction, 16, 20
General linear programming problem, 65
Generalized upper bounding, 221
Gomory's method, 246
Graph, 314
    connected, 315
    directed, 315
    oriented, 315
GUB (generalized upper bounding), 221

H

Half-space
  closed, 46
  open, 46
Hull, *see* Convex hull
Hungarian method, 310
Hyperplane, 46

I

Identity matrix, 10
Implicitly enumerated set, 255
Imputed value, 156, 169
Incidence matrix, 314
Index of summation, 9
Initial basic feasible solution, 101, 125
Initial tableau, 102
Intermediate node, 359
Inverse of a matrix, 24
Invertible matrix, 24

K

Knapsack problem, 228
Krein−Milman theorem, 55

L

Labeling method, 324
Late event time, 354
Leading entry, 16
Line, 50
Line segment, 50
Linear combination, 34
Linear function, 56
Linear programming problem
  canonical form of, 66
  constraint of, 4, 65
  general, 65
  standard form of, 65
Linear system of equations, 10
Linearly dependent set of vectors, 36
Linearly independent set of vectors, 36
Logical restraint, 351
Loop, 283, 314

LP codes, 216
LP systems, 216
LP/360, 217
LU factorization, 219

M

Main diagonal, 8
Marginal value, 157
Mathematical programming, 4
Matrices, equal, 8
Matrix, 7
  augmented, 11
  capacity, 316
  coefficient, 11
  column of, 7
  diagonal of, 8
  element of, 8
  identity, 10
  incidence, 314
  inverse, 24
  invertible, 24
  negative of, 9
  noninvertible, 24
  nonsingular, 24, 38
  partitioned, 12
  product, 9
  row of, 7
  scalar multiple of, 11
  singular, 24
  sparse, 219
  square, 8
  sum, 8
  transpose, 12
  zero, 8
Max flow−min cut theorem, 333
Maximal flow problem, 320
Method
  big M, 136
  branch and bound, 254
  dual simplex, 190
  Gomory's, 246
  Hungarian, 310
  labeling, 324
  revised simplex, 196
  simplex, 99, 109, 169
  two-phase, 125
  Vogel's, 293

Minimum cost flow problem, 359
Minimum cost rule, 282
Mixed integer programming problem, 234, 235
MPS/90, 219
MPSX/370, 217

N

n-space, 29
n-tuple, 29
n-vector, 29
Necessary column, 307, 308
Necessary row, 309
Negative of a matrix, 9
Network, 316
  circularization, 362
Node, 256, 314
  intermediate, 359
  terminal, 257
  transshipment, 359
Nonbasic variable, 93
Noninvertible matrix, 24
Nonsingular matrix, 24, 38

O

Objective function, 4, 65
Objective row, 102
Open half-space, 46
Operations research, 1
Operations research study, phases of, 2
OPHELIE, 217
Optimal solution, 79
Optimality criterion, 103
OR (operations research), 1
Oriented arc, 315
Oriented graph, 315
Out-of-kilter algorithm, 359

P

Parameter, 4
Partitioned matrix, 12
Path, 256, 314
  critical, 355

Penalty cost, 136
PERT, 350
Perturbation, 117
Pivot, 18, 107, 192
Pivotal column, 18, 105, 192
Pivotal row, 105, 109, 192
Pivoting, 107, 192
Polyhedron, see Convex polyhedron
Polytope, see Convex polytope
Price, shadow, 156
Primal problem, 144
Problem
  assignment, 229, 300
  diet, 60, 157
  dual, 144
  equipment replacement, 343
  fixed charge, 233
  general linear programming, 65
  knapsack, 228
  maximal flow, 320
  minimum cost flow, 359
  mixed integer programming, 234, 235
  primal, 144
  pure integer programming, 235
  shortest route, 338
  stock cutting, 232
  transportation, 61, 227, 271
  transshipment, 365
  traveling salesman, 229
  zero-one programming, 228, 255
Product of matrices, 9, 10
Pure integer programming problem, 235

R

Rectangle, 54
Reduced row echelon form, 16
  see also Gauss−Jordan reduction
Reinverting the basis, 224
Replacement value, 157
Restarting, 226
Restraint, see Constraint
  logical, 351
Revised simplex method, 196
Row
  assigned, 303
  of a matrix, 7
  necessary, 309

objective, 102
pivotal, 105, 109
Row sum, 224

S

Scalar multiple, 11
Scaling, 224
Sensitivity analysis, 3, 206
Set
   bounded convex, 54
   closed convex, 55
   convex, 51
   implicitly enumerated, 255
   unbounded convex, 55
Shadow price, 156
Shortest route problem, 338
Simplex method, 99, 109, 169
   dual, 190
   George B. Dantzig, 2
   revised, 196
Singular matrix, 24
Sink, 317, 359
Slack time, 355
Slack variable, 86
Solution
   basic, 93
   basic feasible, 93
   degenerate, 116
   feasible, 79
   optimal, 79
Source, 317, 359
Span, 35
Spanning set, 35
Sparse matrix, 219
Square matrix, 8
Standard form for linear programming
      problem, 65
Stock cutting problem, 232
Submatrix, 12
Subspace, 30
   dimension, 39
   trivial, 31
Sum of matrices, 8
Summation, index of, 9
Supersink, 361
Supersource, 361

T

Tableau, initial, 102
Terminal node, 257
θ-ratio, 105, 109
Transportation algorithm, 282
Transportation problem, 61, 227, 271
Transpose, 12
Transshipment node, 359
Transshipment problem, 365
Traveling salesman problem, 229
Trivial subspace, 31
Two-phase method, 125

U

UKILT 1100, 362
Unbounded convex set, 55
Unconstrained variables, 70

V

Value
   imputed, 156, 169
   marginal, 157
   replacement, 157
Variable
   artificial, 125
   basic, 93
   decision, 4
   departing, 105
   entering, 104
   nonbasic, 93
   slack, 86
   unconstrained, 70
Vector, 30
   components, 29
   coordinate, 40
Vectors, linear combination of, 34
Vertex, 314
Vogel's method, 293

Z

Zero matrix, 8
Zero-one programming problem, 228, 255

Computer Science and Applied Mathematics
A SERIES OF MONOGRAPHS AND TEXTBOOKS

Editor
Werner Rheinboldt
University of Maryland

HANS P. KÜNZI, H. G. TZSCHACH, and C. A. ZEHNDER. Numerical Methods of Mathematical Optimization: With ALGOL and FORTRAN Programs, Corrected and Augmented Edition

AZRIEL ROSENFELD. Picture Processing by Computer

JAMES ORTEGA AND WERNER RHEINBOLDT. Iterative Solution of Nonlinear Equations in Several Variables

AZARIA PAZ. Introduction to Probabilistic Automata

DAVID YOUNG. Iterative Solution of Large Linear Systems

ANN YASUHARA. Recursive Function Theory and Logic

JAMES M. ORTEGA. Numerical Analysis: A Second Course

G. W. STEWART. Introduction to Matrix Computations

CHIN-LIANG CHANG AND RICHARD CHAR-TUNG LEE. Symbolic Logic and Mechanical Theorem Proving

C. C. GOTLIEB AND A. BORODIN. Social Issues in Computing

ERWIN ENGELER. Introduction to the Theory of Computation

F. W. J. OLVER. Asymptotics and Special Functions

DIONYSIOS C. TSICHRITZIS AND PHILIP A. BERNSTEIN. Operating Systems

ROBERT R. KORFHAGE. Discrete Computational Structures

PHILIP J. DAVIS AND PHILIP RABINOWITZ. Methods of Numerical Integration

A. T. BERZTISS. Data Structures: Theory and Practice, Second Edition

N. CHRISTOPHIDES. Graph Theory: An Algorithmic Approach

ALBERT NIJENHUIS AND HERBERT S. WILF. Combinatorial Algorithms

AZRIEL ROSENFELD AND AVINASH C. KAK. Digital Picture Processing

SAKTI P. GHOSH. Data Base Organization for Data Management

DIONYSIOS C. TSICHRITZIS AND FREDERICK H. LOCHOVSKY. Data Base Management Systems

JAMES L. PETERSON. Computer Organization and Assembly Language Programming

WILLIAM F. AMES. Numerical Methods for Partial Differential Equations, Second Edition

ARNOLD O. ALLEN. Probability, Statistics, and Queueing Theory: With Computer Science Applications

ELLIOTT I. ORGANICK, ALEXANDRA I. FORSYTHE, AND ROBERT P. PLUMMER. Programming Language Structures

ALBERT NIJENHUIS AND HERBERT S. WILF. Combinatorial Algorithms. Second edition.

JAMES S. VANDERGRAFT. Introduction to Numerical Computations

AZRIEL ROSENFELD. Picture Languages, Formal Models for Picture Recognition

ISAAC FRIED. Numerical Solution of Differential Equations

ABRAHAM BERMAN AND ROBERT J. PLEMMONS. Nonnegative Matrices in the Mathematical Sciences

BERNARD KOLMAN AND ROBERT E. BECK. Elementary Linear Programming with Applications

CLIVE L. DYM AND ELIZABETH S. IVEY. Principles of Mathematical Modeling

ERNEST L. HALL. Computer Image Processing and Recognition

ALLEN B. TUCKER, JR., Text Processing: Algorithms, Languages, and Applications

In preparation

MARTIN CHARLES GOLUMBIC. Algorithmic Graph Theory and Perfect Graphs

B
C
D
E
F
G
H 3
I 4
J 5